10-12.72

QD   73919
251
.07 Organic reactions, vol. I-
v.19 J. Wiley & sons, inc.; London

Organic Reactions

Organic Reactions

VOLUME 19

EDITORIAL BOARD

WILLIAM G. DAUBEN, *Editor-in-Chief*

JOHN FRIED   BLAINE C. MCKUSICK
ANDREW S. KENDE   JERROLD MEINWALD
JAMES A. MARSHALL   BARRY M. TROST

ROBERT BITTMAN, *Secretary*
Queens College of the City University of New York
Flushing, New York

ADVISORY BOARD

A. H. BLATT   DAVID Y. CURTIN
VIRGIL BOEKELHEIDE   LOUIS F. FIESER
T. L. CAIRNS   JOHN R. JOHNSTON
DONALD J. CRAM   FRANK C. MCGREW
                              HAROLD R. SNYDER

ASSOCIATE EDITORS

J. M. BLATCHLY   GARY H. POSNER
JAY K. KOCHI   MARTIN F. SEMMELHACK
J. F. W. MCOMIE   ROGER A. SHELDON

FORMER MEMBERS OF THE BOARD, NOW DECEASED

ROGER ADAMS   WERNER E. BACHMANN
HOMER ADKINS   ARTHUR C. COPE
                              CARL NIEMANN

JOHN WILEY & SONS, INC.
NEW YORK · LONDON · SYDNEY · TORONTO

Copyright © 1972, by John Wiley & Sons, Inc.

All rights reserved. Published simultaneously in Canada.

No part of this book may be reproduced by any means, nor transmitted, nor translated into a machine language without the written permission of the publisher.

Library of Congress Catalog Card Number: 42-20265

ISBN 0-471-19619-3

Printed in the United States of America.

10 9 8 7 6 5 4 3 2 1

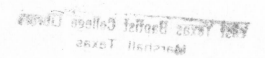

ROGER ADAMS 1889-1971

This volume which, like its eighteen predecessors, so clearly bears the imprint of his leadership and his hand is dedicated to Roger Adams, one of the founders of Organic Reactions.

PREFACE TO THE SERIES

In the course of nearly every program of research in organic chemistry the investigator finds it necessary to use several of the better-known synthetic reactions. To discover the optimum conditions for the application of even the most familiar one to a compound not previously subjected to the reaction often requires an extensive search of the literature; even then a series of experiments may be necessary. When the results of the investigation are published, the synthesis, which may have required months of work, is usually described without comment. The background of knowledge and experience gained in the literature search and experimentation is thus lost to those who subsequently have occasion to apply the general method. The student of preparative organic chemistry faces similar difficulties. The textbooks and laboratory manuals furnish numerous examples of the application of various syntheses, but only rarely do they convey an accurate conception of the scope and usefulness of the processes.

For many years American organic chemists have discussed these problems. The plan of compiling critical discussions of the more important reactions thus was evolved. The volumes of *Organic Reactions* are collections of chapters each devoted to a single reaction, or a definite phase of a reaction, of wide applicability. The authors have had experience with the processes surveyed. The subjects are presented from the preparative viewpoint, and particular attention is given to limitations, interfering influences, effects of structure, and the selection of experimental techniques. Each chapter includes several detailed procedures illustrating the significant modifications of the method. Most of these procedures have been found satisfactory by the author or one of the editors, but unlike those in *Organic Syntheses* they have not been subjected to careful testing in two or more laboratories.

Each chapter contains tables that include all the examples of the reaction under consideration that the author has been able to find. It is inevitable, however, that in the search of the literature some examples will be missed, especially when the reaction is used as one step in an extended synthesis. Nevertheless, the investigator will be able to use the tables and their accompanying bibliographies in place of most or all of the literature search so often required.

Because of the systematic arrangement of the material in the chapters and the entries in the tables, users of the books will be able to find information desired by reference to the table of contents of the appropriate chapter. In the interest of economy the entries in the indices have been kept to a minimum, and, in particular, the compounds listed in the tables are not repeated in the indices.

The success of this publication, which will appear periodically, depends upon the cooperation of organic chemists and their willingness to devote time and effort to the preparation of the chapters. They have manifested their interest already by the almost unanimous acceptance of invitations to contribute to the work. The editors will welcome their continued interest and their suggestions for improvements in *Organic Reactions*.

Chemists who are considering the preparation of a manuscript for submission to Organic Reactions are urged to write the secretary before they begin work.

CONTENTS

		PAGE
1.	CONJUGATE ADDITION REACTIONS OF ORGANOCOPPER REAGENTS *Gary H. Posner*	1
2.	FORMATION OF CARBON–CARBON BONDS VIA π-ALLYLNICKEL COMPOUNDS *Martin F. Semmelhack*	115
3.	THE THIELE–WINTER ACETOXYLATION OF QUINONES *J. F. W. McOmie and J. M. Blatchly*	199
4.	OXIDATIVE DECARBOXYLATION OF ACIDS BY LEAD TETRAACETATE *Roger A. Sheldon and Jay K. Kochi*	279

AUTHOR INDEX, VOLUMES 1–19 423

CHAPTER INDEX, VOLUMES 1–19 425

SUBJECT INDEX, VOLUME 19 431

Organic Reactions

CHAPTER 1

CONJUGATE ADDITION REACTIONS OF ORGANOCOPPER REAGENTS

Gary H. Posner

The Johns Hopkins University, Baltimore, Maryland

CONTENTS

	PAGE
Acknowledgments	2
Introduction	3
Mechanism	6
Scope and Limitations	8
The Organocopper Reagent	8
Nature and Preparation of the Reagent	9
Catalytic Organocopper Reagents	9
Stoichiometric Organocopper Reagents	11
RCu	11
R_2CuLi	12
RCu-Ligand	14
Structural Variation of R	15
Stereochemical Stability of R	16
Effect of Structure on Amount and Stereochemistry of Conjugate Addition	18
Structure of Reagent	18
Structure of R	20
Selection of Reagent for Conjugate Addition	22
The α,β-Unsaturated Carbonyl Substrate	23
α,β-Ethylenic Carbonyl Derivatives	23
Acyclic Enones	24
Acyclic Enoate Esters	25
Optical Isomerism and Diastereoisomerism	31
Cyclic Enones	32
Polyethylenic Esters and Ketones	40
Compatible Functional Groups	42
α,β-Acetylenic Carbonyl Derivatives	43
Miscellaneous Substrates	46

Conjugate Addition of Other Organometallic Compounds . . 48

Synthetic Utility 51

Experimental Factors 54

    Preparation and Handling of Organocopper Reagent 54
    Temperature 55
    Solvent 56
    Workup Procedure 56

Experimental Procedures 57

    Ethyl 2-Cyano-3-isopropyl-3-methylpentanoate 58
    trans-10-Methyl-7α-t-butyl-4-octal-3-one 58
    9α,10α-Dimethyl-1β-isopropenyl-trans-3-decalone 59
    (±)-Eremophil-11-en-3-one 59
    3,5-Dimethylcyclohexanone 60
    3-Methyl-5-phenylcyclohexanone 60
    3-Allylcyclohexanone 61
    Methyl 3-Methyl-trans-2-decenoate 61
    (−)-3-Phenylbutyric Acid 62
    16α-Methylprogesterone 62

Tabular Survey 63

    Table I. Copper-Catalyzed Conjugate Addition of Grignard Reagents to
        α,β-Ethylenic Carbonyl Compounds 64
        A. Acyclic Carbonyl Substrates 64
        B. Cyclic Carbonyl Substrates 78
    Table II. Copper-Catalyzed Conjugate Addition of Grignard Reagents to
        α,β-Acetylenic Carbonyl Compounds 94
    Table III. Conjugate Addition of Organocopper Reagents to α,β-Ethylenic
        Carbonyl Compounds 96
        A. Acyclic Carbonyl Substrates 96
        B. Cyclic Carbonyl Substrates 98
    Table IV. Conjugate Addition of Organocopper Reagents to α,β-Acetylenic
        Carbonyl Compounds 110

References to Tables I–IV 113

ACKNOWLEDGMENTS

The help of Miss Susan A. Vladuchick, of E. I. Dupont de Nemours and Company, in searching the chemical literature for examples of organocopper conjugate addition reactions is gratefully acknowledged. Acknowledgment is made also to Miss Kathleen Offutt for her meticulous typing of this manuscript.

INTRODUCTION

Addition of an organometallic reagent to an α,β-unsaturated carbonyl compound occurs in one or both of two modes: attack at the carbonyl function produces a 1,2 adduct, whereas reaction with the entire conjugated system gives a 1,4 adduct (Eq. 1). Organoalkali metal reagents derived from unstabilized carbanions (*e.g.*, methyllithium, phenylsodium) generally add across the carbonyl group,[1] whereas enolate and other stabilized anions usually undergo Michael (1,4) addition,[2] introducing a carboalkyl group at the β carbon. Addition of Grignard reagents to α,β-unsaturated aldehydes normally gives 1,2 adducts, and Grignard addition to α,β-unsaturated carbonyl species in which the carbonyl group is sterically inaccessible often gives 1,4 adducts exclusively; most unsaturated carbonyl functions, however, react with organomagnesium reagents to give mixtures of 1,2 and 1,4 addition products.[3] Selective 1,4, or conjugate, addition of aliphatic and aromatic groups to various α,β-ethylenic and acetylenic carbonyl units has been achieved with high success using organocopper reagents. The wide scope and effectiveness of such reagents have made conjugate addition of alkyl and aryl groups a useful reaction in organic synthesis, allowing efficient and specific placement of hydrocarbon units β to a carbonyl function.

(Eq. 1)

For twenty-five years, from 1941 to 1966, copper-promoted conjugate additions of hydrocarbon groups were generally effected by allowing an organomagnesium reagent and an α,β-unsaturated carbonyl compound to react in the presence of a catalytic amount of a copper salt. The first recorded example of such a copper-catalyzed conjugate addition is due

---

[1] T. Eicher, in *The Chemistry of Carbonyl Compounds*, S. Patai, Ed., Interscience Publishers, New York, 1966, pp. 624–631, 662–678.

[2] E. D. Bergmann, D. Ginsburg, and R. Pappo, *Org. Reactions*, **10**, 179 (1959).

[3] M. S. Kharasch and O. Reinmuth, *Grignard Reactions of Nonmetallic Substances*, Prentice-Hall, Englewood Cliffs, N.J., 1954, pp. 196–239.

to Kharasch and Tawney.[4]* They found that a catalytic amount of cuprous chloride in a diethyl ether solution of methylmagnesium bromide and isophorone (1) increased the amount of conjugate addition from 1.5 to 82.5%.† Birch and Robinson in 1943 applied the same general procedure to achieve angular methylation of bicyclic enone 2, forming the *cis*-decalone 3 in 60% yield.[5] Steroidal enones with the α,β-ethylenic carbonyl function in various positions in rings A, B, C, or D of the steroid nucleus were shown in several laboratories to undergo position specific and stereospecific conjugate addition of Grignard reagents in the presence of catalytic amounts of copper salts (see p. 40).

The scope of copper-catalyzed conjugate additions was extended to various acyclic α,β-ethylenic (enoate) and acetylenic (ynoate) esters by Munch-Petersen and his students. In a series of papers appearing from 1957 to 1966, they explored cuprous chloride-promoted conjugate addition of Grignard reagents to enoate esters,[6,7] to ethylene-1,2-dicarboxylates,[8,9] to ynoates,[10] to 2,4-alkadienoates,[11] and to 2,4,6-alkatrienoates.[12]

* Undoubtedly the first example of a copper-catalyzed Grignard conjugate addition actually occurred when such reactions were normally performed in copper reaction vessels.

† In this and subsequent reactions of organometallic reagents with α,β-unsaturated carbonyl compounds, hydrolysis of intermediate species will not be designated as a separate step in the reaction.

[4] M. S. Kharasch and P. O. Tawney, *J. Amer. Chem. Soc.*, **63**, 2308 (1941).

[5] A. J. Birch and R. Robinson, *J. Chem. Soc.*, **1943**, 501.

[6] J. Munch-Petersen, *J. Org. Chem.*, **22**, 170 (1957).

[7] T. Kindt-Larsen, V. Bitsch, I. G. K. Andersen, A. Jart, and J. Munch-Petersen, *Acta Chem. Scand.*, **17**, 1426 (1963).

[8] E. Bjerl Nielsen, J. Munch-Petersen, P. Moller Jorgensen, and S. Refn, *Acta Chem. Scand.*, **13**, 1943 (1959).

[9] V. K. Andersen and J. Munch-Petersen, *Acta Chem. Scand.*, **16**, 947 (1962).

[10] C. Bretting, J. Munch-Petersen, P. Moller Jorgensen, and S. Refn, *Acta Chem. Scand.*, 14, 151 (1960).

[11] J. Munch-Petersen, C. Bretting, P. Moller Jorgensen, S. Refn, and V. K. Andersen, *Acta Chem. Scand.*, **15**, 277 (1961).

[12] S. Jacobsen, A. Jart, T. Kindt-Larsen, I. G. K. Andersen, and J. Munch-Petersen, *Acta Chem. Scand.*, **17**, 2423 (1963).

Short reviews of the scope, mechanism, and stereochemistry of copper-catalyzed conjugate additions of Grignard reagents have appeared.[3, 13–15]

In 1966 House, Respess, and Whitesides provided experimental evidence implicating organocopper species as the reactive intermediates in copper-catalyzed Grignard conjugate additions.[16] They showed that preformed methylcopper[17] and lithium dimethylcopper[18] (prepared from methyllithium and cuprous iodide, Eqs. 2a, 2b) react with *trans*-3-penten-2-one (**4**) in diethyl ether to give in high yield 4-methyl-2-pentanone, the same conjugate adduct obtained from enone **4** by copper-catalyzed addition of methylmagnesium bromide. The discovery of this pronounced ability of methylcopper and other organocopper reagents to add 1,4 to α,β-unsaturated carbonyl compounds has revived interest in the chemistry of organocopper compounds.[17]

$$CuI + CH_3Li \longrightarrow CH_3Cu + LiI \qquad (Eq.\ 2a)$$

$$CH_3Cu + CH_3Li \longrightarrow (CH_3)_2CuLi \qquad (Eq.\ 2b)$$

$$trans\text{-}CH_3CH{=}CHCOCH_3 \xrightarrow{CH_3Cu} (CH_3)_2CHCH_2COCH_3$$
$$\mathbf{4}$$

Studies since 1966 have shown that stoichiometric organocopper reagents,* in addition to their broad utility in coupling reactions with organic halides[19–23] and in reactions with other substrates,[24] undergo unusually effective conjugate addition to an extremely wide variety of

---

* Organocopper reagents formed from 0.5 or more equivalents of copper salt per equivalent of organometallic will be denoted "stoichiometric" organocopper reagents, in distinction to "catalytic" organocopper reagents formed from an organometallic and a catalytic amount of a copper salt.

[13] S. Patai and Z. Rappoport, in *The Chemistry of Alkenes*, S. Patai, Ed., Interscience Publishers, New York, 1964, pp. 550 ff.

[14] J. Munch-Petersen, *Bull. Soc. Chim. Fr.*, **1966**, 471, and references cited therein.

[15] K. Thornburg, Organic Chemistry Seminar, University of Illinois, Urbana, Illinois, January 6, 1969.

[16] H. O. House, W. L. Respess, and G. M. Whitesides, *J. Org. Chem.*, **31**, 3128 (1966).

[17] H. Gilman and J. M. Straley, *Rec. Trav. Chim.*, **55**, 821 (1936).

[18] H. Gilman, R. G. Jones, and L. A. Woods, *J. Org. Chem.*, **17**, 1630 (1952).

[19] E. J. Corey and G. H. Posner, *J. Amer. Chem. Soc.*, **89**, 3911 (1967).

[20] E. J. Corey and G. H. Posner, *J. Amer. Chem. Soc.*, **90**, 5615 (1968).

[21] G. M. Whitesides, W. F. Fischer, Jr., J. San Filippo, Jr., R. W. Bashe, and H. O. House, *J. Amer. Chem. Soc.*, **91**, 4871 (1969), and references therein.

[22] O. P. Vig, J. C. Kapur and S. D. Sharma, *J. Indian Chem. Soc.*, **45**, 734 (1968).

[23] O. P. Vig, S. D. Sharma, and J. C. Kapur, *J. Indian Chem. Soc.*, **46**, 167 (1969).

[24] Organocopper reagents have been shown to react with the following substrates, giving the products indicated: (a) ethynylcarbinol acetates → allenes: P. Rona and P. Crabbé, *J. Amer. Chem. Soc.*, **91**, 3289 (1969); (b) allylic acetates → olefins: R. J. Anderson, C. A. Henrick, and J. B. Siddall, *ibid.*, **92**, 735 (1970); (c) epoxides → alcohols: R. W. Herr, D. M. Wieland, and C. R. Johnson, *ibid.*, **92**, 3813 (1970); (d) vinylic epoxides → allylic alcohols: R. J. Anderson, *ibid.*, **92**, 4978 (1970), and R. W. Herr and C. R. Johnson, *ibid.*, **92**, 4979 (1970).

α,β-unsaturated carbonyl substrates. In three of the four cases where direct comparison with copper-catalyzed Grignard conjugate addition is possible for addition of methyl, n-alkyl, and phenyl but not allylic groups, stoichiometric organocopper reagents generally produce conjugate adducts in higher yields and with greater stereoselectivity than do organocopper reagents generated catalytically from Grignard reagents and small amounts of copper salts.

Because selective conjugate addition is often achieved more effectively by stoichiometric than by catalytic organocopper reagents, the emphasis of this chapter is on stoichiometric organocopper reagents. Where appropriate, the two types of organocopper reagents are compared. Consideration is given to possible mechanisms of organocopper conjugate addition, to the scope, limitations, and synthetic utility of this reaction, and to optimal experimental conditions for its application.

MECHANISM

Even the basic notions about the mechanism(s) of organocopper conjugate addition are currently a matter of speculation. Two working hypotheses have been suggested. Both agree that copper catalysis of Grignard reagents causes rapid formation *in situ* of an organocopper species (Eq. 3) which adds rapidly to the β-carbon atom of the α,β-unsaturated substrate, consuming the substrate in most instances before the competing carbonyl addition of the Grignard reagent can occur, and regenerating active organocopper reagent to repeat the process in a catalytic manner. The two hypotheses differ, however, on the details of how the organic group is transferred from the organocopper reagent to the β-carbon atom of the α,β double bond.

$$RMgX + CuY \rightarrow RCu + MgXY \qquad \text{(Eq. 3)}$$
$$X = Cl, Br, I$$
$$Y = Cl, Br, I, CN, (OAc)_2$$

The first mechanistic hypothesis involves complex formation between the copper(I) alkyl and the carbon-carbon double bond of the α,β-ethylenic carbonyl substrate.[3, 8] The β-carbon atom, already electrophilic because it is at the terminus of the polarized conjugated system (*cf.* 5), is thus made more electrophilic by the copper complexation. By a concerted cyclic six-membered transition state involving substrate and two molecules of RCu, conjugate addition occurs with concurrent attachment of copper to the α-carbon atom and liberation of one molecule of RCu. Immediately thereafter (or perhaps simultaneously with the first step) the Grignard reagent reacts with the C-copper enolate to form an O-magnesium enolate and a second molecule of RCu, making the process catalytic in RCu (Eq.

$$-\overset{|}{\underset{|}{C}}=\overset{|}{\underset{|}{C}}-\overset{|}{\underset{|}{C}}=O \longleftrightarrow -\overset{+}{\underset{|}{C}}-\overset{|}{\underset{|}{C}}=\overset{|}{\underset{|}{C}}-O^-$$
5

(Eq. 4)

4).[7, 11, 12, 25] Arguments for and against a cyclic six-membered transition state for conjugate addition have been reviewed elsewhere.[3, 16, 26, 27]

The second currently more accepted mechanistic hypothesis involves neither a complex between organocopper reagent and substrate α,β double bond nor a cyclic six-centered transition state.[16] Copper-double bond complexation is deemed unlikely on the basis of the following two observations: (1) "the presence of either excess tri-n-butylphosphine or pyridine in the reaction mixture had no observed effects although these additives might be expected to compete with the unsaturated ketone for coordination positions on copper;" and (2) no copper(I)-ketone complex was discerned from examining the infrared, nmr, and ultraviolet spectra of ether solutions containing the ketone and copper(I) iodide-tri-n-butylphosphine complex.[16] A cyclic mechanism for lithium dimethylcopper conjugate addition to *trans*-3-penten-2-one, and presumably to most unsaturated carbonyl systems, was shown to be no more than a minor pathway by trapping the intermediate enolate anions, 69% of which had the *trans* configuration (6) which could not have been formed via a cyclic mechanism involving one molecule of enone and one molecule of organocopper reagent.[16]

This second mechanistic hypothesis involves either partial or complete electron transfer from organocuprate, $[RCu(I)Y]^{\ominus}$, to unsaturated substrate forming either a charge-transfer complex or an ion radical. Subsequent transfer of an organic radical from a transient organocopper(II)

---

[25] J. Munch-Petersen and V. K. Andersen, *Acta Chem. Scand.*, **15**, 271 (1961).
[26] E. R. Alexander and G. R. Coraor, *J. Amer. Chem. Soc.*, **73**, 2721 (1951).
[27] J. Klein, *Tetrahedron*, **20**, 465 (1964).

species or collapse of the charge-transfer complex would complete the addition sequence and would generate copper(I)-Y, as in Eq. 5.[16, 28] The apparent requirement of a net negative charge on the copper complex for efficient conjugate addition has recently been confirmed.[29, 30]

$$[\text{R:Cu(I)Y}]^- \qquad \text{R:Cu(II)Y} \qquad \text{Cu(I)Y}$$

$$\underset{|\ \ |}{-\text{C}=\text{C}-\overset{\overset{\displaystyle \text{O}}{\|}}{\text{C}}-} \longrightarrow \underset{|}{-\overset{}{\text{C}}-\text{C}=\overset{\overset{\displaystyle \text{O}^\ominus}{|}}{\text{C}}-} \longrightarrow \underset{|}{-\overset{\overset{\displaystyle \text{R}}{|}}{\text{C}}-\text{C}=\overset{\overset{\displaystyle \text{O}^\ominus}{|}}{\text{C}}-} \qquad \text{(Eq. 5)}$$

Y = R, halogen, CN

Concerning the second hypothesis of an electron transfer from copper reagent to substrate, it should be noted that excess isoprene does not retard organocopper conjugate addition—an indication that this reaction does not involve organic free radicals which are reactive toward isoprene.[31]

The mechanism(s) of organocopper conjugate addition to polyethylenic and to α,β-acetylenic carbonyl compounds is equally unclear.

Experimental clarification of the mechanism(s) operating in organocopper conjugate addition would provide much needed answers to such questions as: (1) What purpose does the normally required two- or threefold excess of organocopper reagent serve? (2) Generally, how do the spatial and electronic requirements of the organocopper reagent correlate with the spatial and electronic demands of the substrate (*i.e.*, structure-reactivity relationships)? (3) What factors govern the relative reactivity of *cis* and *trans* geometrical isomers? (4) What factors control the normally highly stereoselective organocopper conjugate addition to acetylenic carbonyl substrates, and how can addition be made completely stereoselective? (5) When should organocopper conjugate addition be expected to result in asymmetric 1,4 adducts?

As the pieces of the mechanistic jigsaw puzzle are slowly put together and a clear picture of the mechanism of organocopper conjugate addition is formed, there will emerge a more thorough understanding of why copper among the transition metals promotes conjugate addition so effectively.

SCOPE AND LIMITATIONS

The Organocopper Reagent

The scope of conjugate addition as a function of the organocopper reagent is discussed in detail in this section. Of the by-products formed in

[28] H. O. House and W. F. Fischer, Jr., *J. Org. Chem.*, **33**, 949 (1968).
[29] N. T. Luong-Thi and H. Rivière, *Tetrahedron Lett.*, **1970**, 1579.
[30] N. T. Luong-Thi and H. Rivière, *Tetrahedron Lett.*, **1970**, 1583.
[31] H. O. House, D. D. Traficante, and R. A. Evans, *J. Org. Chem.*, **28**, 348 (1963).

these organocopper reactions, two kinds are typical: 1,2 adducts arising from carbonyl addition, and symmetrical dimers $(R-R)^{32}$ arising from the organic group (R) in the organocopper reagent itself. The limitations imposed on conjugate addition by these side products can usually be minimized by proper selection of organocopper reagent and experimental conditions (see pp. 54–57).

Nature and Preparation of the Reagent

Throughout this chapter a distinction is made between catalytic and stoichiometric organocopper reagents; a catalytic organocopper reagent is formed from an organometallic reagent (usually Grignard) and a *catalytic amount* of a copper salt, whereas a stoichiometric organocopper reagent is formed from 0.5 or more equivalents of copper salt per equivalent of organometallic reagent (usually organolithium). This distinction between catalytic and stoichiometric organocopper reagents is based primarily on the repeatedly observed difference in yield and stereochemistry of conjugate addition when one or the other type of reagent is used, and is not based on any clearly established difference in nature between these two types of organocopper reagents. When both types of organocopper reagent are being discussed, "organocopper" will be used with no descriptive adjective.

Catalytic Organocopper Reagents.
Much of the early experimental work is difficult to interpret for three reasons: (1) the investigators generally did not realize the importance of trace impurities in the magnesium used to prepare the Grignard reagents;[33-35] (2) too often only desired products were isolated and identified, leaving substantial amounts of by-products uncharacterized; and (3) analytical techniques lacked present-day precision. Furthermore the synthetic utility of even recent conjugate additions using catalytic organocopper reagents is often difficult to evaluate and is probably not at its maximum; relatively low yields of conjugate adducts may be due to an inherently poor conjugate addition of the particular catalytic organocopper reagent *or* to the thermal instability of the reagent at the often arbitrarily selected reaction temperature (see p. 55).

Two aspects of copper-catalyzed Grignard conjugate additions, however, are most impressive: first, the wide range of generally successful conjugate additions achieved in this way (see Tables I and II); and, second, the variability of success as a function, for example, of the copper

---

[32] J. A. Katzenellenbogen, Ph.D. Thesis, Harvard University, Cambridge, Mass., 1969; *Diss. Abstr.*, **31**, 1826-B (1970).
[33] J. Hilden and J. Munch-Petersen, *Acta Chem. Scand.*, **21**, 1370 (1967).
[34] J. A. Marshall and H. Roebke, *J. Org. Chem.*, **31**, 3109 (1966).
[35] G. Costa, A. Camus, L. Gatti, and N. Marsich, *J. Organometal. Chem.*, **5**, 568 (1966).

salt and of the order of adding the reagents. (Variability of success in terms of the organocopper reagent itself is discussed on p. 19.)

For many years after Kharasch and Tawney's report in 1941,[4] catalytic organocopper reagents were formed in ether from Grignard reagents and cuprous halides, cuprous chloride being the favorite. Cuprous chloride, however, is virtually insoluble in ether. In 1962 Birch and Smith[36] showed that cupric acetate, soluble in tetrahydrofuran and readily reduced to cuprous ions by the excess of Grignard reagents,[37] allows more effective Grignard conjugate addition than does cuprous chloride in ether.[38] Similarly, Prout and Abdulslam in 1966 demonstrated that cuprous cyanide, being more soluble in ether than the cuprous halides, promotes better Grignard conjugate addition.[39] For example, conjugate addition of isopropylmagnesium bromide to cyanoacetate 7 varies in yield with the copper catalyst: CuCl (29–48% yield); CuBr (46%); CuI (56%); CuCN (65%).[39] Efforts to solubilize the copper salt to make organocopper formation more efficient have led recently to the use of cuprous halides complexed with a suitable ligand, for example a phosphine or sulfide[16] (cf. p. 14).

$$C_2H_5C(CH_3)=C(CN)CO_2C_2H_5 \xrightarrow[CuX]{i\text{-}C_3H_7MgBr} C_2H_5\underset{\underset{C_3H_7\text{-}i}{|}}{C}(CH_3)CH(CN)CO_2C_2H_5$$

7

The order of adding reagents can affect the course of conjugate addition substantially. Normal addition can be performed in two ways: (a) by adding the unsaturated substrate *to* a mixture of Grignard reagent *and* copper salt or (b) by adding unsaturated substrate *and* copper salt *to* the Grignard reagent (*i.e.*, stepwise addition of copper salt). The former is by far the more often used procedure. Stepwise, or portionwise, addition of copper catalyst was first shown by Munch-Petersen to cause a 10–30% increase in yield of conjugate adducts;[25] this method has been used sporadically, however, to diminish the amount of undesired 1,2 addition. Apparently each fresh portion of copper salt produces more organocopper reagent which enters the catalytic cycle.

Inverse addition involves adding the Grignard reagent *to* a mixture of unsaturated substrate *and* copper catalyst.[8] In many reactions inverse addition has led to a dramatic increase in yield of 1,4 adducts. Ethyl isopropylidene cyanoacetate, for example, reacts with *n*-butylmagnesium bromide and cuprous chloride to give conjugate adduct 8 in 42% yield by normal addition and in 89% yield by inverse addition.[14]

[36] A. J. Birch and M. Smith, *Proc. Chem. Soc.*, **1962**, 356.
[37] G. E. Coates and F. Glockling, *Organometallic Chemistry*, H. Zeiss, Ed., Reinhold Publishing Corp., New York, 1960, p. 446.
[38] W. E. Parham and L. J. Czuba, *J. Org. Chem.*, **34**, 1899 (1969).
[39] F. S. Prout and M. M. E. Abdulslam, *J. Chem. Eng. Data*, **11**, 616 (1966).

$$(CH_3)_2C\!\!=\!\!C(CN)CO_2C_2H_5 \xrightarrow[CuCl]{n\text{-}C_4H_9MgBr} (CH_3)_2\underset{\underset{8}{C_4H_9\text{-}n}}{C}CH(CN)CO_2C_2H_5$$

The nature of catalytic organocopper reagents has not been studied; their short lifetime and low concentration relative to Grignard reagent make examination difficult. Several kinds of catalytic organocopper reagents may exist, depending on the catalyst, the Grignard reagent, the solvent, and the reaction conditions used.

Stoichiometric Organocopper Reagents. *RCu.* The nature of organocopper RCu species depends strongly on the method and the reaction conditions used for its preparation; a good discussion of the difficulties in preparing pure stoichiometric RCu reagents is available.[35] Gilman and Straley have compiled a summary of the relative reactivities of some RCu species and have related them to other organometallic compounds.[17]

Two methods, the first general and the second of limited generality, have been used to form these RCu reagents: (1) metathetical reaction of an organometallic reagent with a salt of another metal (Eq. 6), and (2) thermal decarboxylation of cuprous carboxylates (Eq. 7).[40, 41] Pure RCu species have been prepared only via Eq. 6, although in principle they probably could also be prepared by cuprous carboxylate decarboxylation, which has so far not been widely used and may in practice be limited to preparation of arylcopper species only.

$$\text{R—Met} + \text{CuX} \longrightarrow \text{RCu} + \text{Met—X} \qquad (\text{Eq. 6})$$

$$\text{RCO}_2\text{Cu} \xrightarrow{\text{Solvent}} \text{RCuSolvent} + \text{CO}_2 \qquad (\text{Eq. 7})$$

For the metathesis represented in Eq. 6 to proceed, copper must be below Met in the electromotive series.[42] When Met is lead,[43] zinc,[44] magnesium, or lithium, this condition is fulfilled and RCu is formed as indicated along with Met—X. RCu free of zinc[44] or magnesium[29, 45] salts has been produced and isolated, and in general is a colored, explosive solid which is easily hydrolyzed and air-oxidized.[17, 45] Studies of the state of aggregation of several fluorinated arylcopper species have been performed.[45] Such salt-free RCu, however, does not undergo conjugate addition.[28–30]

---

[40] A. Cairncross, J. R. Roland, R. M. Henderson, and W. A. Sheppard, *J. Amer. Chem. Soc.*, **92**, 3187 (1970).
[41] T. Cohen and R. A. Schambach, *J. Amer. Chem. Soc.*, **92**, 3189 (1970).
[42] R. G. Jones and H. Gilman, *Chem. Rev.*, **54**, 847 (1954).
[43] H. Gilman and L. A. Woods, *J. Amer. Chem. Soc.*, **65**, 435 (1943).
[44] K. H. Thiele and J. Köhler, *J. Organometal. Chem.*, **12**, 225 (1968).
[45] A. Cairncross and W. A. Sheppard, *J. Amer. Chem. Soc.*, **90**, 2186 (1968).

Adding inorganic salts (*e.g.*, LiI,[28] LiBr[46]) to solutions of pure RCu usually restores to the organocopper reagent its ability to undergo conjugate addition. This effect has been interpreted in terms of the second mechanistic hypothesis discussed earlier: an organocuprate[21, 47] species (Eq. 8), because of the net negative charge on copper, can transfer an electron to the substrate more easily than can neutral RCu[28]; recent chemical[29, 30] and spectroscopic[28] evidence appears to confirm the presence of a mixed cuprate species such as **9** which undergoes effective conjugate addition.

$$\text{RCu} + \text{Met—X} \rightarrow [\text{RCuX}]^-\text{Met}^+ \qquad \text{(Eq. 8)}$$
$$\mathbf{9}$$

R_2CuLi. If, instead of a salt, RLi is added to RCu, then a lithium diorganocopper species represented as R_2CuLi is formed (Eq. 9).[16, 18] It should be noted that the R_2CuLi species can be formed either without (Eq. 9) or with (Eq. 10) a lithium salt. Although the presence of lithium salts has been shown to influence the coupling of lithium dimethylcopper with organic halides,[21] no such effect has been observed in conjugate addition reactions.[28] Since successful conjugate addition generally occurs with R_2CuLi in the presence of LiX, and since the reaction represented by Eq. 10 is easier to carry out experimentally than that represented by Eq. 9 (*i.e.*, salt-free organocopper must be prepared in a separate step), R_2CuLi species for conjugate additions have usually been prepared according to Eq. 10 from 2 equivalents of organolithium reagent and 1 equivalent of the appropriate copper halide. Details on the preparation and handling of lithium diorganocopper reagents, which have variable thermal stability depending on R, are given in the section "Experimental Factors."

$$\text{RCu} + \text{RLi} \rightarrow \text{R}_2\text{CuLi} \qquad \text{(Eq. 9)}$$
$$2\,\text{RLi} + \text{CuX} \rightarrow \text{R}_2\text{CuLi} + \text{LiX} \qquad \text{(Eq. 10)}$$

Generally, ethereal solutions of lithium diorganocopper reagents are easily prepared under nitrogen or argon at 0° (R = CH$_3$) or below —20° (R $\neq$ CH$_3$) from organolithium compounds and cuprous iodide or bromide in a 2:1 molar ratio. As the second equivalent of organolithium is added, the reaction mixture becomes essentially homogeneous. Frequently color development due to undesirable side reactions (*e.g.*, oxidation or thermal

---

[46] N. T. Luong-Thi and H. Rivière, *Compt. Rend.*, **267**, 776 (1968).

[47] W. Tochtermann [*Angew. Chem. Int. Ed. Engl.*, **5**, 351 (1966)] and G. Wittig [*Quart. Rev.* (London), **20**, 191 (1966)] have provided reviews of the properties and reactions of ate complexes.

decomposition)[48] accompanies formation of lithium diorganocuprates. Such colored solutions of organocopper reagents nevertheless have been routinely used with success in conjugate addition reactions. Formation of the lithium diorganocopper reagent is complete in less than 5 minutes at 0°, as shown for example by a negative Gilman test with Michler's ketone.[49]

Lithium diorganocopper reagents are substantially less basic and less nucleophilic than the corresponding organolithium species. Methyllithium partially metalates toluene to form benzyllithium,[50] but lithium dimethylcopper does not attack toluene and in fact can be formed in this solvent.[51] Methyllithium completely consumes saturated aliphatic ketones by carbonyl addition in less than 1 second at 25° in ether, whereas under the same conditions 2 minutes are needed for only 31% ketone consumption by lithium dimethylcopper.[16,52]

The nature of the diorganocopper reagent represented as R_2CuLi is unclear; the R_2CuLi notation itself was introduced merely to indicate the stoichiometry of the reaction represented in Eq. 10 and was not intended to convey structural information.[16] Neither the degree of aggregation nor the extent of solvation of these stoichiometric diorganocopper reagents is known. Nuclear magnetic resonance study of lithium dimethylcopper shows that "the methyl signal for the dimethylcopper anion remains sharp (half-bandwidth 1.8 cps) even at relatively low temperatures. From this observation and the shift in both the methyl signal and the ^{13}CH coupling constant as excess methyllithium is added, [it was concluded] that the equilibrium [represented in Eq. 10] lies far to the right; the sharpness of the signal also argues against the presence of substantial concentrations of paramagnetic copper(II) species in these solutions."[16]

An attempt to isolate lithium dimethylcopper by low-temperature evaporative distillation of the ether solvent gave initially a white solid which rapidly decomposed.[53]

The major organic side product in conjugate addition reactions with R_2CuLi reagents is often the symmetrical dimer RR.[32] This product does not usually constitute a serious limitation on the use of R_2CuLi species, for the dimer RR is often a gas and, when a liquid or solid, is usually easily separable from the polar carbonyl-containing conjugate adduct.

[48] E. J. Corey and I. Kuwajima, in unpublished results, have found that the dark color of ethereal lithium dibutylcopper, for example, is due to decomposition of the reagent; low-temperature centrifugation gave a colorless supernatant solution containing active organocopper reagent; cf. G. M. Whitesides, J. San Filippo, Jr., C. P. Casey, and E. J. Panek, J. Amer. Chem. Soc., **89,** 5302 (1967).
[49] H. Gilman and F. Schulze, J. Amer. Chem. Soc., **47,** 2002 (1925).
[50] J. M. Mallan and R. L. Bebb, Chem. Rev., **69,** 716 (1969).
[51] Gary Gaston de Grande, B.S. Thesis, University of Illinois, Urbana, Illinois, 1970.
[52] E. J. Corey and I. Kuwajima, J. Amer. Chem. Soc., **92,** 395 (1970).
[53] E. J. Corey and G. H. Posner, unpublished results.

RCu-Ligand. To solubilize the copper salt used to prepare the organocopper reagent and to stabilize this reagent once formed, a variety of heteroatom ligands have been used to complex or chelate with copper: amines,[44, 54] phosphines,[16, 54] phosphites,[28] and sulfides.[55] The detailed structure of these RCu-ligand species in solution is not known;[56] but ebullioscopic[57] and preliminary X-ray[58] measurements of the related tri-n-butylphosphine-cuprous iodide [$(n$-$C_4H_9)_3$PCuI][59] show this complex to be a tetramer, and recent X-ray work shows the triphenylphosphine-cuprous chloride complex to be a chlorine-bridged dimer.[60]

Also, recently, Luong-Thi and Rivière have suggested and attempted to show that the reactive entities in conjugate addition of stoichiometric organocopper reagents derived from XCu-ligand and RLi (Eq. 11) are mixed cuprates like 10 and not RCu-ligand species.[29, 30]

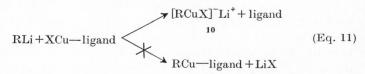

(Eq. 11)

In most instances of conjugate addition with a stoichiometric organocopper reagent, the RCu-ligand reagent is less useful than the R_2CuLi reagent mainly because the ligand is often released during the reaction, thus making product isolation more difficult.[28, 55] To some extent this difficulty is overcome with a phosphite ligand chosen to be partially water soluble.[28] Copper(I) sulfide complexes are nicely crystalline and thus may be easily purified.[55] Conjugate addition of R may be more economical with RCu-ligand reagents which allow complete utilization of R than with R_2CuLi reagents which transfer only one of the two R groups.[32, 54]

Of the one catalytic and three stoichiometric organocopper reagents just discussed, only catalytic organocopper reagents were used for conjugate addition between 1941 and 1966. In 1966 the ability of the stoichiometric organocopper reagents to undergo conjugate addition was first conclusively demonstrated,[61] and since then both types of organocopper

[54] J. B. Siddall, M. Biskup, and J. H. Fried, *J. Amer. Chem. Soc.*, **91**, 1853 (1969).
[55] H. O. House and W. F. Fischer, Jr., *J. Org. Chem.*, **34**, 3615 (1969).
[56] G. Costa, A. Camus, N. Marsich, and L. Gatti, *J. Organometal. Chem.*, **8**, 339 (1967).
[57] F. G. Mann, D. Purdie, and A. F. Wells, *J. Chem. Soc.*, **1936**, 1503.
[58] A. F. Wells, *Z. Kristollogr.*, **94**, 447 (1936).
[59] G. B. Kauffman and L. A. Teter, *Inorg. Syn.*, **7**, 9 (1963).
[60] D. F. Lewis, S. J. Lippard, and P. S. Welcker, *J. Amer. Chem. Soc.*, **92**, 3805 (1970).
[61] The only exception to this statement is the isolated observation by H. Gilman, R. G. Jones, and L. A. Woods [*J. Org. Chem.*, **17**, 1630 (1952)] that phenylcopper adds in a conjugate manner to benzalacetophenone.

reagents have been used. Continually more emphasis is being given now to the stoichiometric organocopper reagents, presumably because they give conjugate adducts in more reproducible fashion and often in higher yields and with greater stereoselectivity than do the corresponding catalytic organocopper reagents.

Structural Variation of R

The aim in this section is to enumerate all of the kinds of R groups that have undergone organocopper conjugate addition. The structural types of R will be considered in terms of the formal hybridization of the carbon bound to copper: sp^3, sp^2, sp, and special cases (e.g., allylic, homoallylic, and fluorinated R). References to only the more unusual examples will be given. In later sections both the effect of the structure of R on the course of conjugate addition and the choice of an organocopper reagent (e.g., catalytic or stoichiometric) for conjugate addition of a particular type of R are discussed.

sp^3-Hybridized primary R from methyl virtually through n-dodecyl and including isobutyl[62] and isopentyl[63] have all undergone conjugate addition. Of these, only methyl,[16, 51, 64] ethyl,[65] n-butyl,[65] and n-heptyl[32, 64] have been added via stoichiometric organocopper reagents; this limited variety of R having undergone *stoichiometric* organocopper conjugate addition is due presumably to the novelty of these reagents and not to any fundamental limitation in their reactivity.

Isopropyl,[34, 66] cyclopentyl,[65] and t-butyl[34] are the only examples of secondary and tertiary sp^3-hybridized R having undergone conjugate addition; catalytic organocopper reagents have been used most often in these cases. Conjugate addition of t-butyl and cyclopentyl via stoichiometric organocopper reagents has just recently been achieved.[65]

Phenyl[46, 67] and various substituted phenyls[68, 69] have also all undergone organocopper conjugate addition. Vinylic[70-72] R has thus far undergone mainly catalytic organocopper conjugate addition; lithium

[62] L. Mandell and J. M. Brodmann, *J. Org. Chem.*, **31**, 591 (1966).
[63] W. H. Hook and R. Robinson, *J. Chem. Soc.*, **1944**, 152.
[64] E. J. Corey and J. A. Katzenellenbogen, *J. Amer. Chem. Soc.*, **91**, 1851 (1969).
[65] C. A. Henrick and J. B. Siddall, unpublished results.
[66] H. Rivière and J. Tostain, *Bull. Soc. Chim. Fr.*, **1969**, 568.
[67] J. Klein and R. M. Turkel, *J. Amer. Chem. Soc.*, **91**, 6186 (1969).
[68] R. Filler and Y. S. Rao, *J. Org. Chem.*, **27**, 3348 (1962).
[69] H. O. House and R. W. Bashe, *J. Org. Chem.*, **32**, 784 (1967).
[70] H. O. House, R. A. Latham, and C. D. Slater, *J. Org. Chem.*, **31**, 2667 (1966).
[71] E. Piers, W. de Waal, and R. W. Britton, *Can. J. Chem.*, **47**, 4299 (1969).
[72] E. Piers, R. W. Britton, and W. de Waal, *Can. J. Chem.*, **47**, 4307 (1969).

dipropenylcopper has only recently been shown to be useful in 1,4 addition to 2-cyclohexenone.[73]

Neither cuprous acetylides[74] nor lithium acetylenic-copper reagents[55] add in a conjugate manner to unsaturated carbonyl derivatives.[74a]

Allylic and benzylic R generally do not undergo more effective conjugate addition with organocopper than with Grignard reagents,[55, 62] but homoallylic RCu[32] and RCu-phosphine species do.[54, 65] Although perfluoroalkyl-[75] and perfluoroaryl-copper[45] compounds are known (and are more stable than the corresponding hydrocarbon species), conjugate addition of these stoichiometric organocopper reagents has not been studied.

Stereochemical Stability of R

Only one example of an sp^3-hybridized R capable of existing as one or both of two epimers has been reported; endo-2-norbornylcopper-tri-n-butylphosphine complex 11 is configurationally stable at $-78°$ in ether and undergoes highly stereoselective coupling reactions.[76] Conjugate addition has been achieved recently with this stoichiometric RCu-ligand reagent.[76b, c]

CuP(C$_4$H$_9$-n)$_3$
11

Several examples are known of sp^2-hybridized R in which R is configurationably stable. Tri-n-butylphosphine complexes of cis- and trans-1-propenylcopper and cis- and trans-2-butenylcopper undergo stereospecific thermal dimerizations[77, 78] and stereospecific reduction with cuprous hydride,[76] but they have not been treated with α,β-unsaturated carbonyl compounds. Conjugate addition of lithium di-cis- and di-trans-1-propenylcopper to 2-cyclohexenone has recently been shown to be stereospecific.[73]

[73] (a) C. P. Casey and R. A. Boggs, Tetrahedron Lett., 1971, 2455; (b) see also conjugate addition of divinylcopperlithium tri-n-butylphosphine complexes: J. Hooz and R. B. Layton, Can. J. Chem., 48, 1626 (1970); (c) Conjugate addition of divinylcopperlithium itself has been reported also: E. J. Corey and R. L. Carney, J. Amer. Chem. Soc., 93, 7318 (1971).

[74] W. S. Johnson, N. P. Jensen, J. Hooz, and E. J. Leopold, J. Amer. Chem. Soc., 90, 5872 (1968).

[74] (a) Conjugate addition of acetylene units has recently been achieved with organoaluminum reagents: J. Hooz and R. B. Layton, J. Amer. Chem. Soc., 93, 7320 (1971).

[75] V. C. R. McLoughlin and J. Thrower, Tetrahedron, 25, 5921 (1969).

[76] (a) G. M. Whitesides, J. San Filippo, Jr., E. R. Stedronsky, and C. P. Casey, J. Amer. Chem. Soc., 91, 6542 (1969); (b) G. M. Whitesides, 162nd Meeting of the American Chemical Society, Div. Inorg. Chem., Paper No. 50, Sept. 1971, Washington, D.C.; (c) G. M. Whitesides and P. E. Kendall, J. Amer. Chem. Soc., in press.

[77] (a) G. M. Whitesides and C. P. Casey, J. Amer. Chem. Soc., 88, 4541 (1966); (b) G. M. Whitesides, C. P. Casey, and J. K. Krieger, ibid., 93, 1379 (1971).

[78] T. Kauffman and W. Sahm, Angew. Chem. Int. Ed. Engl., 6, 85 (1967).

α-Carbalkoxy and α-carboxy vinylic R (*e.g.*, **12**) have been prepared *in situ* from conjugate addition of organocopper reagents to acetylenic esters and acids.[54, 64, 67] The configurational stability of these presumably C-copper enolates is inferred from the high stereochemical purity (*cis*) of the products formed upon hydrolysis, upon deuterium oxide quenching,[54] upon iodination,[64] and upon alkylation[65] (see p. 61). Time, temperature, solvent, complexing additives, and the nature of the organocopper reagent used to prepare vinylic copper species **12** are key factors in controlling the stability of these vinylic intermediates.

$$RCu + R'C \equiv CCO_2CH_3 \longrightarrow \underset{\underset{\mathbf{12}}{}}{\overset{R'}{\underset{R}{>}}C=C\overset{CO_2CH_3}{\underset{Cu}{<}}} \qquad \text{(Eq. 12)}$$

a, $R = CH_3$, $R' = C_2H_5$
b, $R = n\text{-}C_7H_{15}$, $R' = CH_3$
c, $R = CH_3$, $R' = n\text{-}C_7H_{15}$

In ether at $-78°$ for 2 minutes, vinylic copper species **12a**, for example, can be hydrolyzed to give a 92:8 *cis:trans* ratio of 2-pentenoates, whereas under the same conditions for 60 minutes the 2-pentenoate *cis:trans* ratio upon hydrolysis is only 85:15; this result was interpreted to mean that C-copper enolate **12** slowly equilibrates with its geometrical isomer.[54] A similar result and interpretation have been reported for copper enolate **12b**.[32]

The configurational stability of vinylic copper species **12c** in tetrahydrofuran has been studied at various temperatures ranging from -78 to $0°$. Only when the temperature is allowed to rise above $-30°$ is there a small amount of isomerization; at $0°$ for 10 minutes a total of 29% isomerization has occurred.[32] Furthermore 2-pentenoate **13** is formed in ether from lithium dimethylcopper and methyl 2-pentynoate in a 92:8 *cis:trans* ratio at $-78°$ and in a 97:3 *cis:trans* ratio at $-100°$.[54]

$$(CH_3)_2CuLi + C_2H_5C \equiv CCO_2CH_3 \rightarrow \underset{\mathbf{13}}{C_2H_5C(CH_3){=}CHCO_2CH_3}$$

The configurational stability of the intermediate sp^2-hybridized organocopper species decreases as the solvent is changed from tetrahydrofuran, to toluene, to diethyl ether. Thus the ratio **12c:12b** is 99.8:0.2 in tetrahydrofuran, 92.5:7.5 in toluene, and 24:76 in ether, as determined by gas chromatography after a 2–3 hour reaction time.[32, 64]

Complexing additives such as tetramethylethylenediamine (TMED), trimethyl phosphite, and pyrrolidine tend to stabilize vinylic copper species. Thus after 15 minutes the ratio **12c:12b** is 35:65 in ether, 92:8 in ether containing TMED, and 79:21 in ether containing trimethyl phosphite.[32,64]

Finally, in one study the nature of the organocopper reagent which adds to the acetylenic carbonyl compound has apparently been found to influence the configurational stability of the intermediate vinylic copper species. Phenylpropiolic acid reacts with methylcopper to give *cis* adduct 14 but with lithium dimethylcopper to give *trans* adduct 14. Furthermore addition of methyllithium to the reaction mixture of methylcopper and phenylpropiolic acid gives *trans* adduct 14.[67] Both organocopper reagents are assumed to generate the same vinylic copper species initially, which isomerizes only when lithium dimethylcopper is used.[67]

$$C_6H_5C(CH_3)\!=\!CHCO_2H \qquad (cis = CH_3 \text{ and } H \text{ adjacent})$$
$$\mathbf{14}$$

No explicit reports have appeared on conjugate addition of α-carbalkoxy or α-carboxy vinylic R.

The positional stability of copper or lithium enolates formed via organocopper conjugate addition to enones has been studied. Such structurally specific enolates react with acetyl chloride, acetic anhydride[16] (Eq. 13), or diethyl phosphorochloridate (Eq. 14)[79] to give pure enol derivatives not contaminated by double bond isomers.

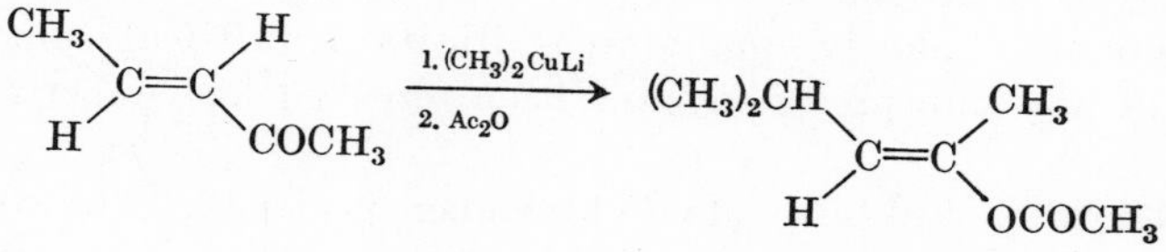

(Eq. 13)[16]

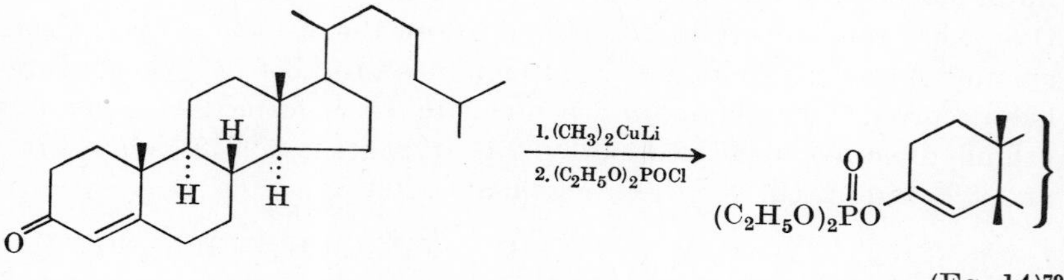

(Eq. 14)[79]

Effect of Structure on Amount and Stereochemistry of Conjugate Addition

Structure of Reagent. Catalytic and stoichiometric organocopper reagents often undergo conjugate addition to different extents and with different stereochemical results. In three of the four cases where direct comparison is possible—for methyl, *n*-alkyl, and phenyl but not allylic

[79] R. E. Ireland and G. Pfister, *Tetrahedron Lett.*, **1969**, 2145; *cf.* D. C. Muchmore, *Org. Syntheses*, **51**, procedure no. 1734 (1971).

groups—stoichiometric organocopper reagents generally produce 1,4 adducts in higher yields (*i.e.*, less or no 1,2 addition) and with greater stereoselectivity than do catalytic organocopper reagents.

Thus octalone 15 reacts with lithium dimethylcopper [$(CH_3)_2CuLi$] at 0° in ether to give *cis*-decalone 16 in 77% yield, whereas octalone 15 with cupric acetate and methylmagnesium iodide gives essentially no 1,4 adduct 16.[80] A similar result was observed for octalone 17; reaction with methylmagnesium iodide and cuprous acetate gives *cis*-decalone 18 in 16–21% yield, whereas the yield of this 1,4 adduct 18 is doubled by using lithium dimethylcopper.[81] α,β-Unsaturated enone 19 is converted by methylmagnesium iodide and cuprous acetate to a 56:34 *mixture* of a tertiary alcohol (1,2 adduct) and β-methylated ketone 20, whereas enone 19 reacts with lithium dimethylcopper to give ketone 20 in 100% yield.[38] α,β-Acetylenic ester 21 is converted to its 1,4 adduct 22 in 73% yield by cuprous chloride catalysis of *n*-butylmagnesium bromide,[10] and ester 23 to its conjugate adduct 24 by lithium di-*n*-heptylcopper in 90% yield.[64]

15— R = CH_3, R' = H —16
17— R = H, R' = CH_3 —18

$CH_3C\equiv CCO_2C_4H_9$-*s* $\xrightarrow{\substack{n\text{-}C_4H_9MgBr \\ CuCl}}$ $CH_3C=CHCO_2C_4H_9$-*s*
                                                              |
                                                       C_4H_9-*n*

21   22

$CH_3C\equiv CCO_2CH_3$ $\xrightarrow{(n\text{-}C_7H_{15})_2CuLi}$ $CH_3C=CHCO_2CH_3$
                                                           |
                                                    C_7H_{15}-*n*

23   24

Concerning stereoselectivity of conjugate addition, 5-methyl-2-cyclohexenone (25) reacts with lithium dimethylcopper to give 3,5-di-methylcyclohexanone (26) having a *trans*:*cis* ratio of 98:2 (98% yield); reaction of the same enone 25 with methylmagnesium bromide and 1 mol%

[80] E. Piers and R. J. Keziere, *Tetrahedron Lett.*, **1968**, 583.
[81] J. A. Marshall and H. Roebke, *J. Org. Chem.*, **33**, 840 (1968).

cuprous iodide gives conjugate adduct **26** having a *trans:cis* ratio of 93:7 (>90% yield).[28]

<img: reaction of 25 with (CH₃)₂CuLi giving 26 trans + 26 cis>

A more striking example of the higher stereoselectivity achieved by stoichiometric organocopper reagents appears in the conjugate addition of phenyl to 4-methyl-2-cyclohexenone (**27**).[46] Cuprous chloride-promoted reaction of phenylmagnesium bromide with enone **27** gives 4-methyl-3-phenylcyclohexanone (**28**, R = C_6H_5) having a *trans:cis* ratio of 87:13 (75% yield); reacting with the same substrate **27**, phenylcopper (prepared from phenyllithium and cuprous iodide) apparently formed ketone **28**, R = C_6H_5, with a *trans:cis* ratio of 96:4 (100% yield).[46]

<img: reaction of 27 with RCu or R₂CuLi giving 28>

Within the class of stoichiometric organocopper reagents also, product yields and stereochemistry often depend on the nature of the reagent used: RCu, R₂CuLi, or RCu-ligand. The yield of conjugate adduct **20**, for example, is 100% with lithium dimethylcopper but only 82% with the tri-*n*-butylphosphine complex of methylcopper,[38] and 1,4 adduct **28**, R = CH_3, is formed also in 100% yield with lithium dimethylcopper but in 80% yield with methylcopper.[46]

The stereoselectivity of conjugate addition to acetylenic carbonyl compounds as a function of organocopper reagent has not been thoroughly studied. Generally each of the three stoichiometric organocopper reagents adds to alkylacetylenic esters in high yield to give predominantly overall *cis* addition of R and Cu.[54, 64, 65] Exceptions have been noted for β-phenylacetylenic esters and acids which form 1,4 adducts the stereochemistry of which depends on the organocopper reagent used.[67] (See p. 18).

No explanation or rationalization has appeared to account for the generally greater stereoselectivity of R₂CuLi reagents over RCu and RCu-ligand species in conjugate addition reactions.

Structure of R. Within the class of stoichiometric organocopper reagents, relatively little information is available about the effect of the structure of R on the yield and stereochemistry of conjugate addition. Perhaps the most valid generalization is that allylic[55, 82] and benzylic[83] R groups tend to cause the organocopper reagent to react much like the corresponding Grignard reagent, RMgX.

In contrast, several comprehensive studies have been done on the extent to which the structure of R in catalytic organocopper reagents affects the yield and particularly the stereochemistry of conjugate addition. The structure of R often does affect the amount and the stereochemistry of 1,4 addition to unsaturated carbonyl compounds. Octalone **2**, for example, undergoes copper catalytic conjugate addition of methyl in 60–80% yield[5, 36, 84] and of more bulky phenyl in 25% yield.[85] Likewise, 10-methylhexalone **29** produces conjugate adducts **30** in diminishing yields generally as the size of R increases: R = ethyl (55–65%), R = isopropyl (58%), R = t-butyl (40%), and R = phenyl (0).[34] Furthermore, as the size of R increases, the $\alpha:\beta$ ratio of R in 1,6 adduct **30** increases from 93:7 for R = methyl to 98:2 for R = ethyl, and to 100:0 for R = t-butyl or isopropyl.[34] Likewise, 4-methyl-2-cyclohexenone **(27)** reacts with RMgBr in the presence of cuprous chloride to give *trans*- and *cis*-3,4-dialkylcyclohexanones; as the size of R increases, the *trans*:*cis* ratio also increases: from 72:28 for R = methyl, to 78:22 for R = ethyl, to 88:12 for R = isopropyl.[66]

The ability of R to influence the course of catalytic conjugate addition has been analyzed in terms of steric and electronic factors[86, 87] and in terms of ground state and transition state considerations.[34, 66, 88] Since the role of

[82] R. E. Ireland and D. R. Marshall, unpublished results.
[83] R. E. Ireland and S. W. Baldwin, unpublished results.
[84] J. A. Marshall, W. I. Fanta, and H. Roebke, *J. Org. Chem.*, **31**, 1016 (1966).
[85] S. M. McElvain and D. C. Remy, *J. Amer. Chem. Soc.*, **82**, 3960 (1960).
[86] E. Toromanoff, *Bull. Soc. Chim. Fr.*, **1962**, 707.
[87] N. L. Allinger and C. K. Riew, *Tetrahedron Lett.*, **1966**, 1269.
[88] J. A. Marshall and N. H. Andersen, *J. Org. Chem.*, **31**, 667 (1966).

copper in these reactions is unclear and since copper may complex with the unsaturated substrate, conformational analyses of pure chair or boat structures may not be relevant. In any case, no complete explanation of the effect of R on the course of organocopper conjugate addition has been reported.

Selection of Reagent for Conjugate Addition

Choosing a reagent for any chemical transformation requires evaluation of several basic characteristics of the product of the reaction and of the reagent. Ideally the desired product should be formed in high yield and with good stereochemical purity, and it should not react further with the reagent, nor should it be formed along with side products (*e.g.*, ligand from RCu-ligand) which make product isolation difficult. The reagent of choice should be easy to prepare and to handle, and it should give reproducible results. On the basis of these considerations, selection of an organocopper reagent for conjugate addition of a particular type of R to α,β-unsaturated carbonyl substrates can be discussed. Organocopper reagents having sp^3- and sp^2-hybridized R as well as allylic and benzylic R are treated first; then the discussion turns to choosing an organocopper reagent for conjugate addition specifically to sterically hindered substrates.

Conjugate addition of sp^3-hybridized primary R is generally best achieved using lithium diorganocopper reagents. These stoichiometric R_2CuLi reagents meet most of the criteria (yield, stereochemistry, absence of contaminating side products, ease of preparation and handling, and reproducibility) better than any of the other organocopper reagents.

Secondary and tertiary sp^3-hybridized R have been added in a conjugate manner mainly via catalytic organocopper reagents. Only recently have cyclopentyl and *t*-butyl R in stoichiometric organocopper reagents (RCu-ligand and RR'CuLi) undergone highly stereoselective 1,4 addition to acetylenic esters in good yield.[65] The success of these conjugate additions suggests that, when yield or stereochemistry is important, 1,4 addition of secondary and tertiary sp^3-hybridized R be attempted first with stoichiometric organocopper reagents and, if these fail, then with catalytic organocopper reagents. Copper catalysis of Grignard conjugate addition can best be achieved using either cuprous chloride which is added in portions during the reaction,[14, 25] or inverse addition of Grignard reagent to a soluble copper salt such as cuprous cyanide in ether[39] or cupric acetate in tetrahydrofuran.[36, 38]

Aryl R, like sp^3-hybridized primary R, generally undergoes most effective conjugate addition as a lithium diarylcopper reagent.[46, 67] Vinylic R, however, has undergone mainly catalytic organocopper conjugate addition with cuprous chloride,[71, 72] cupric acetate,[89] or cuprous iodide[70] catalyst in tetrahydrofuran. (Compare ref. 73.)

For conjugate addition of allylic and benzylic R, organocopper reagents generally offer no advantage over the corresponding Grignard reagents.[55, 90–92] The only two reported exceptions to this generalization are the conjugate addition of lithium diallylcopper to cyclohexenone[55] and copper-catalyzed conjugate addition of m-methoxybenzylmagnesium chloride to methyl vinyl ketone.[93]

Homoallylic R has undergone effective conjugate addition only via stoichiometric RCu[32] and RCu-ligand reagents.[54, 65]

Conjugate addition to sterically crowded α,β-unsaturated carbonyl groups generally proceeds much better with lithium diorganocopper (R_2CuLi) reagents than with other stoichiometric or with catalytic organocopper reagents (see p. 19).

The α,β-Unsaturated Carbonyl Substrate

In this section the scope and limitations of conjugate addition are discussed in terms of the α,β-unsaturated carbonyl substrate. α,β-Unsaturated carbonyl compounds can be classified by the degree of unsaturation: α,β-ethylenic or α,β-acetylenic. The discussion is focused first on α,β-ethylenic carbonyl derivatives: acyclic enones, acyclic enoate esters, cyclic enones, and polyethylenic esters and ketones. For each class of substrate discussion is centered on a parent carbon skeleton and the way substitution on that skeleton affects the rate and course of conjugate addition. After discussion of α,β-acetylenic carbonyl species, several miscellaneous classes of unsaturated substrates are treated.

α,β-Ethylenic Carbonyl Derivatives

Within the class of α,β-ethylenic carbonyl derivatives no data have been reported for either catalytic or stoichiometric organocopper conjugate addition to ethylenic anhydrides or amides. α,β-Ethylenic carboxylic acids generally do not undergo copper-catalyzed or stoichiometric organocopper conjugate addition; α-bromocrotonic acid (**31**) reacts with phenylmagnesium bromide to give a 1:1 mixture of *erythro*- and *threo*-acid **32** in 70% yield, and a catalytic amount of cuprous chloride does not change

---

[89] H. Roebke, Ph.D. Thesis, Northwestern University, Evanston, Illinois, 1969; *Diss. Abstr.*, **30**, 3104-B (1970).
[90] M. Kawana and S. Emoto, *Bull. Chem. Soc. Jap.*, **39**, 910 (1966).
[91] J. Munch-Petersen, P. Moller Jorgensen, and S. Refn, *Acta Chem. Scand.*, **13**, 1955 (1959).
[92] J. W. McFarland and D. N. Buchanan, *J. Org. Chem.*, **30**, 2003 (1965).
[93] S. Danishefsky and B. H. Migdalof, *Chem. Commun.*, **1969**, 1107.

the yield or the distribution of products.[94] Furthermore, the β,β-disubstituted acid **33** fails to react with lithium dimethylcopper.[16]

$$trans\text{-}CH_3CH=C(Br)CO_2H \xrightarrow[(CuCl)]{C_6H_5MgBr} CH_3CHCH(Br)CO_2H$$
$$\qquad\qquad\qquad\qquad\qquad\qquad\qquad\qquad\qquad\;\;|$$
$$\qquad\qquad\qquad\qquad\qquad\qquad\qquad\qquad\;\;C_6H_5$$
$$\quad\;\;\textbf{31}\qquad\qquad\qquad\qquad\qquad\qquad\qquad\qquad\textbf{32}$$

$$t\text{-}C_4H_9-\!\!\left\langle\;\;\right\rangle\!\!=CHCO_2H$$
$$\textbf{33}$$

Ethylenic aldehydes have not been widely studied, presumably because of their well-established tendency to undergo 1,2 addition with most reactive organometallic reagents.[1] Nevertheless, the α,β-disubstituted acrolein **34** reacts with lithium dimethylcopper in toluene at $-78°$ to give 1,4 adduct **35** only slightly contaminated by the expected 1,2 addition product.[51]

$$trans\text{-}n\text{-}C_6H_{13}CH=C(CH_3)CHO \xrightarrow{(CH_3)_2CuLi} n\text{-}C_6H_{13}CH(CH_3)CH(CH_3)CHO$$
$$\qquad\qquad\qquad\textbf{34}\qquad\qquad\qquad\qquad\qquad\qquad\qquad\qquad\textbf{35}$$

By far the most thoroughly studied classes of reactive ethylenic substrates are ethylenic ketones and esters. These classes are discussed separately and then compared with each other.

A summary is given below of the various ethylenic carbonyl substrates (**36**) in terms of (**a**) successful conjugate addition, (**b**) unsuccessful conjugate addition, and (**c**) no data reported for organocopper or copper-catalyzed conjugate addition.

$$-C=C-C=O$$
$$\;\;|\;\;\;\;|\;\;\;\;\;|$$
$$\qquad\;\;\;\;Y$$
$$\textbf{36}$$

a, Y = H, C, OR
b, Y = OH
c, Y = NR$_2$, OCOR

Acyclic Enones. The four-carbon skeleton of methyl vinyl ketone (**37**) is taken as the parent structure for discussion of acyclic enones. These enones react rapidly with organocopper reagents, conjugate addition being complete in less than 0.1 second at 25°.[16] Through forthcoming examples we shall see that alkyl substituents at the α, α' or β positions of parent enone **37** may cause only mild alteration in the rate or course of conjugate addition. If lower reaction temperatures, shorter reaction times, and more highly substituted substrates were used, perhaps structural effects on reactivity would be revealed more clearly.

[94] J. Klein and S. Zitrin, *J. Org. Chem.*, **35**, 666 (1970).

Methyl vinyl ketone (37) reacts with m-methoxybenzylmagnesium chloride in the presence of cuprous chloride to give 1,4 adduct 38.[93] β-Methyl-substituted enone 4 is transformed into 4-methyl-2-pentanone by copper catalysis of methylmagnesium bromide (73–82 %)[16, 31] and by stoichiometric methylcopper reagents (85–96 %).[16] Complete α' substitution does not substantially retard conjugate addition; lithium dimethylcopper adds to t-butyl enone 39 to give 1,4 adduct 40 in 79 % yield.[55]

$$\overset{\beta}{C}H_2=\overset{\alpha}{C}HCO\overset{\alpha'}{C}H_3 \xrightarrow[\text{CuCl}]{m\text{-CH}_3OC_6H_4CH_2MgCl} m\text{-CH}_3OC_6H_4(CH_2)_3COCH_3$$
$$\qquad\quad 37 \qquad\qquad\qquad\qquad\qquad\qquad\qquad 38$$

$$trans\text{-CH}_3CH=CHCOCH_3 \xrightarrow[\text{CuX}]{CH_3MgBr} (CH_3)_2CHCH_2COCH_3$$
$$\qquad\qquad 4$$

$$trans\text{-}t\text{-}C_4H_9CH=CHCOC_4H_9\text{-}t \xrightarrow{(CH_3)_2CuLi} t\text{-}C_4H_9CH(CH_3)CH_2COC_4H_9\text{-}t$$
$$\qquad\qquad\qquad 39 \qquad\qquad\qquad\qquad\qquad\qquad 40$$

Two β-alkyl substituents appear to hinder slightly some organocopper conjugate addition to enones. Lithium dimethylcopper addition to enone 41a gave conjugate adduct 42a (96 %);[16] in the copper-catalyzed methyl Grignard addition to the same substrate, however, addition was 74 % 1,4 and 26 % 1,2.[16] Incursion of the 1,2 addition mode, typical of Grignard addition to this substrate,[16] suggests that usually rapid organocopper conjugate addition was retarded by the two β substituents, allowing normal Grignard 1,2 addition to compete to the extent of 26 %.

41
a, R = CH$_3$
b, R = OC$_2$H$_5$

42
a, R = CH$_3$
b, R = OC$_2$H$_5$

Acyclic Enoate Esters. As might be expected from consideration of the lower electrophilicity of enoates relative to enones, enoates are substantially less reactive than enones toward nucleophilic organocopper reagents. Enoate 43b, for example, reacts with lithium dimethylcopper in ether at 0° for 44 hours only to the extent of 8 %, whereas the corresponding enone 43a is completely consumed under the same conditions.[51] Furthermore, although enoate 41b is inert to lithium dimethylcopper, structurally similar enone 41a reacts to give saturated ketone 42a in 96 % yield.[16] The substantially higher reactivity of ethylenic ketones relative to ethylenic esters might possibly suggest that these two substrates react with organocopper reagents by two different mechanisms, as has been proposed for conjugate additions of Grignard reagents to enones and enoates.[95]

$$n\text{-}C_6H_{13}CH{=}C(CH_3)COR$$
43a, R = CH_3
43b, R = OC_2H_5

The lower reactivity of enoates compared to that of enones makes possible a better elucidation of the basic structure-activity relationships in enoates. The same or similar factors presumably operate also in enones whose inherently higher reactivity partially masks these relationships. The lower reactivity of enoates, however, limits their synthetic usefulness because the more highly substituted esters do not undergo conjugate addition even with the more effective organocopper reagents, e.g., $(CH_3)_2CuLi$.

Acrylate **44** will be taken as the parent system. The course of reaction between an organometallic reagent and an enoate ester is often controlled by changing the structure of R, the alcohol portion of the ester; large R groups tend to diminish carbonyl addition, which is known to be acutely subject to steric hindrance.[96] Munch-Petersen has made extensive use of sec-butyl groups in enoate esters to guide Grignard reagents effectively toward 1,4 addition; ethyl in place of sec-butyl causes 1,2 addition to predominate. When ethyl enoates are treated with methyl Grignard reagents in the presence of a copper catalyst or with stoichiometric methylcopper reagents, however, mainly 1,4 addition takes place. Thus β-substituted ethyl enoate **45** is converted by methylmagnesium bromide in the presence of cuprous chloride to 1,4 adduct **46** in good yield,[13, 97] and ethyl enoate **47** undergoes only 1,4 addition with lithium dimethylcopper.[51]

$$\overset{\beta}{C}H_2{=}\overset{\alpha}{C}HCO_2R$$
44

$$trans\text{-}CH_3CH{=}CHCO_2C_2H_5 \xrightarrow[\text{CuCl}]{CH_3MgBr} (CH_3)_2CHCH_2CO_2C_2H_5$$
45 **46**

$$trans\text{-}n\text{-}C_6H_{13}CH{=}CHCO_2C_2H_5 \xrightarrow{(CH_3)_2CuLi} n\text{-}C_6H_{13}CH(CH_3)CH_2CO_2C_2H_5$$
47

An attempt to achieve conjugate addition of n-alkyl groups by cuprous chloride catalysis of n-alkyl Grignard reagents succeeded for ethyl crotonate but not for ethyl isocrotonate (cis isomer) or for ethyl cinnamate.[98] No explanation of these results has been given, nor has a comparable study been reported with stoichiometric n-alkylcopper reagents.

[95] T. Holm, *Acta Chem. Scand.*, **19**, 1824 (1965).
[96] M. S. Newman in *Steric Effects in Organic Chemistry*, M. S. Newman, Ed., John Wiley & Sons, Inc., New York, 1956, Ch. 4.
[97] J. Munch-Petersen, *Acta Chem. Scand.*, **12**, 2007 (1958).
[98] J. Munch-Petersen, *Acta Chem. Scand.*, **12**, 2046 (1958).

β,β Disubstitution prevents conjugate addition of methyl to enoate esters;[24b] addition of other alkyl groups has not been explored. Thus enoate **41b** is inert to lithium dimethylcopper,[16] and β,β-dimethyl enoate **48** does not undergo 1,4 addition with methylmagnesium bromide and cuprous chloride.[97] An organocopper reagent formed *in situ* from a Grignard reagent and a catalytic amount of a cuprous salt competes with the Grignard reagent for the unsaturated ester substrate; hence, when conjugate addition by the organocopper reagent is sterically hindered by β,β disubstitution, normal Grignard 1,2 addition predominates. The 1,2 addition product, however, is an α,β-unsaturated ketone which, as discussed earlier, is readily consumed by organocopper reagent. The overall reaction produces a saturated ketone, a usual side product in copper-catalyzed Grignard additions to β,β-disubstituted enoate esters (*e.g.*, Eq. 15).

$$(CH_3)_2C\!=\!CHCO_2C_4H_9\text{-}s \xrightarrow[CuCl]{CH_3MgBr} [(CH_3)_2C\!=\!CHCOCH_3]$$
$$\text{48} \qquad\qquad\qquad\qquad \downarrow$$
$$(CH_3)_3CCH_2COCH_3 \qquad \text{(Eq. 15)}$$

α,β Disubstitution prevents conjugate addition to some but not all enoate esters. α,β-Dialkyl substitution retards methyl conjugate addition in both copper-catalyzed Grignard and stoichiometric organocopper reactions with ethyl and *sec*-butyl enoate esters (Eqs. 16, 17).

$$trans\text{-}C_6H_{13}CH\!=\!C(CH_3)CO_2C_2H_5 \xrightarrow[0°, 44\text{ hr}]{(CH_3)_2CuLi} 8\%\text{ reaction} \qquad \text{(Eq. 16)}^{51}$$
43b

$$trans\text{-}CH_3CH\!=\!C(CH_3)CO_2R \xrightarrow[CuCl]{CH_3MgBr} (CH_3)_2CHCH(CH_3)COCH_3$$
$$\qquad\qquad\qquad\qquad\qquad\qquad\qquad\qquad\qquad \text{(Eq. 17)}^{7,\ 97}$$
$R = C_2H_5, s\text{-}C_4H_9 \qquad\qquad (1,2 + 1,4\text{-addition})$

Conjugate addition of other alkyl groups to α,β-dialkyl-substituted enoates has not been studied using stoichiometric organocopper reagents; through copper catalysis of *n*-butylmagnesium bromide, ethyl enoate **49a**[98] undergoes 1,2 addition but *sec*-butyl enoate **49b**[7] undergoes 1,4 addition. On the basis of these results, *n*-alkylcopper reagents, being more reactive than methylcopper reagents, will probably be found to add 1,4 to α,β-dialkyl-substituted enoate esters.

$$trans\text{-}CH_3CH\!=\!C(CH_3)CO_2R$$
49a, $R = C_2H_5$
49b, $R = C_4H_9\text{-}s$

Since β-monosubstituted enoates undergo methyl conjugate addition but α,β-dialkyl-substituted enoates do not, the α substituent must contribute to this inactivity. The relative importance of the steric and electronic effects of the α substituent in retarding conjugate addition has not been clearly ascertained. Munch-Petersen attempted to resolve this problem by examining the course of cuprous chloride promoted Grignard conjugate addition to α,β-dialkyl substituted enoates in which the α substituent was varied from, for example, n-butyl to isobutyl to sec-butyl (Eq. 18).[14] The n- and iso-butyl esters gave conjugate adducts in high yields, but the α-sec-butyl enoate **50c** gave mainly a saturated aldehyde, the product presumably of 1,2 reduction and then 1,4 addition. Munch-Petersen explained these results in terms of steric effects. He assumed that, when the α substituent is secondary, both 1,2 and 1,4 addition are sufficiently impeded to make carbonyl reduction, which is less sterically demanding than carbonyl addition, predominate. Unfortunately none of these α-substituted crotonates was treated with a stoichiometric organocopper reagent, and therefore it is not clear what the reactive species is in Munch-Petersen's work; moreover, the results of copper catalysis closely parallel the results when the Grignard reagent itself is used. A complete explanation for the effects of substituents on reactivity is still lacking. α Substituents with constant steric and varying electronic effects might make possible a better understanding of the relative importance of these two factors; p-trifluoromethyl- and p-methoxy-phenyl and phenyl are suggested.

$$trans\text{-}CH_3CH{=}C(R)CO_2C_4H_9\text{-}s \xrightarrow[\text{CuCl}]{n\text{-}C_4H_9MgBr} CH_3CHCH(R)CO_2C_4H_9\text{-}s \quad \text{(Eq. 18)}$$
$$\underset{C_4H_9\text{-}n}{|}$$

50

a, $R = n\text{-}C_4H_9$ (84%)
b, $R = i\text{-}C_4H_9$ (86%)
c, $R = s\text{-}C_4H_9$ (12%)

α-Bromo-β-methyl-disubstituted enoate **51** reacts with phenylmagnesium bromide, with *or without* cuprous chloride, to give conjugate adduct **52** in 70% yield,[94] and α-bromo-β-phenyl-disubstituted enoate **53** undergoes mainly 1,2 addition with methylmagnesium bromide and cuprous chloride.[94] On the basis of these results and comparable ones with the corresponding α-bromo enoic acids, Klein and Zitrin conclude that the absence of catalytic effect of added cuprous chloride on the proportion of conjugate addition indicates that methylcopper (formed *in situ*) is not sufficiently reactive in the 1,4 addition mode to be able to compete with the 1,2 addition mode of the Grignard reagent, and therefore that the α-bromo substituent hinders the usual 1,4 addition of methylcopper.[94] Furthermore, since bromine stabilizes negative charge α to it[99] and would therefore be expected to favor conjugate addition on electronic grounds, α bromine appears to retard 1,4 addition by means other than electronic.

[99] G. Köbrich and R. H. Fischer, *Chem. Ber.*, **101**, 3208 (1968).

$$trans\text{-}CH_3CH=C(Br)CO_2CH_3 \xrightarrow[\text{CuCl}]{C_6H_5MgBr} \underset{\underset{C_6H_5}{|}}{CH_3CHCH(Br)CO_2CH_3}$$
$$\phantom{trans\text{-}}\mathbf{51}\mathbf{52}$$

$$trans\text{-}C_6H_5CH=C(Br)CO_2CH_3 \xrightarrow[\text{CuCl}]{CH_3MgBr} trans\text{-}C_6H_5CH=C(Br)COCH_3$$
$$\phantom{trans\text{-}}\mathbf{53}$$

Stoichiometric organocopper reagents also fail to undergo conjugate addition with α-halo-β,β-disubstituted enoate esters; alkylation and metal-halogen exchange predominate.[32]

Completely alkyl-substituted enoate esters presumably would not undergo conjugate addition, but no data are available to verify this expectation. When the α substituent is a strongly electron-withdrawing group capable of delocalizing an adjacent negative charge, such as cyano or alkoxycarbonyl, conjugate addition is facilitated. α-Carbethoxy-β,β-dimethyl enoate **54** undergoes copper-catalyzed conjugate addition of methyl-,[100] n-butyl-,[8, 14] and phenyl-magnesium bromides,[101] and a wide variety of α-cyano-β,β-dialkyl enoates such as **55** undergo cuprous salt-promoted conjugate addition of various isoalkyl[39, 102] and n-alkyl[8, 63, 103] Grignard reagents and of lithium dimethylcopper (Eq. 19).[104]

$$(CH_3)_2C=C(CO_2C_2H_5)_2 \xrightarrow[\text{CuCl}]{RMgBr} (CH_3)_2C(R)CH(CO_2C_2H_5)_2$$
$$\mathbf{54} R = CH_3, n\text{-}C_4H_9, C_6H_5$$

$$(CH_3)_2C=C(CN)CO_2C_2H_5 \xrightarrow[\text{CuX}]{RMgBr} (CH_3)_2C(R)CH(CN)CO_2C_2H_5$$
$$\mathbf{55}$$
$$ R = i\text{-alkyl}, n\text{-alkyl}$$
$$ X = Cl, I, CN$$

[structure] =C(CN)CO$_2$C$_2$H$_5$ $\xrightarrow{(CH_3)_2CuLi}$ [structure] CH(CN)CO$_2$C$_2$H$_5$ (Eq. 19)

When in these alkylidene cyanoacetates the accessibility of the β-carbon atom is diminished by a large β substituent as in the arylidene cyanoacetate **56**,[105] or by a substituent spatially proximate to the β-carbon atom as in the 10-methyldecalin derivative **57b**,[106] then conjugate addition is retarded and less sterically demanding reduction often occurs.

---

[100] E. L. Eliel and M. C. Knoeber, *J. Amer. Chem. Soc.*, **90**, 3444 (1968).
[101] M. S. Newman, S. Mladenovic, and L. K. Lala, *J. Amer. Chem. Soc.*, **90**, 747 (1968).
[102] A. Brandstrom and I. Forsblad, *Arkiv Kemi*, **6**, 561 (1954).
[103] N. Rabjohn, L. V. Phillips, and R. J. DeFeo, *J. Org. Chem.*, **24**, 1964 (1959).
[104] R. R. Sobti and S. Dev, *Tetrahedron Lett.*, **1967**, 2893; *Tetrahedron*, **26**, 649 (1970).
[105] R. C. Pandey and S. Dev, *Tetrahedron*, **24**, 3829 (1968).
[106] U. Ghatak, N. N. Saha, and P. C. Dutta, *J. Amer. Chem. Soc.*, **79**, 4487 (1957).

$trans$-m-CH$_3$C$_6$H$_4$C(CH$_3$)=C(CN)CO$_2$C$_2$H$_5$ $\xrightarrow{\text{CH}_3\text{MgI}}_{\text{CuCl}}$
56

m-CH$_3$C$_6$H$_4$CH(CH$_3$)CH(CN)CO$_2$C$_2$H$_5$

[Structure **57** with C(CN)CO$_2$C$_2$H$_5$ and R substituents] $\xrightarrow{\text{CH}_3\text{MgI}}_{\text{CuI}}$ [Structure with R′, CH(CN)CO$_2$C$_2$H$_5$ and R substituents]

57
a, R = H
b, R = CH$_3$

a, R = H, R′ = CH$_3$
b, R = CH$_3$, R′ = H

Substrate stereochemistry often affects the course of conjugate addition but, despite a rationalization of this phenomenon on the basis of a cyclic mechanism[7] or in terms of a copper–double bond complex,[51] no fully satisfactory explanation has yet appeared. Thus geometrical isomers may undergo 1,4 addition but in substantially different yields and presumably at very different rates. The available data summarized in Table A indicate that $trans$-crotonates and -cinnamates are more reactive toward catalytic organocopper reagents than cis-crotonates and -cinnamates and that the reverse is true for maleates and fumarates.

TABLE A. REACTIVITY OF $cis/trans$ ISOMER PAIRS TOWARD CATALYTIC ORGANOCOPPER REAGENTS (CUPROUS CHLORIDE CATALYST)

		% 1,4 Adduct from	
Substrate	Grignard	cis-Substrate	$trans$-Substrate
CH$_3$CH=CHCO$_2$C$_4$H$_9$-s	n-C$_4$H$_9$MgBr	51	80
CH$_3$CH=C(CH$_3$)CO$_2$C$_4$H$_9$-s	n-C$_4$H$_9$MgBr	15–20	55
C$_6$H$_5$CH=C(Br)CO$_2$CH$_3$	CH$_3$MgBr	43	62
s-C$_4$H$_9$O$_2$CCH=CHCO$_2$C$_4$H$_9$-s	n-C$_4$H$_9$MgBr	57–67	38
s-C$_4$H$_9$O$_2$CC(CH$_3$)=CHCO$_2$C$_4$H$_9$-s	n-C$_4$H$_9$MgBr	40	21

The 1,4 adducts from α-methyl-maleate and -fumarate are mixtures of α,α (**58**) and α,α′ (**59**) succinates in which the former predominates (90% from the maleate and 70% from the fumarate). The predominance of the α,α-disubstituted succinate **58** is incompatible with a steric directing factor due to the original α-methyl group but might be explained on the basis of the positive inductive effect of this methyl group which would cause intermediate **60a** to be less stable than intermediate **60b**.

There are no data concerning conjugate addition of stoichiometric organocopper reagents for any of the isomers just cited. The only such

s-$C_4H_9O_2CC(CH_3)CH_2CO_2C_4H_9$-$s$
 |
 C_4H_9-n
 58

s-$C_4H_9O_2CCH(CH_3)CHCO_2C_4H_9$-$s$
 |
 C_4H_9-n
 59

Met
 |
$RO_2CC(CH_3)CCO_2R$
 |
 C_4H_9-n
 60a

Met
 |
$RO_2CC(CH_3)CHCO_2R$
 |
 C_4H_9-n
 60b

experiment was performed on *cis*- and *trans*-α,β-ethylenic aldehydes **61**; the *cis*, i.e., *Z*, isomer undergoes conjugate addition with lithium dimethylcopper at least three times as fast as the *trans*, i.e., *E*, isomer.[51] Obviously more information is needed before the effect of substrate stereochemistry on organocopper conjugate addition can be fully understood.

$$n\text{-}C_6H_{13}CH{=}C(CH_3)CHO \xrightarrow{(CH_3)_2CuLi} n\text{-}C_6H_{13}CH(CH_3)CH(CH_3)CHO$$
61

Optical Isomerism and Diastereoisomerism. Formation of optical and diastereoisomers by organocopper conjugate addition to α,β-unsaturated carbonyl substrates has been examined for only two types of substrates, acyclic β-methyl-monosubstituted enoates **62** and acyclic α,β-disubstituted carbonyl derivatives **63**.

$CH_3CH{=}CHCO_2R$

62
R = menthyl, xylose, or glucose derivative

$RCH{=}CH(R')CR''$
         ‖
         O

63
R = CH_3, n-C_6H_{13}, $CO_2C_4H_9$-s
R' = CH_3, Br
R'' = H, OH, OCH_3, OC_4H_9-s

The first report of partial asymmetric synthesis by organocopper conjugate addition deals with the copper-catalyzed phenyl Grignard addition to the acyclic β-methyl enoate, (−)-menthyl crotonate **(62)**.[107] Whereas phenyl Grignard alone gives (+)-3-phenylbutyric acid after hydrolysis, the copper-catalyzed conjugate addition produces (−)-3-phenylbutyric acid. This result was later corroborated and extended to crotonate esters of sugar derivatives.[90] In these cases as well, with R racemic but containing many oxygen atoms on the asymmetric carbons of a glucose or xylose derivative, cuprous chloride causes phenyl Grignard conjugate addition to produce 1,4 adducts (3-phenylbutyric acids) whose optical rotations have the opposite signs compared to the conjugate adducts formed without copper catalyst; the optical yields in the catalyzed conjugate additions are in the range 5–74%, depending on the nature of alcoholic R in enoate **62**. The ability of a copper catalyst to influence the preferred direction of attack on the enoates has been interpreted in terms

[107] Y. Inouye and H. M. Walborsky, *J. Org. Chem.*, **27**, 2706 (1962).

of an organocopper complex with the double bond of the enoate.[90] The extent to which this complex mechanism contributes to the overall mechanism of organocopper conjugate addition as well as the effect on asymmetric synthesis of such variables as size of organocopper reagent (e.g., C_6H_5Cu vs. CH_3Cu), size of alcoholic R, and size of the β substituent have yet to be clarified.

Organocopper conjugate addition to acyclic α,β-disubstituted carbonyl derivatives such as **63** can in principle produce one or both of two diastereomers, depending on the stereochemistry of conjugate addition (*cis* or *trans*) and on the configurational stability of the intermediate copper (or lithium or magnesium) enolate. In most reactions involving such acceptor substrates as **63**, however, no determinations of diastereomer ratios have been achieved[8, 108] (*cf.* ref 94).

Cyclic Enones. Although no planned study has been made of cycloalkenone ring size as a controlling factor in conjugate addition, Stoll and Commarmont report that 2-cyclopentadecenone reacts with methylmagnesium bromide and cuprous chloride to give mainly 1,2 addition.[109] On the basis of this result they conclude that the catalytic effect of cuprous chloride in promoting conjugate addition of Grignard reagents is less pronounced in macrocyclic enones than in cyclohexenone. 2-Cycloheptenone and 2-cyclooctenone undergo copper-catalyzed phenyl Grignard conjugate addition in modest yields.[110]

Unlike acyclic enones whose reactivity is affected mainly by α, α' and β substituents, cyclic enones are also often affected in reactions with organocopper reagents by substituents in other parts of the molecule. These substituents, which are not directly connected to the reactive portion of the molecule, may nevertheless influence the stereochemistry and amount of conjugate addition by their spatial proximity to the reaction site.

As a basis for discussion, cyclohexenone is taken as the parent cyclic enone.

In the cyclohexenone system **64** the effect of varying the 4-alkyl group on the stereochemistry of copper-catalyzed methyl Grignard conjugate addition has been studied.[66] From each substrate (**64**, $R = CH_3, C_2H_5$,

[108] J. Munch-Petersen and V. K. Andersen, *Acta Chem. Scand.*, **15**, 293 (1961).

[109] M. Stoll and A. Commarmont, *Helv. Chim. Acta*, **31**, 554 (1948); *cf.* B. D. Mookharjee, R .W. Trenkle, and R. R. Patel, *J. Org. Chem.*, **36**, 3266 (1971) and B. D. Mookharjee, R. R. Patel, and W. O. Ledrig, *J. Org. Chem.* **36**, 4124 (1971).

[110] A. C. Cope and S. S. Hecht, *J. Amer. Chem. Soc.*, **89**, 6920 (1967).

i-C_3H_7), two isomers of 1,4 adduct **65** are formed, the *trans* always predominating over the *cis*. As the bulk of the 4-alkyl group is increased from methyl to ethyl to isopropyl, however, the *trans*:*cis* ratio is also increased, from 72:28 to 77:23 to 89:11. A similar trend is observed for phenyl Grignard conjugate addition to these 4-alkylcyclohexenones **64**. Thus the bulk of the 4-alkyl substituent in this flexible system affects the stereochemistry of conjugate addition, with larger R causing higher stereoselectivity. Rivière and Tostain[66] conclude from these results that steric interactions between reagent and ground-state reactant are of major importance (steric approach control[111]) and that steric interactions in the transition state are of minor importance.

α′,α′-Dimethyl and probably α′,α′-dialkyl substitution do not decrease the amount but may affect the stereochemistry of conjugate addition of Grignard reagents catalyzed by copper salts. Thus cyclohexenone itself undergoes 1,4 addition of n-propyl,[55] isopropenyl,[70] allyl,[55] and isopentyl[62] groups in 63–90% yields (Eq. 20), and the α′,α′-dimethyloctalone **66** forms 1,4-methyl, -isopropyl, and -phenyl adducts in 84–97% yields.[88] Furthermore α′-alkylated cyclopentenone **68** reacts with methylmagnesium bromide to form 1,4 adduct **69**.[112] (Equations on p. 34.)

The stereochemistry of conjugate addition has been studied as a function of α′ substitution in octalones such as **70**. Thus octalone **70**, R = CH_3,[71] and steroidal analogs,[113, 114] as well as octalone **71**,[73] all undergo only axial alkyl conjugate addition, but α′,α′-dimethyloctalone **66**[88] gives 1,4 adduct **67** in which the newly introduced isopropyl group is 54% axial and 46% equatorial. A similar but less dramatic result is observed with methyl conjugate addition to α′,α′-dimethyloctalone **66**: 82% axial and 17% equatorial methyl. It should be noted that axial and equatorial conjugate addition to octalone **66** might be due to the absence of an angular methyl group rather than to the presence of α′,α′-dimethyl groups.

---

[111] *Cf.* W. G. Dauben, G. J. Fonken, and D. S. Noyce, *J. Amer. Chem. Soc.*, **78**, 2579 (1956).
[112] (a) T. Sakan and K. Abe, *Tetrahedron Lett.*, **1968**, 2471; (b) G. Stork, G. L. Nelson, F. Rouessac, and O. Gringore, *J. Amer. Chem. Soc.*, **93**, 3091 (1971).
[113] W. J. Wechter, G. Slomp, F. A. MacKellar, R. Wiechert, and U. Kerb, *Tetrahedron*, **21**, 1625 (1965).
[114] D. Bertin and J. Perronnet, *Compt. Rend.*, **257**, 1946 (1963).

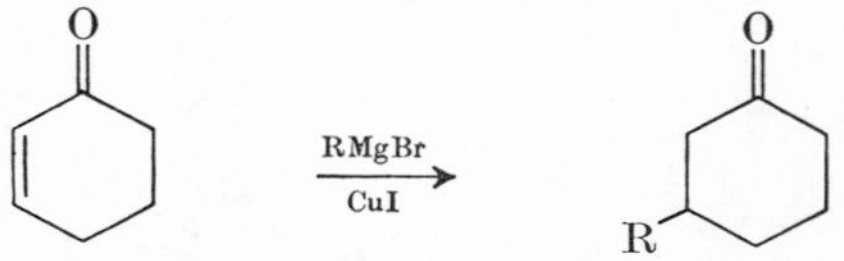

(Eq. 20)

R = n-C$_3$H$_7$, CH$_2$=C(CH$_3$), (CH$_3$)$_2$CH(CH$_2$)$_2$

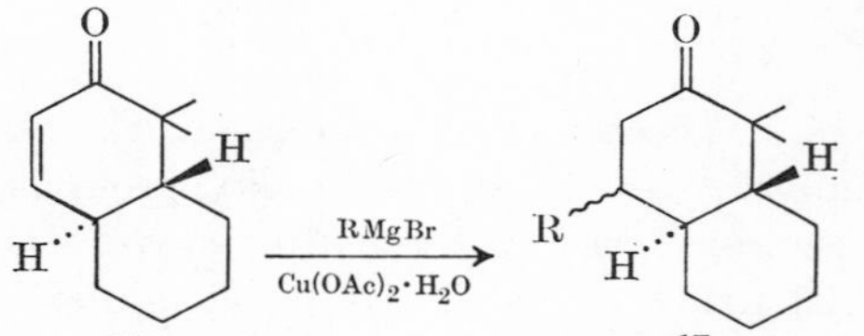

66 R = CH$_3$, (CH$_3$)$_2$CH, C$_6$H$_5$ **67**

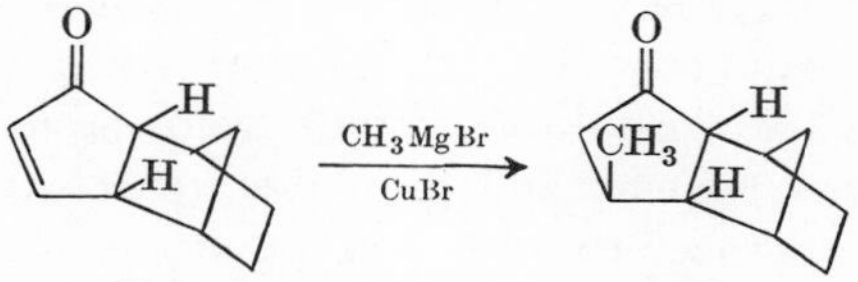

68 **69**

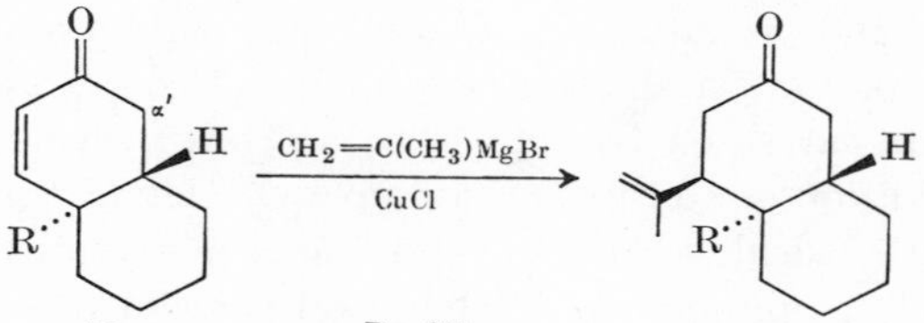

70 R = CH$_3$

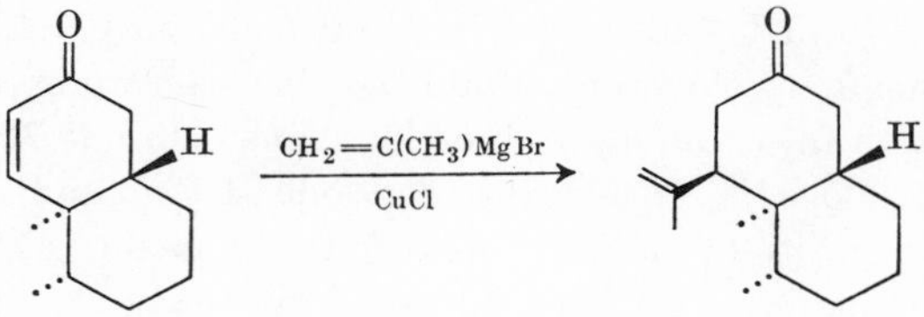

71

The only recorded example of an α',β-dialkyl-substituted system is the tetracyclic enone **72** which reacts with methylmagnesium iodide and cuprous bromide to give only 1,2 adduct **73**.[115]

β-Methyl substitution alone is found not to diminish the extent of either copper-catalyzed or stoichiometric organocopper methyl conjugate addition (1,4 addition of other alkyl groups has not been attempted). β-Methylcyclohexenones **74**[116] and **1**[4] react with methyl Grignard and

[115] A. J. Birch and R. R. Robinson, *J. Chem. Soc.*, **1944**, 503.
[116] G. Büchi, O. Jeger, and L. Ruzicka, *Helv. Chim. Acta*, **31**, 241 (1948).

cuprous chloride, giving 1,4 adducts in 60–83% yields, and enone **74** reacts with various methylcopper reagents to give conjugate adducts in 88–97% yields.[28]

β-Alkyl substitution, examined only in 4-octal-3-one (**2**, steroid numbering) and various derivatives, also does not in general adversely affect methyl conjugate addition which proceeds with copper catalysis of methyl Grignard in 60–83% yields.[5, 36, 68, 84] Similarly, copper-catalyzed methyl conjugate addition to substituted octalones **75**[117] and **76**[36] and to the bicyclic enone **77**[36] (p. 36) proceeds in good yield. Although copper-catalytic conjugate addition of bulky (*e.g.*, phenyl) Grignard reagents proceeds satisfactorily for unsubstituted cyclohexenones,[46, 66] such 1,4 addition is severely hampered by β-alkyl substitution.[85]

[117] J. A. Settepani, M. Torigoe, and J. Fishman, *Tetrahedron*, **21**, 3661 (1965).

77

With octalone **78a** as the parent compound, the effects of an angular methyl substituent (*i.e.*, **78b**), of an α-methyl group (*i.e.*, **78c**), and of both angular and α-methyl substituents (*i.e.*, **78d**), on conjugate addition can be examined and compared.

78a, R = H
78b, R = CH$_3$

78c, R = H
78d, R = CH$_3$

10-Methyloctalone **78b**,[84, 118] and analogous 10-methyl steroidal enones,[36, 119] undergo copper-catalyzed methyl Grignard conjugate addition only to the extent of 20–40%, compared with 60–83% for unsubstituted octalone **78a**. The angular methyl group, although not directly attached but nevertheless close in space to the enone chromophore, sterically hinders copper-catalyzed conjugate addition. This retarding effect of angular methyl was effectively used to perform selective catalytic conjugate addition to the ring-D enone portion of steroidal bisenone **79** in 90% yield.[120] The 10-methyl group only slightly retards conjugate addition of stoichiometric organocopper reagents; lithium dimethylcopper adds 1,4 to 9-tetrahydropyranyloxy-10-methyloctalone **80** in 78% yield.[121]

Marshall and his students have examined the amount and the stereochemistry of organocopper conjugate addition as functions of axial and equatorial C-7 substituents in rigid octalones **81**.[81, 84] They find that parent octalone **81a** as well as 7β-isopropenyl- and 7β-isopropyl-octalones **81b** and **81c** undergo effective organocopper conjugate addition to form only *cis*-decalone **82**, but that 7α-isopropenyloctalone **83** gives no conjugate adducts with either stoichiometric or catalytic organocopper reagents. Thus "the orientation of an alkyl substituent at C-7 in octalones

---

[118] R. E. Ireland, M. I. Dawson, J. Bordner, and R. E. Dickerson, *J. Amer. Chem. Soc.*, **92** 2568 (1970).

[119] M. Torigoe and J. Fishman, *Tetrahedron Lett.*, **1963**, 1251.

[120] K. Heusler, J. Kebrle, C. Meystre, H. Ueberwasser, P. Wieland, G. Anner, and A. Wettstein, *Helv. Chim. Acta*, **42**, 2043 (1959).

[121] E. Piers, R. W. Britton, and W. de Waal, *Chem. Commun.*, **1969**, 1069; E. Piers, W. de Waal, and R. W. Britton, *J. Amer. Chem. Soc.*, **93**, 5113 (1971).

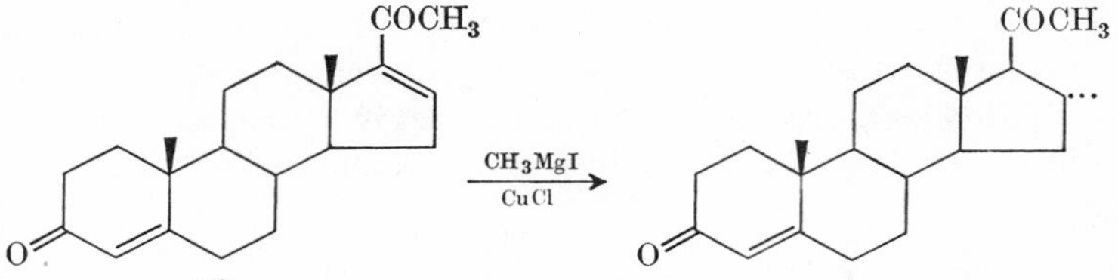

(Eq. 21)

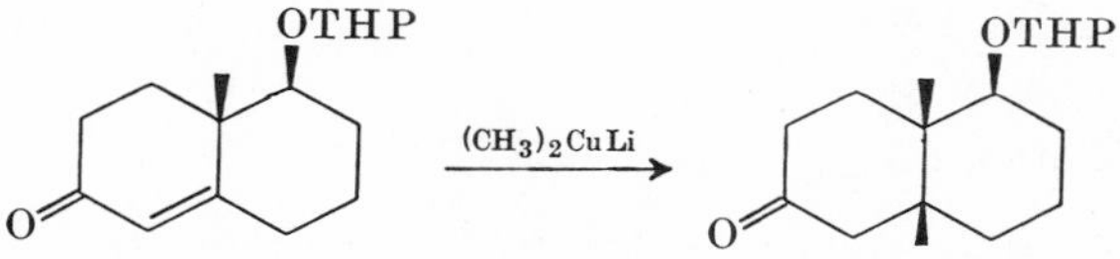

(Eq. 22)

such as [81] and [83] controls the conjugate methylation of these compounds. Axial groups block the 1,4 addition and equatorial groups exert a negligible effect."[81] These results are interpreted to mean that steric interactions between reagent and ground-state reactant are of lesser importance than steric interactions in the transition state.[81]

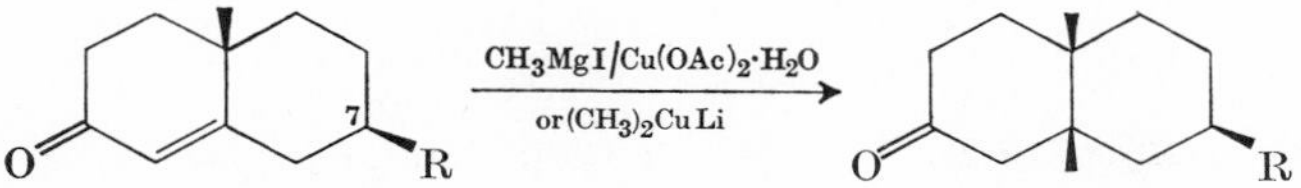

81
a, R = H
b, R = C(CH₃)=CH₂
c, R = (CH₃)₂CH

α-Methyloctalone **78c** has not been subjected to organocopper reagents, but substituted α-methyloctalone **84** has; it reacts with methylmagnesium iodide and cupric acetate monohydrate to give *cis*-decalone **85** (R = C(CH₃)=CH₂) in 10–25% yield, and with lithium dimethylcopper to give the same 1,4 adduct in 77% yield.[122] Thus comparison of 7β-isopropenyl-4-methyloctalone **84** with 7β-isopropenyl-10-methyloctalone **81b** shows that copper-catalyzed Grignard conjugate addition proceeds equally poorly in both cases, and that lithium dimethylcopper conjugate addition

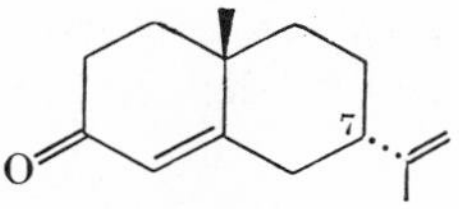

[122] E. Piers and R. J. Keziere, *Can. J. Chem.*, **47**, 137 (1969).

proceeds equally well in both cases. No determination then can be made as to which of the two substituents, α-methyl or angular methyl, influences conjugate addition to a greater extent. Larger alkyl substituents on the octalone nucleus would probably produce more pronounced effects.

$$\text{78c, R = H} \quad \text{84, R = C(CH}_3\text{)=CH}_2 \quad \xrightarrow{\text{CH}_3\text{MgI/Cu(OAc)}_2\cdot\text{H}_2\text{O} \text{ or } (\text{CH}_3)_2\text{CuLi}} \quad \text{85}$$

4-Alkyl-10-methyloctalones **86** and **87** are inert to lithium dimethylcopper,[123] indicating that the combination of angular and α substituents in one octalone molecule is sufficient to prevent methyl conjugate addition.

86 R = CH$_3$, m-CH$_3$OC$_6$H$_4$(CH$_2$)$_2$

87 R = CH$_3$, m-CH$_3$OC$_6$H$_4$(CH$_2$)$_2$

Although the vast majority of cyclic enones react with organocopper reagents to form conjugate adducts with the newly introduced group axial, prediction of the stereochemistry of conjugate addition to substituted enones may be hazardous even when based on close analogy (see Eqs. 23 and 24).[124] Any mechanistic interpretation of these results, however, is and will remain speculative until the interaction between organocopper reagent and unsaturated substrate is defined in detail.

$$\xrightarrow{(\text{CH}_3)_2\text{CuLi}} \quad (90\%) \quad \text{(Eq. 23)}$$

$$\xrightarrow{(\text{CH}_3)_2\text{CuLi}} \quad \text{(Eq. 24)}$$

1α (20%)
1β (80%)

[123] R. E. Ireland, S. C. Welch, and C. Kowalski, unpublished results.
[124] J. A. Marshall and S. F. Brady, *Tetrahedron Lett.*, **1969**, 1387; *J. Org. Chem.*, **35**, 4068 (1970).

Cisoid enone α-methylenecyclohexanone (**88**) represents another structural variation within the class of cyclic α,β-ethylenic ketones that has been subjected to organocopper reagents. Thus bicyclic enone **89** reacts with lithium dimethyl- and diphenyl-copper reagents to give methyl[82] and phenyl[83] conjugate adducts **90** in good yields, and octalone **91**[69, 125] and an analogous steroidal enone[126] give moderate amounts of conjugate addition via copper catalysis of Grignard reagents. β-Phenyl enone **92** is converted to 1,4 adduct **93** by copper catalysis of phenyl Grignard[127] and,

[125] H. O. House and H. W. Thompson, *J. Org. Chem.*, **28**, 360 (1963).
[126] H. Mori, *Chem. Pharm. Bull. (Tokyo)*, **12**, 1224 (1964) [*C.A.* **62**, 1710h (1965)].
[127] S. O. Winthrop and L. G. Humber, *J. Org. Chem.*, **26**, 2834 (1961).

finally, fully β-substituted cyclobutanone **94** is converted to conjugate adduct **95** by cuprous chloride catalysis of methylmagnesium iodide.[128]

That the vast majority of cyclic enones which react with organocopper reagents form conjugate adducts with the newly introduced group axial is typical of kinetically controlled 1,4 additions, reactions which are subject to the stereoelectronic requirement that the reagent approach the α,β-unsaturated substrate in a plane perpendicular to the α,β double bond.[86, 87, 125] This tendency of organocopper reagents to introduce hydrocarbon groups axially has been used with consistent success in methyl conjugate addition to various steroidal enones; axial methyl has thus been placed at the following positions of the steroid nucleus **96**: 1,[113, 114, 129] 4,[130, 131] 5,[36, 79, 129, 132] 6,[130] 7,[133, 134] 10,[117, 119] and 16.[120, 135]

Polyethylenic Esters and Ketones. Copper-catalyzed Grignard addition to polyethylenic carbonyl derivatives has been studied with respect to the relative amounts of 1,4 or 1,6 or 1,8 conjugate addition. Acyclic dienoate ester **97** undergoes only 1,6 addition with n-butylmagnesium bromide and cuprous chloride[10] and, under the same conditions, acyclic trienoate ester **98** produces a 1,8 adduct in 20–25% yield as the only identifiable product.[11]

$$CH_3(CH=CH)_2CO_2C_4H_9\text{-}s \xrightarrow[\text{CuCl}]{n\text{-}C_4H_9MgBr} CH_3\underset{\underset{C_4H_9\text{-}n}{|}}{CH}CH_2CH=CHCO_2C_4H_9\text{-}s$$
$\qquad\quad$ **97**

$$CH_3(CH=CH)_3CO_2C_4H_9\text{-}s \xrightarrow[\text{2. H}_2/\text{Pd}/\text{BaSO}_4]{1.\ n\text{-}C_4H_9MgBr/\text{CuCl}} CH_3\underset{\underset{C_4H_9\text{-}n}{|}}{CH}(CH_2)_5CO_2C_4H_9\text{-}s$$
$\qquad\quad$ **98**

---

[128] J. Salaün and J. M. Conia, *Bull. Soc. Chim. Fr.*, **1968**, 3730.
[129] M. Fetizon and J. Gramain, *Bull. Soc. Chim. Fr.*, **1968**, 3301.
[130] J. R. Bull, *J. Chem. Soc. (C)*, **1969**, 1128.
[131] H. Mori, *Chem. Pharm. Bull. (Tokyo)*, **12**, 1224 (1964) [*C.A.* **62**, 1710h (1965)].
[132] P. François and J. Levisalles, *Bull. Soc. Chim. Fr.*, **1968**, 318.
[133] J. A. Campbell and J. C. Babcock, *J. Amer. Chem. Soc.*, **81**, 4069 (1959).
[134] N. W. Atwater, R. H. Bible, Jr., E. A. Brown, R. R. Burtner, J. S. Mihina, L. N. Nysted, and P. B. Sollman, *J. Org. Chem.*, **26**, 3077 (1961).
[135] R. D. Hoffsommer, H. L. Slates, D. Taub, and N. L. Wendler, *J. Org. Chem.*, **24**, 1617 (1959).

Cyclic polyenones have also been examined. Dienone **99**[36] reacts with methylmagnesium iodide and cupric acetate in tetrahydrofuran to give predominantly a 1,6 adduct in 54% yield, but under similar conditions steroidal dienones **100**[133] afford 1,6 adducts in only 25–40% yields. Highly functionalized steroidal dienone **102**[134] and trienone **103**[135a] undergo conjugate addition in high yield with methyl Grignard and cuprous chloride. (Equations on p. 42.) This result might suggest that, when an α,β-enone and an $\alpha,\beta,\gamma,\delta$-dienone having generally the same steric environments compete for an organocopper reagent, addition to the β-carbon atom of the α,β-enone is energetically more favorable than addition to the δ-carbon atom of the $\alpha,\beta,\gamma,\delta$-dienone.[135b]

One additional example of the way in which the structure of a cyclic enone may affect the stereochemistry of conjugate addition involves the steroidal dienone **100**. The dienone **100a** undergoes "normal" copper-catalyzed Grignard conjugate addition to give mainly 7α-methyl adduct **101a**, but 11β-hydroxy dienone **100b** reacts under the same conditions to give mainly 7β-methyl adduct **101b**.[133] No data are available for stoichiometric organocopper conjugate addition to the epimeric 11α-hydroxy dienones.

Despite a lengthy discussion of probable factors governing conjugate addition to polyethylenic ketones and consideration of a cyclic or a carbanion mechanism for such reactions,[12] no definitive experiments have been done to establish clearly the mechanism by which copper salts

[134a] R. Wiechert, U. Kerb, and K. Kieslich, *Chem. Ber.*, **96**, 2765 (1963).

[135a] Regioselectivity of lithium dimethylcopper conjugate addition to several cyclic dienones has recently been reported: J. A. Marshall, R. A. Ruden, L. K. Hirsch, and M. Phillippe, *Tetrahedron Lett.*, **1971**, 3795.

[Structures: 102 → product with CH₃MgBr/CuCl; 103 → product with CH₃MgI/CuCl]

promote addition of Grignard reagents to the terminal carbon atom of a conjugated system. Furthermore, no such polyethylenic substrates have been treated with stoichiometric organocopper reagents.

Compatible Functional Groups. Organocopper reagents are relatively weak nucleophiles, as evidenced for example by their inability to add across the carbonyl group of Michler's ketone (*i.e.*, negative Gilman test[49]). Especially when organocopper conjugate addition to an α,β-unsaturated carbonyl substrate is rapid under mild conditions, a wide variety of functional groups on the substrate but not directly bonded to the enone chromophore can survive the reaction unscathed. The following are the examples of hydroxyl derivatives which are usually not destroyed by organocopper reagents:* hydroxyl and allylic hydroxyl,[126] tetrahydropyranyloxyl,[121] and acetoxyl and allylic acetoxyl;[133] in addition to the ethylene ketal carbonyl derivative,[113] saturated ketones[113] and lactones[134] and sterically hindered α,β-unsaturated ketones[120] can survive organocopper conjugate addition. Examples of substrates carrying these substituents come mainly from steroidal enones; for more details see Tables IB and IIIB. Just two of the more interesting cases are shown in Eqs. 25 and 26 (see also Eqs. 21 and 22).

* Undoubtedly the hydroxyl group reacts with organocopper reagent to generate an alkoxide, but hydrolytic workup regenerates the hydroxyl group.

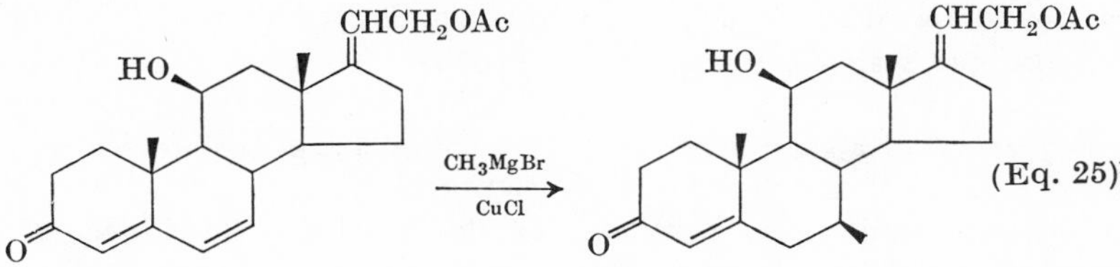

(Eq. 25)[133]

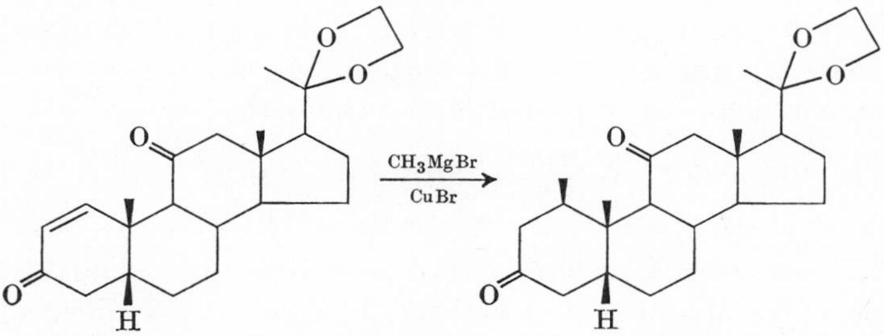

(Eq. 26)[113]

No α,β-unsaturated carbonyl substrate bearing a halogen or an epoxide elsewhere in the molecule has yet been treated with organocopper reagents.

α,β-Acetylenic Carbonyl Derivatives

No data are recorded for either copper-catalyzed or stoichiometric organocopper conjugate addition to α,β-acetylenic aldehydes or anhydrides. Amides, possibly because of their low electrophilicity, are inert to methylcopper; thus phenylpropiolic amides **104** fail to react with methylcopper.[136] More reactive organocopper reagents, such as lithium diethylcopper, however, do add in a 1,4 manner to acetylenic amides.[65] Unexpectedly the tetra-n-butylammonium carboxylate **105** is converted by methylcopper in tetrahydrofuran to 1,4 adduct **106** in 80% yield.[136]

$$C_6H_5C{\equiv}CCONRR'$$
104

$$R = H, \quad R' = H$$
$$H, CH_3$$
$$CH_3 CH_3$$

$$C_6H_5C{\equiv}CCO_2^-N^+(C_4H_9\text{-}n)_4 \xrightarrow[\text{THF}]{CH_3Cu} trans\text{-}C_6H_5C(CH_3){=}CHCO_2^-N^+(C_4H_9\text{-}n)_4$$
105   **106**

α,β-Acetylenic ketones have received very limited attention. The only relevant literature report shows that ynone **107** reacts with lithium dimethylcopper in tetrahydrofuran at −78° to give a 1:1 *cis:trans* mixture of enone **108**.[32] No attempts have yet been made to control the stereochemistry of this reaction by decreasing reaction time, by lowering

[136] J. Klein and N. Aminadav, *J. Chem. Soc.* (C), **1970**, 1380.

reaction temperature, by using chelating ligands,[32] or by using less reactive organocopper reagents (*e.g.*, methylcopper).

$$n\text{-}C_7H_{15}C\equiv CCOCH_3 \xrightarrow{(CH_3)_2CuLi} n\text{-}C_7H_{15}C(CH_3)=CHCOCH_3$$
$$\phantom{n\text{-}C_7H_{15}C\equiv CCOCH_3 \xrightarrow{(CH_3)_2CuLi}}\ 107 108$$

Conjugate addition to acetylenic nitriles has been the subject of one investigation. 2-Butynonitrile (**109**) reacts with ethylmagnesium bromide in the presence of 1 mol% cuprous chloride to give a 1:1 *cis:trans* mixture of 1,4 adduct **110** in 80% yield.[137] It is not clear, however, that the reactive species in this conjugate addition is an organocopper reagent, for essentially the same results were obtained without cuprous chloride.

$$CH_3C\equiv CCN \xrightarrow[CuCl]{C_2H_5MgBr} C_2H_5C(CH_3)=CHCN$$
$$\ 109 110$$

Ynoic acids and esters have been studied intensively in several laboratories. No direct comparison of the reactivity of an α,β-acetylenic acid and its esters has been reported.

α,β-Acetylenic carbonyl substrates are summarized below in terms of (a) successful conjugate addition, (b) unsuccessful conjugate addition, and (c) no data reported for organocopper or copper-catalyzed conjugate addition.

$$-C\equiv C-C=O$$
$$|$$
$$Y$$
$$111$$

a. Y = C, OH, OR, O⁻N⁺R₄, NR₂
b. Y = NR₂
c. Y = H, OCOR

α,β-Ethylenic esters react with organocopper reagents much more slowly than do analogous acetylenic esters. When ethyl 2-nonenoate (**47**)[51] and methyl 2-decynoate (**112**)[32] are treated for 3 hours with lithium dimethylcopper in toluene at −78°, the former is unscathed while the latter has reacted to the extent of 47%. A more dramatic illustration of this difference in reactivity is found when tetrahydrofuran is used as solvent: enoate **47** is stable toward lithium dimethylcopper for 3 hours at 0° in this solvent, whereas ynoate **112** is completely consumed after 2.5 hours *at −78°!* The higher reactivity of ynoate relative to enoate esters toward organocopper reagents may have mechanistic implications, for generally most nucleophilic additions to unsaturated systems proceed more readily with acetylenic than with ethylenic substrates.[138]

$$trans\text{-}n\text{-}C_6H_{13}CH=CHCO_2C_2H_5 n\text{-}C_7H_{15}C\equiv CCO_2CH_3$$
$$\phantom{trans\text{-}n\text{-}C_6H_{13}CH=CHCO_2C_2H_5 xx}47 112$$

[137] G. Boularand and R. Vessiere, *Bull. Soc. Chim. Fr.*, **1967**, 1706.
[138] J. March, *Advanced Organic Chemistry: Reactions, Mechanisms, and Structure*, McGraw-Hill Book Company, New York, N.Y., 1968, p. 575.

Similarly, enoic acids are substantially less reactive than ynoic acids toward organocopper reagents. Enoic acid **33** is unaffected by lithium dimethylcopper,[16] but phenylpropiolic acid **(113a)** and tetrolic acid **(113b)** undergo effective conjugate addition with a variety of organocopper reagents (*e.g.*, methylcopper, phenylcopper, lithium dimethylcopper).[65, 67]

$$t\text{-}C_4H_9-\underset{33}{\underset{}{\bigcirc}}=CHCO_2H \qquad \underset{\underset{\substack{a,\ R = C_6H_5 \\ b,\ R = CH_3}}{113}}{R-C\equiv C-CO_2H}$$

The most interesting and useful information about acetylenic carbonyl systems has come from studies of the stereochemistry of conjugate addition.

Acetylenic acids in which R (structure **114**) is methyl,[67, 136, 139] *n*-butyl,[139] or phenyl,[67, 136, 139] undergo organocopper conjugate addition in varying yields. Acetylenic esters with various alcoholic R′ groups (*e.g.*, CH_3, C_2H_5, sec-C_4H_9), and in which R is hydrogen,[65] methyl,[10, 32, 64, 67, 136] ethyl,[54] *n*-heptyl,[32, 64] or phenyl,[10, 67, 136] also undergo organocopper conjugate addition generally in good yields. As noted previously, large alcoholic R′ groups tend to retard Grignard carbonyl addition, allowing organocopper conjugate addition to compete effectively; ethyl phenylpropiolate reacts with *n*-butylmagnesium bromide and cuprous chloride to give only tar, but the corresponding sec-butyl ester **115** reacts under the same conditions to give conjugate adduct **116** in 55% yield.[10] Available data are insufficient, however, to make possible a general correlation of the structure of R and R′ with reactivity of the acetylenic carbonyl system toward either catalytic or stoichiometric organocopper reagents.

$$\underset{114}{R-C\equiv C-CO_2R'}$$

$$\underset{115}{C_6H_5C\equiv CCO_2C_4H_9\text{-}s} \xrightarrow[\text{CuCl}]{n\text{-}C_4H_9MgBr} \underset{116}{\underset{\underset{n}{\overset{|}{C_4H_9\text{-}n}}}{C_6H_5C=CHCO_2C_4H_9\text{-}s}}$$

In contrast to the *trans* addition of most organometallic reagents to acetylenes,[140] α,β-acetylenic acids[67, 136, 139] and esters[14, 32, 54, 64] generally undergo conjugate *cis* addition with organocopper reagents (Eq. 27), the stereospecificity of the addition often approaching 100%. Although only an enlightened guess can be made concerning the mechanism of this

---

[139] I. Iwai and T. Konotsune, *Yakugaku Zasshi*, **82**, 601 (1962) [*C.A.*, **58**, 1392a (1963)].

[140] J. M. Landesberg and D. Kellner, *J. Org. Chem.*, **33**, 3374 (1968), and references cited therein.

transformation,[32] the overall *cis* addition of R″ and Y (see **117**) across the triple bond has been used to prepare stereospecifically geometrical isomers of various tri- and tetra-substituted olefins.

$$RC{\equiv}CCO_2R' \xrightarrow[\substack{-78 \text{ to } -100° \\ THF}]{R_2''CuLi} \underset{\underset{\textbf{117}}{}}{\overset{R}{\underset{R''}{>}}C{=}C\overset{CO_2R'}{\underset{Y}{<}}} \quad \text{(Eq. 27)}$$

114

In terms of organic synthesis, it should be noted that in principle and often in practice[64] the geometric isomer of unsaturated ester **117** can be formed by allowing R_2CuLi to react with $R''C{\equiv}CCO_2R'$ (Eq. 28). Since the product stereochemistry and yield are high when R and R″ are primary alkyl,[64] the only major limitation appears to be the availability of both reactants.

$$R''C{\equiv}CCO_2R' \xrightarrow{R_2CuLi} \overset{R''}{\underset{R}{>}}C{=}C\overset{CO_2R'}{\underset{Y}{<}} \quad \text{(Eq. 28)}$$

According to current thinking,[32, 54, 64, 67, 136] the organocopper conjugate addition proceeds through the intermediacy of a vinylcopper species (**117**, Y = Cu). Most research done on organocopper conjugate additions to acetylenic acids and esters has been directed at finding optimum conditions of time, temperature, and solvent (pp. 55, 56) for controlling the configurational stability of the vinylcopper intermediate and thus for maximizing stereoselectivity. The reactive intermediate has been treated with acid to generate isoprenoid trisubstituted olefins (**117**, Y = H),[32, 54, 64] with deuterium oxide to form specifically an isomerically pure deutero-olefin (**117**, Y = D),[54] with iodine to produce vinyl iodides stereospecifically,[32, 64] with oxygen to cause stereoselective dimerization to a butadiene derivative (**117**,

$$Y = \left(\overset{R}{\underset{R''}{>}}C{=}C\overset{CO_2R}{<} \right)^{32,64}$$

and with allylic halides to form stereoselectively 1,4 dienes (**117**, Y = $CH_2CH{=}CR_2$).[65] Although only a few reports appear in the literature through 1970 concerning organocopper conjugate addition to acetylenic carbonyl derivatives, the synthetic utility of *cis* addition across a triple bond should stimulate substantial interest in this reaction.

Miscellaneous Substrates

Several classes of unsaturated substrates have been shown to react in a conjugate manner with organocopper reagents. The vinylogous amide **118** is transformed by methylmagnesium iodide and cuprous bromide into 1,4 adduct **119**,[141] and *p*-quinone diimine **120** reacts with methylcopper to give a mixture of methylated and reduced products.[142] Various aryl azlactones **121** react with aryl Grignard reagents and cuprous chloride to give 1,4 adducts **122** in yields ranging from 25 to 75%,[68] and α,β-unsaturated phosphine sulfides **123** and oxide **124** undergo copper-promoted Grignard conjugate addition.[143,143a]

$$\underset{\textbf{118}}{\text{[structure]}} \xrightarrow{\underset{\text{CuBr}}{\text{CH}_3\text{MgI}}} \underset{\textbf{119}}{\text{[structure]}}$$

$$\underset{\textbf{120}}{\text{C}_6\text{H}_5\text{N}=\!\!=\!\!\text{NC}_6\text{H}_5} \xrightarrow{\text{CH}_3\text{Cu}} \underset{\text{R}=\text{H, CH}_3}{\text{C}_6\text{H}_5\text{NH}\text{—}\text{N(R)C}_6\text{H}_5}$$

$$\underset{\textbf{121}}{\text{[azlactone with CHAr]}} \xrightarrow{\underset{\text{CuCl}}{\text{Ar'MgBr}}} \underset{\textbf{122}}{\text{[azlactone with CHArAr']}}$$

$$\underset{\textbf{123}}{\overset{\text{S}}{\underset{\|}{\text{R}_2\text{PC}\!\equiv\!\text{CR}'}}} \xrightarrow{\underset{\text{CuCl}}{\text{R''MgX}}} \overset{\text{S}}{\underset{\|}{\text{R}_2\text{PCH}\!=\!\text{CR}'\text{R}''}}$$

X = Br, I
R = CH$_3$, C$_2$H$_5$, C$_6$H$_5$
R' = CH$_3$, C$_2$H$_5$, C$_6$H$_5$
R'' = CH$_3$, C$_2$H$_5$, C$_6$H$_5$

$$\underset{\textbf{124}}{\overset{\text{O}}{\underset{\|}{(\text{C}_6\text{H}_5)_2\text{PC}\!\equiv\!\text{CC}_2\text{H}_5}}} \xrightarrow{\underset{\text{CuCl}}{\text{C}_2\text{H}_5\text{MgBr}}} \overset{\text{O}}{\underset{\|}{(\text{C}_6\text{H}_5)_2\text{PCH}\!=\!\text{C}(\text{C}_2\text{H}_5)_2}}$$

The scope of organocopper conjugate additions with respect to classes of reactive substrates has certainly not been fully explored. Substrates of the general forms —C=C—Z and —C≡C—Z have been carefully examined

---

[141] Z. Horii, K. Morikawa, and I. Ninomiya, *Chem. Pharm. Bull.* (*Tokyo*), **17**, 846 (1969).
[142] J. Honzl and M. Metalová, *Tetrahedron*, **25**, 3641 (1969).
[143] A. M. Aguiar and J. R. S. Irelan, *J. Org. Chem.*, **34**, 4030 (1969).
[143a] Lithium dimethylcopper also adds in a 1,4 manner to allenic phosphine oxides: J. Berlan, M. Capmau, and W. Chodkiewicz, *Compt. Rend.*, **273**, 295 (1971).

only where Z = carbonyl; compounds with a wide variety of electron-withdrawing groups Z might be studied, *e.g.*, Z = NO_2, SR, SO_2R, SO_2OR, PR_2, POR_2,[143] or PSR_2.[143]

CONJUGATE ADDITION OF OTHER ORGANOMETALLIC COMPOUNDS

Conjugate addition of organocopper reagents should probably be distinguished mechanistically from Michael addition of stabilized anions,[2] from 1,4 addition of Grignard reagents to aromatic systems,[144] and from conjugate addition of organoalkali reagents to systems in which steric hindrance prevents 1,2 addition.[1] Although several organic derivatives of group II metals (Be, Zn, Cd), of group III metals (B, Al), and of transition metals (Mn, Co, Ni) have undergone conjugate addition to α,β-unsaturated carbonyl substrates,[1] none of these organometallic compounds approaches organocopper reagents in general utility for conjugate addition.

Several organic derivatives of beryllium, zinc, and cadmium of the forms RMX, R_2M, and R_3MLi have been treated with a limited variety of unsaturated carbonyl compounds. The organometallic halide (RMX) and the diorganometallic (R_2M) species undergo conjugate addition in fair to good yields; isopropylcadmium chloride reacts with diethyl maleate to give conjugate adduct **125** in 29% yield,[145] and diphenylberyllium and diphenylzinc react in a 1,4 manner with benzalacetophenone (90–91% yields, Eq. 29).[146] α-Naphthylmethylcadmium chloride undergoes more effective conjugate addition to alkylidene malonates **126** than does the corresponding Grignard reagent.[147] The triorganometallic complexes (R_3MLi), however, usually undergo substantial 1,2 addition.[148] A reasonable mechanism proposed for the conjugate additions of these Lewis acid organometallic species to α,β-unsaturated carbonyl compounds involves oxygen coordination to metal, forming an "ate" complex, followed by transfer of R to the electrophilic β-carbon atom.[1] Comparison of the mechanism(s) of organocopper and group II organometallic conjugate addition is not possible, for the mechanism of neither reaction has been established, but it is likely that organocopper conjugate addition proceeds by an entirely different mechanism.

Triaryl-boron and -aluminum species undergo 1,4 addition to benzalacetophenone, as do aryl Grignard reagents.[1, 146] Generally, however, organoboron and organoaluminum compounds have been used mainly for

---

[144] R. C. Fuson, *Adv. Organmetal. Chem.*, **1**, 221 (1964).
[145] C. S. Marvel, R. L. Myers, and J. H. Saunders, *J. Amer. Chem. Soc.*, **70**, 1694 (1948).
[146] H. Gilman and R. H. Kirby, *J. Amer. Chem. Soc.*, **63**, 2046 (1941).
[147] B. Riegel, S. Siegel, and W. M. Lilienfeld, *J. Amer. Chem. Soc.*, **68**, 984 (1946).
[148] G. Wittig, F. J. Meyer, and G. Lange, *Ann.*, **571**, 167 (1951).

cis-$C_2H_5O_2CCH=CHCO_2C_2H_5$ $\xrightarrow{i\text{-}C_3H_7CdCl}$ $C_2H_5O_2CCHCH_2CO_2C_2H_5$
$\ \ |$
$\ \ C_3H_7\text{-}i$
 **125**

$C_6H_5CH=CHCOC_6H_5$ $\xrightarrow[M=Be, Zn]{(C_6H_5)_2M}$ $(C_6H_5)_2CHCH_2COC_6H_5$ (Eq. 29)

$RCH=C(CO_2C_2H_5)_2$ $\xrightarrow[R=CH_3, C_2H_5, i\text{-}C_3H_7]{CH_2CdCl}$ $\underset{}{RCHCH(CO_2C_2H_5)_2}$
126

carbonyl addition and reduction.[1, 149] Recently, trialkylboranes have been shown to undergo conjugate addition of primary and secondary alkyl to acetylacetylene, producing a mixture of cis- and $trans$-β-alkyl α,β-unsaturated ketones.[150] Trialkylboranes also add in a conjugate manner to β-unsubstituted[151–153] and β-alkyl[154, 155] vinylic aldehydes and ketones. The *β-unsubstituted* enals and enones react rapidly with trialkylboranes, but only one of the three alkyl groups is transferred to the β-carbon atom of the substrate. Trialkylborane conjugate addition to *β-alkyl* enals and enones uses all three of the alkyl groups on boron but must be induced by diaryl peroxides,[154] by light,[154] or by oxygen.[155] Despite these limitations, trialkylborane conjugate addition to vinylic aldehydes and ketones may be as effective as, or even more effective in a few instances (*e.g.*, α-bromoacrolein) than, 1,4 addition of organocopper reagents. Thus the only examples of organocopper conjugate addition to methyl vinyl ketone involve a lithium divinylcuprate[73b] and cuprous chloride catalysis of *m*-methoxybenzylmagnesium chloride,[93] but variously substituted enones have undergone successful organocopper conjugate addition (see Tables I and III).* No example has been reported of organocopper conjugate

---

* Conjugate addition of secondary and tertiary alkyl groups to methyl vinyl ketone has recently been accomplished with B-alkylboracyclanes; H. C. Brown and E. Negishi, *J. Amer. Chem. Soc.*, **93**, 3777 (1971).

[149] For 1,2 addition of triethylaluminum to unsaturated carbonyl systems, see Y. Baba, *Bull. Chem. Soc. Jap.*, **41**, 928 (1968).

[150] A. Suzuki, S. Nozawa, M. Itoh, H. C. Brown, G. W. Kabalka, and G. H. Holland, *J. Amer. Chem. Soc.*, **92**, 3503 (1970).

[151] A. Suzuki, A. Arase, H. Matsumoto, M. Itoh, H. C. Brown, M. M. Rogic, and M. W. Rathke, *J. Amer. Chem. Soc.*, **89**, 5708 (1967).

[152] H. C. Brown, M. M. Rogic, M. W. Rathke, and G. W. Kabalka, *J. Amer. Chem. Soc.*, **89**, 5709 (1967).

[153] H. C. Brown, G. W. Kabalka, M. W. Rathke, and M. M. Rogic, *J. Amer. Chem. Soc.*, **90**, 4165 (1968).

[154] H. C. Brown and G. W. Kabalka, *J. Amer. Chem. Soc.*, **92**, 712 (1970).

[155] H. C. Brown and G. W. Kabalka, *J. Amer. Chem. Soc.*, **92**, 714 (1970).

addition to β-unsubstituted acroleins, but the β-alkyl-α-methylacrolein **61** is effectively methylated with lithium dimethylcopper.[51] Copper catalysis apparently has no influence on the course of Grignard addition to α-bromocrotonic acid or esters,[94] nor do stoichiometric organocopper reagents add in a conjugate manner to α-halo β,β-disubstituted enoate esters.[32] A significant advantage of the trialkylboranes for conjugate addition is that they are easily prepared in high yield by hydroboration of the corresponding olefins.[153]

$$n\text{-}C_6H_{13}CH\!=\!C(CH_3)CHO$$
$$\mathbf{61}$$

Although organocopper reagents are of general utility for conjugate addition, unsaturated carbonyl substrates bearing specific functional groups (*e.g.*, α-bromoacrolein just discussed) or having specific steric interactions may undergo more successful conjugate addition with other organometallic reagents. Thus conjugate addition of cyanide using diethylaluminum cyanide affords an indirect method for methyl conjugate addition, as indicated by the accompanying scheme.[156] This method has recently been used effectively on enones inert to lithium dimethylcopper.[157]

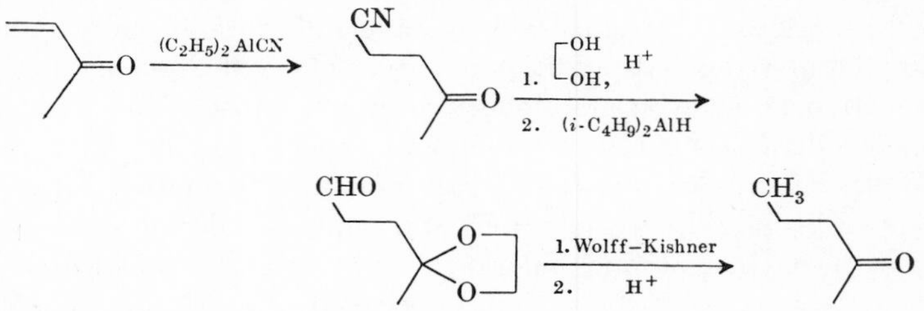

Several organo-transition metal compounds other than organocopper reagents undergo limited conjugate addition. Phenylmanganese iodide adds in a 1,4 manner to benzalacetophenone,[146] but a mixture of 1,4 and 1,2 adducts is obtained from cyclohexenone and lithium trimethylmanganese.[53, 158] It is presently unclear whether there is any relationship between organocopper conjugate and anionic cobalt(I) conjugate addition.[159] Recently 1,4 addition of acyl groups to α,β-ethylenic nitriles,[160]

---

[156] W. Nagata, M. Narisada, and T. Sugasawa, *J. Chem. Soc.* (C), **1967**, 648 and references therein.
[157] See O. R. Rodig and N. J. Johnston, *J. Org. Chem.*, **34**, 1942 (1969).
[158] E. J. Corey and G. H. Posner, *Tetrahedron Lett.*, **1970**, 315.
[159] G. N. Schrauzer, *Accts. Chem. Res.*, **1**, 97 (1968).
[160] E. Yoshisato, M. Ryang, and S. Tsutsumi, *J. Org. Chem.*, **34**, 1500 (1969).

ketones, and esters[161] has been achieved with acylnickel reagents; allylnickel species also undergo conjugate addition to α,β-ethylenic nitriles and esters.[162]

SYNTHETIC UTILITY

Attaching a hydrocarbon group selectively at a carbon atom β to a carbonyl group has continually been a serious challenge to synthetic organic chemists. Typically the β carbon is electrophilic or, in relatively few cases, nucleophilic (see Eq. 30). As discussed in the Introduction, the electrophilic β carbon atom of an α,β-unsaturated carbonyl system can undergo carbon-carbon bond formation with nucleophilic enolate anions (Michael reaction)[2] to form a 1,5-dicarbonyl species (Eq. 31), or the substrate can undergo carbonyl carbon addition with various organoalkali metal reagents derived from unstabilized carbanions. Reaction with Grignard reagents usually affords a *mixture* of 1,2 and 1,4 addition products.[3] Organocopper reagents are generally the reagents of choice for efficient and specific placement of hydrocarbon groups β to a carbonyl function.

$$\text{(Eq. 30)}^{163}$$

$$\underset{\|}{\overset{O}{C}}-\bar{C}R_2 + CH_2{=}CHCOCH_3 \rightarrow \underset{\|}{\overset{O}{C}}-CR_2-CH_2CH_2COCH_3 \quad \text{(Eq. 31)}$$

Other useful methods of limited scope are available for attachment of alkyl groups to the β carbon atom of an α,β-unsaturated carbonyl species. Trialkylboranes (p. 49) and diethylaluminum cyanide (p. 50) have already been mentioned. Conversion of an α,β enone to an α,β-cyclopropyl ketone followed by lithium in ammonia reduction allows overall addition of a

$$\text{(Eq. 32)}$$

---

[161] E. J. Corey and L. S. Hegedus, *J. Amer. Chem. Soc.*, **91**, 4926 (1969).

[162] G. P. Chiusoli and L. Cassar, *Angew. Chem. Int. Ed. Eng.*, **6**, 124 (1967) and references cited therein.

[163] G. Stork, P. Rosen, N. Goldman, R. V. Coombs, and J. Tsuji, *J. Amer. Chem. Soc.*, **87**, 275 (1965); *cf.* R. G. Carlson and R. G. Blecke, *Chem. Commun.*, **1969**, 93.

methyl group to the β carbon atom of the original α,β-unsaturated ketone (Eq. 32).[71, 164]

In exploring the scope and limitations of organocopper conjugate addition to α,β-unsaturated carbonyl substrates, investigators have synthesized a large variety of β-alkyl and β-aryl carbonyl compounds, themselves often interesting and useful molecules as discussed throughout this review. A substantial number of β-substituted carbonyl species, however, have been prepared by organocopper conjugate addition *specifically as synthetic intermediates* whose structural elaboration has led to a variety of natural products. Enumeration of these naturally occurring substances provides an indication of the broad synthetic utility of organocopper reagents.

Methylcopper reagents have been used to introduce axial methyl groups in various positions of the steroid nucleus (p. 40), to generate structurally specific olefins as exemplified by the synthesis of 5-methyl-3-coprostene[79],* (Eq. 33), and to prepare stereospecifically the diterpene ($\pm$)-methyl vinhaticoate (**127**)[165] and the terpene d,l-verbenalol (**128**).[112] Lithium dimethylcopper has been used to prepare ($\pm$)-isolongifolene (**129**)[104] and to place an axial methyl group on various cyclic enones, leading stereoselectively to a general synthesis of hydroazulenes such as

(Eq. 33)

---

* The methyl group indicated as "CH$_3$" is the one introduced by the organocopper reagent.

[164] (a) W. G. Dauben and E. J. Deviny, *J. Org. Chem.*, **31**, 3794 (1966); (b) for a recent example of the utility of this method, see R. E. Ireland and S. C. Welch, *J. Amer. Chem. Soc.*, **92**, 7232 (1970).

[165] T. A. Spencer, R. M. Villarica, D. L. Storm, T. D. Weaver, R. J. Friary, J. Posler, and P. R. Shafer, *J. Amer. Chem. Soc.*, **89**, 5497 (1967); *cf.* T. A. Spencer, R. A. J. Smith, D. L. Storm, and R. M. Villarica, *J. Amer. Chem. Soc.*, **93**, 4856 (1971).

130[166] and to such sesquiterpenes as (±)-nootkatone **(131)**,[167] (±)-eremophil-3,11-diene **(132)**,[80] hinesol **(133)**,[124] and (±)-seychellene **(134)**.[121]

The isopropenyl group has also been introduced axially via an organocopper reagent in a stereoselective preparation of the degradation product **135** of the sesquiterpene aristolone,[72] and homoallylcopper species have been applied to the synthesis of isoprenoid systems, leading to a total synthesis of the insect juvenile hormone **136**.[32, 54]

[166] J. A. Marshall, N. H. Andersen, and P. C. Johnson, *J. Org. Chem.*, **35**, 186 (1970).
[167] M. Pesaro, G. Bozzato, and P. Schudel, *Chem. Commun.*, **1968**, 1152.

                            135   136

Clearly, all of the possibilities for synthetically useful organocopper conjugate additions have not yet been explored. Within the next ten years, advances in experimental technique (*e.g.*, control of temperature, solvent, complexing ligands), development of different organocopper reagents (*e.g.*, XCH_2Cu, $RR'CuLi$, $ROCu$), and use of diverse types of substrates (*e.g.*, α,β-cyclopropyl ketones,[167a,b] vinyl phosphines and sulfides) should substantially increase the general synthetic utility of organocopper reagents.[167c]

EXPERIMENTAL FACTORS

Preparation and Handling of Organocopper Reagents

Because of their high reactivity and low thermal stability, both catalytic and stoichiometric organocopper reagents are prepared *in situ* and used immediately. Air and moisture must be rigorously excluded; reactions are generally run in an atmosphere of argon or prepurified nitrogen, and liquid reagents are best transferred by dry hypodermic syringes and introduced into the reaction mixture through a rubber septum-capped side arm of the reaction flask. Solid reagents should generally be added through a funnel with its stem extending into the neck of the reaction flask out of which a rapid and constant flow of inert gas is maintained.

Complete formation of stoichiometric organocopper reagent can often be judged by a negative Gilman test with Michler's ketone[49] or, less accurately, by visually following the dissolution of copper salt. Alternatively, an aliquot quenched at low temperature with benzoyl chloride,[17] for example, might indicate whether organocopper formation is complete (*i.e.*, if any alcoholic product is produced, then organocopper formation is probably incomplete).[168]

Since most of the copper salts used to prepare the organocopper reagents are not hygroscopic, glove-bag or dry-box procedures are unnecessary; highly effective stoichiometric organocopper reagents have been prepared from carefully purified and dried[16] as well as from commercial samples (*e.g.*, Fisher Chemical Co.)[64] of cuprous iodide. Similarly, commercially

---

[167] (a) Organocopper conjugate addition to cyclopropyl enones has recently been reported: C. Frejaville and R. Jullien, *Tetrahedron Lett.*, **1971**, 2039; (b) J. A. Marshall and R. A. Ruden, *Tetrahedron Lett.*, **1971**, 2875; (c) P. E. Eaton and R. H. Mueller, *J. Amer. Chem. Soc.*, **94**, 1015 (1972) and P. E. Eaton, G. F. Copper, R. C. Johnson, and R. H. Mueller, *J. Org. Chem.*, in press.

[168] G. H. Posner and C. E. Whitten, *Tetrahedron Lett.*, **1970**, 4647.

available organolithium reagents are usually satisfactory, a limitation occasionally being the solvent in which the reagent is prepared; this solvent becomes part of the reaction mixture and may influence the course of conjugate addition.

Temperature

The effects of reaction temperature, solvent, and workup procedure on *catalytic* organocopper conjugate additions have not been studied in detail. The sensitivity of these catalytic organocopper reactions to the order of mixing reagents and to the solubility of the copper catalyst[38] has been noted (p. 10).

Stoichiometric organocopper reagents are often highly sensitive to such variables as reaction temperature, solvent, and workup procedure; their sensitivity to inorganic salts[28-30] and to complexing additives has already been discussed (see pp. 12 and 14).

Temperature variation has been observed to affect both the rate of formation of organocopper reagents and their stability once they have been formed. Lithium di-n-butylcopper, for example, is formed in ether from n-butyllithium and cuprous iodide rapidly ($\ll 1$ minute) at $0°$ but slowly (>10 minutes) at $-40°$;[32] optimum conditions for formation of such organocopper reagents, therefore, must involve careful control of reaction temperature (see p. 61).

The thermal stability of an organocopper reagent depends on the nature of the reagent and on the structure of the organic group. Phenylcopper, for example, is stable below $80°$, but different complexes of phenylcopper [*e.g.*, $(C_6H_5Cu)_4C_6H_5Li\cdot n(C_2H_5)_2O$ or $(C_6H_5Cu)_2(C_6H_5)_2Mg\cdot n(THF)$ or $C_6H_5CuP(C_6H_5)_3$] have different thermal stabilities.[35, 56]

The structure of R strongly influences the stability and reactivity of organocopper reagents. Thus, whereas phenylcopper is stable below $80°$, methylcopper decomposes at room temperature and ethylcopper at $-18°$ in ether.[17, 18] This trend in organocopper thermal stability generally follows the thermal stability of the corresponding organolithium reagents (stability: $C_6H_5Li > CH_3Li > C_2H_5Li$).[169] Similarly, lithium dimethylcopper in ether solution under an inert atmosphere is stable for hours at $0°$, but the more reactive lithium di-n-alkylcopper and secondary and tertiary organocopper reagents rapidly decompose in ether above $-20°$. Thus, for most effective preparation and use of these less stable, more reactive organocopper reagents, the reaction temperature should be carefully controlled. Usually stoichiometric n-alkyl, secondary, and tertiary organocopper reagents are prepared either below $-20°$ for a sufficient amount of time (usually >5 minutes) to allow complete formation of

[169] T. L. Brown, *Advan. Organometal. Chem.*, **3**, 365 (1965).

reagent[21] or at 0° for several minutes (rapid reagent formation) followed immediately by cooling to below −20°[32] (see methyl 3-methyl-*trans*-2-decenoate, p. 61). The reaction temperature used for conjugate addition generally should be the lowest temperature which gives an acceptable reaction rate.

Solvent

Only two kinds of solvent have been widely used for organocopper conjugate additions: aromatic hydrocarbons (benzene and toluene) and ethers (tetrahydrofuran and diethyl ether). Not only does the solvent affect the configurational stability of vinylic copper species formed by organocopper conjugate addition to acetylenic carbonyl substrates (see p. 17), but the solvent also effects the rate of conjugate addition to these acetylenic substrates. Thus at −78° in 5 minutes conjugate addition of lithium dimethylcopper to methyl 2-decynoate proceeds to the extent of 7% in tetrahydrofuran, 47% in toluene, and 100% in ether.[32] Despite the slow rate of reaction in tetrahydrofuran, it is the solvent of choice because with it the most stereoselective conjugate addition to acetylenic substrate occurs.

The choice of solvent for stoichiometric organocopper conjugate additions to α,β-ethylenic carbonyl compounds is less clear, although diethyl ether has been widely used with success; toluene has recently been shown to be useful also, but additions in tetrahydrofuran may be too sluggish to be practicable.[51] For catalytic organocopper reagents, ether or tetrahydrofuran is most often used, but selection of one over the other to optimize yields of 1,4 adducts should usually be made on the basis of small-scale preliminary experiments with each solvent and different copper catalysts; tetrahydrofuran has been used to increase conjugate addition by solubilizing the copper catalyst (usually cupric acetate monohydrate),[36] but tetrahydrofuran has also been observed to decrease the ratio of 1,4 to 1,2 adducts relative to the ratio achieved in ether.[16, 84] The solvent of choice should normally be freshly distilled from a drying agent (*e.g.*, lithium aluminum hydride) before use to remove traces of water.

Workup Procedure

The workup procedure may be critical in some cases. The major difficulty has been formation of a (1,4 plus 1,2) di-adduct, arising presumably by hydrolysis of the intermediate enolate formed from conjugate addition followed by 1,2 addition of residual unhydrolyzed organometallic species.[16] Two procedures have been developed to eliminate diadduct (saturated alcohol) formation. The first involves pouring the reaction mixture slowly into aqueous ammonium chloride with vigorous stirring,[16] and the second involves pouring the reaction mixture slowly into cold, rapidly stirred dilute hydrochloric acid.[122] For substrates not sensitive to cold hydrochloric acid for a short time, the second procedure should be used because it eliminates di-addition more effectively than does the first procedure.[122]

The effect of organocopper *concentration* on conjugate addition has been noted in only a few isolated instances. For both catalytic and stoichiometric organocopper reagents higher yields of conjugate addition were obtained on one occasion in more dilute solutions.[34, 51]

Organocopper reagents will undoubtedly continue to be used for effective conjugate addition reactions. To systematize reporting of experimental results in this area, the following comments are made. Inorganic salts[16] and trace impurities[33, 34] in the reagents used to prepare organocopper species[21] have been found on occasion to alter the extent of conjugate addition; future authors are strongly urged, therefore, to indicate explicitly the source and/or method of purification of all reagents. Furthermore, since the reactivity of many organometallic reagents depends on their state of aggregation[170, 171] which may change as concentration of reagent is varied, publications dealing with organocopper reagents should specify the concentration of any organocopper reagent used.

EXPERIMENTAL PROCEDURES

All of the preceding discussion has been directed at an *analysis* of the most important aspects of organocopper conjugate addition. In this section, examples are given to show how these various aspects can be put together to allow effective organocopper conjugate addition. The examples have been carefully chosen to illustrate useful and general experimental procedures, all performed under inert atmosphere, and have been organized into three categories: catalytic organocopper reagents, stoichiometric organocopper reagents, and special types of α,β-unsaturated substrates.

Catalytic organocopper reagents are illustrated by copper-catalyzed Grignard conjugate addition of secondary R (isopropyl addition to ethyl *sec*-butylidenecyanoacetate), of tertiary R (*t*-butyl addition to 10-methyl-4,6-hexal-3-one), and of vinylic R (isopropenyl addition to $9\alpha,10\alpha$-dimethyl-*trans*-1-octal-3-one). Stoichiometric organocopper reagents are exemplified by conjugate addition of methyl R (lithium dimethylcopper addition to $10\beta H$-7β-isopropenyl-4-methyl-4-octal-3-one and methylcopper-trimethylphosphite addition to 5-methyl-2-cyclohexenone), of phenyl R (lithium diphenylcopper addition to 5-methyl-2-cyclohexenone), and of allylic R (lithium diallylcopper addition to cyclohexenone).

[170] W. H. Glaze and C. H. Freeman, *J. Amer. Chem. Soc.*, **91**, 7198 (1969).
[171] T. L. Brown, *J. Organometal. Chem.*, **5**, 191 (1966).

Conjugate addition of organocopper reagents to α,β-unsaturated substrates requiring special experimental procedures are illustrated by lithium di-n-heptylcopper stereospecific addition to methyl 2-butynoate, by copper-catalyzed asymmetric synthesis of (—)-phenylbutyric acid, and by copper-catalyzed selective conjugate addition to the less hindered (i.e., D-ring enone) of the two enone groups in 16-dehydroprogesterone.

Except for the conjugate addition of n-heptyl, these experimental procedures do not incorporate all measures for optimizing yields and minimizing side-product formation. The precautionary measures outlined in the text, especially careful control of reaction temperature, would probably result generally in higher yields of 1,4 adducts than those reported.

Ethyl 2-Cyano-3-isopropyl-3-methylpentanoate

(Conjugate addition of secondary R using cuprous cyanide catalyst.)[39, 103] To 125 g (0.75 mol) of ethyl sec-butylidenecyanoacetate and 3.75 g (0.04 mol, 4 mol%) of cuprous cyanide in about 500 ml of ether below 20° was added dropwise 0.90 mol of ethereal isopropylmagnesium bromide. After completion of Grignard addition, the reaction mixture was refluxed for 1 hour and let stand at room temperature for 20 hours. Product isolation involved pouring the mixture into 200 ml of concentrated hydrochloric acid and 500 g of crushed ice with rapid stirring, extracting with ether, washing the ether extracts with water and 10% aqueous sodium bicarbonate, and drying the extracts over anhydrous sodium sulfate. Removal of ether under reduced pressure and distillation through a 60-cm heated Vigreux column gave 103.7 g (65.5%) of ethyl 2-cyano-3-isopropyl-3-methylpentanoate, bp 127–130° (6 mm).

Hydrolysis and decarboxylation of the product (0.25 mol) using 1.8 mol of potassium hydroxide in 350 ml of ethylene glycol (50 hours reflux) gave, after distillation, 3-isopropyl-3-methylpentanoic acid (94.1% yield from cyanoester), bp 133–134° (14 mm), n^{25}D 1.4463.

trans-10-Methyl-7α-t-butyl-4-octal-3-one

(Conjugate addition of t-alkyl R; inverse addition of Grignard reagent to substrate.)[34] A solution containing 1.00 g (6.2 mmols) of 10-methyl-4,6-hexal-3-one and 400 mg (2.0 mmols, 9 mol%) of cupric acetate monohydrate in 34 ml of dry tetrahydrofuran was cooled to —15° in a dry ice-acetone bath. The reaction mixture was maintained at —10 to —20° and efficiently stirred while 46 ml of an ethereal solution of 0.5 M (23.0 mmols) t-butylmagnesium chloride was added over 0.5 hour. The mixture was allowed to reach room temperature over a 2-hour period and heated to reflux for 15 minutes.

Product isolation involved addition of aqueous ammonium chloride to the cooled reaction mixture, extraction with ether, washing the extracts with saturated brine, and drying them over anhydrous magnesium sulfate, followed by solvent removal under reduced pressure and then basic alumina chromatography (100 g, Merck). The combined fractions from elution with benzene and 5% ether in benzene were distilled (0.1 mm, 65–85° bath temperature) to give 0.55 g (40%) of trans-10-methyl-7α-t-butyl-4-octal-3-one, bp 85° (0.1 mm), forming a 2,4-dinitrophenylhydrazone, mp 166–167.5°.

9α,10α-Dimethyl-1β-isopropenyl-*trans*-3-decalone

(Conjugate addition of vinylic R.)[72] To a stirred solution of isopropenylmagnesium bromide (0.29 g, 2 mmols) in 2 ml of dry tetrahydrofuran was added approximately 6 mg (3 mol%) of anhydrous cuprous chloride, and the resulting mixture was cooled to 0°. A solution of 9α,10α-dimethyl-*trans*-1-octal-3-one (0.1 g, 0.56 mmol) in 4 ml of tetrahydrofuran was added via hypodermic syringe, and the reaction mixture was stirred under nitrogen at 0° for 15 minutes, and then refluxed for 1 hour. The cooled reaction mixture was poured slowly into rapidly stirred, ice-cold dilute hydrochloric acid, and the product was isolated by extraction with ether. Distillation of the crude product gave 110 mg (89%) of a clear oil indicated by gas-liquid chromatography (10 ft × 0.375 in. column of 30% SE-30 on 60-80 Chromosorb W, 245°) to contain 80–85% of 9α,10α-dimethyl-1β-isopropenyl-*trans*-3-decalone. Preparative glc gave the pure product with n^{20}_D 1.5147.

(±)-Eremophil-11-en-3-one

(Conjugate addition of methyl R using lithium dimethylcopper and acidic workup.)[122] To a stirred slurry of cuprous iodide (6.0 g, 31.5 mmols) in 120 ml of anhydrous ether at 0° in a nitrogen atmosphere was added by syringe 39.4 ml of 1.59 M (62.5 mmols) ethereal methyllithium. The resulting solution of tan lithium dimethylcopper was stirred at 0° for 5 minutes, and then a solution of 2.0 g (10.0 mmols) of 10βH-7β-isopropenyl-4-methyl-4-octal-3-one (steroid numbering) in 80 ml of anhydrous ether was added dropwise over a period of 20 minutes. Stirring was continued for an additional 2 hours at 0°, and then the reaction mixture was poured slowly into 800 ml of vigorously stirred 1.2 N hydrochloric acid. Ether extraction, drying the ethereal extracts, and removing solvent under reduced pressure produced 1.9 g of a clear colorless liquid. Analytical gas chromatography (column of 20% FFAP on 60-80 Chromosorb W, 215°) indicated that the material consisted of 5% octalone starting material, 82% of (±)-eremophil-11-en-3-one, and a number of minor, unidentified components. Preparative gas chromatography (same column) allowed isolation of an analytical sample of (±)-eremophil-11-en-3-one; n^{20}D 1.5041; infrared (film), λ_{max} 5.86, 6.10, 11.27 μ; p-toluenesulfonylhydrazone (mp 159–161°); semicarbazone (mp 191.5–193.5°).

3,5-Dimethylcyclohexanone

(Conjugate addition of methyl using an RCu-phosphite complex).[28] To 1.40 g (7.3 mmols) of purified[59] cuprous iodide and 1.808 g (14.5 mmols) of trimethyl phosphite in 15 ml of ether was added 6 ml of ethereal methyllithium (7.2 mmols), causing precipitation of bright yellow methylcopper which dissolved immediately upon addition of an additional 904 mg (7.2 mmols) of trimethyl phosphite. To a cold (0°) solution of 9.8 mmols of 0.47 M tris(trimethyl phosphite) complex of methylcopper, $[(CH_3O)_3P]_3CuCH_3$, was added with stirring a solution (7–15 ml) containing a weighed amount of n-butylbenzene (vpc internal standard) and enough 5-methyl-2-cyclohexenone to make the concentration of the ketone in the reaction mixture 0.30–0.43 M. The resulting solution was stirred at 0° for 5–10 minutes and then poured into saturated aqueous ammonium chloride adjusted to pH 8 by addition of ammonia. Extraction with ether, drying the ethereal extracts, and vpc analysis (column packed with 1,2,3-tris-(β-cyanoethoxy)propane on Chromosorb P was used) indicated 3,5-dimethylcyclohexanone to be present in 90–91% yield (98% *trans*, 2% *cis*). Preparative gas chromatography gave pure 3,5-dimethylcyclohexanone from which was prepared a 2,4-dinitrophenylhydrazone (mp 108.4–110°).

3-Methyl-5-phenylcyclohexanone

(Conjugate addition of phenyl R using lithium diphenylcopper.)[172] To a cold (0°) solution of lithium diphenylcopper prepared from 124 mmols of phenyllithium and 8.90 g (62.0 mmols) of cuprous bromide in 118 ml of ether was added a solution of 6.64 g (60.3 mmols) of 5-methyl-2-cyclohexenone in 50 ml of ether. After the resulting mixture had been stirred for 15 minutes at 0°, an excess of an aqueous solution (pH 8) prepared from ammonium chloride and ammonia was added and air was passed through for a short time to complete the oxidation of insoluble Cu(I) species to the soluble Cu(II)-NH_3 complex.

The organic layer was separated, combined with the ether extract of the aqueous phase, and then washed with aqueous ammonium chloride, dried, and concentrated. The residual yellow oil (12.66 g) was distilled in a short-path still (0.09 mm and 96–110°) to separate 7.61 g of crude *trans*-3-methyl-5-phenylcyclohexanone. An ether solution of this material was

[172] H. O. House, R. W. Giese, K. Kronberger, J. P. Kaplan, and J. F. Simeone, *J. Amer. Chem. Soc.*, **92**, 2800 (1970).

washed with aqueous 5% NaOH to remove traces of phenol. The *trans* conjugate adduct which constituted >90% of the distilled material was purified via its oxime (mp 92–93°), giving on hydrolysis of the oxime (1 hour reflux with aqueous 30% oxalic acid) and short-path distillation (0.05 mm and 90–95°) *trans*-3-methyl-5-phenylcyclohexanone as a colorless liquid, n^{23}D 1.5356–1.5360; infrared (CCl$_4$), 5.81 μ (C=O).

3-Allylcyclohexanone

(Conjugate addition of allylic R using lithium diallylcopper.)[55] To a cold (−78°) solution of lithium diallylcopper [prepared from 1.45 g (3.00 mmols) of the bis-(di-*n*-butyl sulfide) complex of cuprous iodide, [(*n*-C$_4$H$_9$)$_2$S]$_2$CuI, and 6.00 mmols of allyllithium (prepared from tetraallyltin and *n*-butyllithium)] in 11.2 ml of ether was added 2.5 ml of ethereal 2-cyclohexenone (2.50 mmols) and a known amount of naphthalene (internal standard). During the addition, a red precipitate [presumably allylcopper(I)][21] separated. The mixture was allowed to warm to 0° [during which time any unchanged Cu(I) reagent decomposed with separation of metallic Cu] and then was partitioned between ether and an aqueous solution (pH 8) of ammonia and ammonium chloride. Pure 3-allylcyclohexanone (90–94% yield by analytical gas chromatography) was separated from di-*n*-butyl sulfide and naphthalene by preparative gas chromatography (column packed with SE-30 suspended on Chromosorb P was used).

Methyl 3-Methyl-*trans*-2-decenoate

(Stereospecific conjugate addition to acetylenic ester; use of lithium di-*n*-alkylcopper and low temperature.)[32, 64] Lithium di-*n*-heptylcopper was prepared by the addition of 2.0 mmols of *n*-heptyllithium [prepared in hexane solution from 3.45 g (500 mmols) of lithium wire and 30.8 ml (26.9 g, 200 mmols) of *n*-heptyl chloride and separated by syringe from the lithium salts] to 210 mg (1.1 mmols) of cuprous iodide (Fisher Chemical Company) in 12 ml of tetrahydrofuran at 0°. After 2 minutes at this temperature the dark reagent was cooled to −78°, and 100 mg (1.0 mmol) of methyl 2-butynoate was added. Stirring was continued for 3 hours at −78°, and then the reaction mixture was poured directly into 5 ml of well-stirred methanol cooled in a dry ice-acetone bath.

The product was isolated by adding a saturated salt solution and ether, filtering through a pad of Hyflo Super Cel, extracting with ether, drying (magnesium sulfate), and concentrating under reduced pressure to give 316 mg of crude product. This material contained a considerable amount of tetradecane (produced by Wurtz coupling during the *n*-heptyllithium preparation), and preparative thin-layer chromatography (9:1 hexane: ether) was used to obtain a pure sample (176 mg, 90%), which showed a single spot of R_f 0.49 upon tlc analysis (7:1 hexane:ether); infrared (film), λ_{max} 5.80 (C=O), 6.10 μ (C=C); nmr (CCl$_4$), 2.13 δ (doublet, $J = 1.3$ Hz, 3H, CH$_3$—C=). Glpc analysis (16 ft × 0.125 in. 5% LAC-446 on 80–100 Diatoport S column, 190°, 60 ml/min) showed a peak at 7.1 minutes (98.6%), corresponding to methyl 3-methyl-*trans*-2-decenoate, and a small peak at 5.7 minutes (1.4%), corresponding to the *cis* isomer.

The product from another reaction run on a smaller scale with shorter reaction time showed no trace (<0.1%) of the *cis* isomer.

(−)-3-Phenylbutyric Acid

(Asymmetric induction; stepwise addition of copper catalyst.)[90] To 60 ml of a solution of phenylmagnesium bromide [prepared from 1.82 g (75 mg atoms) of magnesium and 11.8 g (75 mmols) of bromobenzene] at −17 to −15° was added first 27 mg (0.35 mol%) of cuprous chloride and then, dropwise during 1 hour, 30 mmols of the 1,2:5,6-di-0-isopropylidene-D-glucose ester of crotonic acid in 60 ml of ether. After each 20-minute interval during the addition another 27 mg of cuprous chloride was added. The last portion was added just after completion of ester addition, making a total of 108 mg (1.4 mol%) of cuprous chloride. After addition of the ester had been completed, stirring was continued at −17 to −15° for 0.5 hour and then at room temperature for 0.5 hour. Product isolation in-involved the following sequential steps: pouring the reaction mixture into aqueous ammonium chloride (250 ml) with vigorous shaking; extracting with ether and drying and evaporating the ether; hydrolyzing the residual oil (until ester carbonyl infrared absorption disappeared) by refluxing for 5 hours in ethanolic potassium hydroxide; removing the ethanol and adding water and ether; and separating aqueous from organic phases.

The aqueous phase was acidified and extracted with ether to give, after drying and removing the ether, (−)-3-phenylbutyric acid (44% synthetic yield, 70% optical yield), $[\alpha]_D$ −36°, bp 104–106° (0.5–1.0 mm), identified also by infrared comparison with an authentic sample.

The organic phase was washed with water, dried, and evaporated to a residue which was recrystallized from ligroin, giving 1,2:5,6-di-*O*-isopropylidene-D-glucose in 60–85% yield.

16α-Methylprogesterone

(Use of only twofold Grignard excess over substrate and short reaction time to allow selective conjugate addition to the less hindered of two enone groups in the same molecule.)[120] Ether-free methylmagnesium iodide was prepared by adding 30 ml of dry peroxide-free tetrahydrofuran and toluene to 10 ml of a 2.14 M ethereal solution of methylmagnesium iodide (prepared from 4 g of magnesium and 3.2 ml of methyl iodide) and then removing the ether by distillation under nitrogen.

Aliquots indicated that the reaction mixture contained 9.47 mmols of Grignard complex. At 25°, 100 mg (1.0 mmol, 1 mol%) of dry cuprous chloride was added and, after 5 minutes, the reaction mixture was cooled to −10 to 0°. A solution of 1.6 g (5.0 mmols) of 16-dehydroprogesterone in 30 ml of toluene was added rapidly and the reaction mixture allowed to stay at −10 to +12° for 30 minutes.

Product isolation involved successive treatment with aqueous ammonium chloride and sodium thiosulfate, extraction with benzene, drying the extracts and then removing the solvent under reduced pressure. Upon trituration with ether, 1.2 g of 16α-methylprogesterone crystallized; chromatography of the mother liquor gave another 0.25 g (total yield, 90%). Recrystallization from n-hexane gave 16α-methylprogesterone with the following properties: mp 137–138°, $\varepsilon_{241\ nm} = 14,400$, ir: 5.87 μ (20-ketone) and 5.98 μ + 6.17 μ (Δ^4-3-ketone).

TABULAR SURVEY

An attempt has been made to include in the tables all *copper-catalyzed* and *stoichiometric* organocopper conjugate addition reactions reported through December 1970; there are some references through December 1971. Tables I and II cover *copper-catalyzed* Grignard conjugate additions to α,β-ethylenic and α,β-acetylenic carbonyl compounds, respectively. Tables III and IV refer to stoichiometric organocopper reagent conjugate additions to the same two general classes of unsaturated carbonyl substrates. Included in the class of *cyclic* α,β-ethylenic carbonyl compounds (Tables IB and IIIB) are those molecules in which the double bond or the carbonyl of the conjugated system or both are part of a ring.

Within each table the substrates are listed in order of increasing number of carbon atoms, subdivided in order of increasing number of hydrogen atoms, and isomers are arranged according to increasing substituent number (*e.g.*, 3-methyl-2-cyclohexenone before 4-methyl-2-cyclohexenone). α,β-Unsaturated carboxylic acid derivatives are listed by the number of carbon atoms in the corresponding *acid*, the methyl ester for example being followed by the ethyl ester, and so on. Derivatives of alcohols are listed by the number of carbon atoms in the alcohol. When a number of different catalysts or reagents (or reaction conditions) have been used for the same compound, they are listed mainly in order of increasing complexity (*e.g.*, CH_3Cu before C_2H_5Cu).

When there is more than one reference, the experimental data are taken from the first reference, and the remaining references are arranged in numerical order.

TABLE I. Copper-Catalyzed Conjugate Addition of Grignard Reagents to α,β-Ethylenic Carbonyl Compounds

Number of Carbon Atoms	Carbonyl Substrate	Grignard Reagent	Catalyst (Reaction Conditions)[a]	Product(s)[b] and Yield(s)(%)	Refs.
			A. Acyclic Carbonyl Substrates		

4	cis-$C_2H_5O_2CCH$=$CHCO_2C_2H_5$	i-C_3H_7MgBr	CuCl (stepwise addition, $-10°$)	$C_2H_5O_2CCHCH_2CO_2C_2H_5$ (I)
				    |
				    R
				R = i-C_3H_7 (81)
					9

		t-C_4H_9MgCl	CuCl (stepwise addition, $-10°$)	I, R = t-C_4H_9 (70)	8

	s-$C_4H_9O_2CCH$=$CHCO_2C_4H_9$-s			s-$C_4H_9O_2CCHCH_2CO_2C_4H_9$-$s$ (I)
				    |
				    R	

	$trans$-	n-C_4H_9MgBr	CuCl	I, R = n-C_4H_9 (38)	8
	cis-	C_2H_5MgBr	CuCl (stepwise addition, $-10°$)	I, R = C_2H_5 (57, 43^c)	14, 9
		n-C_3H_7MgBr	CuCl (stepwise addition, $-10°$)	I, R = n-C_3H_7 (63)	9
		i-C_3H_7MgBr	CuCl (stepwise addition, $-10°$)	I, R = i-C_3H_7 (81^c)	14, 9
		n-C_4H_9MgBr	CuCl (stepwise addition, $-10°$)	I, R = n-C_4H_9 (57, 67^c)	8, 9
		t-C_4H_9MgBr	CuCl (stepwise addition, $-10°$)	I, R = t-C_4H_9 (74, 78^c)	14, 9
	$CH_3C(Cl)$=$CHCO_2C_2H_5$ $trans$- cis-	C_2H_5MgBr C_2H_5MgBr	CuCl (3 mol %) CuCl (3 mol %)^d	$C_2H_5C(CH_3)$=$CHCO_2C_2H_5$ (I) I (36 $trans$,^{e} 42 cis) I (38 $trans$,^{e} 7 cis)	173
	$CH_3C(Cl)$=$CHCO_2C_4H_9$-s $trans$- cis-	C_2H_5MgBr C_2H_5MgBr	CuCl (3 mol %) CuCl (3 mol %)	$C_2H_5C(CH_3)$=$CHCO_2C_4H_9$-s (I) I (34 $trans$,^{e} 43 cis) I (77 $trans$,^{e} 12 cis)	

	$trans$-CH_3CH=$C(Br)CO_2H$	C_6H_5MgBr	CuCl (2 mol %)	$CH_3CHCH(Br)CO_2H$
				    |
				    C_6H_5
					94

Substrate	Grignard	Catalyst	Products (yields)	Ref.
trans-$CH_3CH=C(Br)CO_2CH_3$	C_6H_5MgBr	CuCl (2 mol %)	erythro- (35), threo- (35) trans-$CH_3CH=CHCOC_6H_5$ (11) $CH_3CHCH(Br)CO_2CH_3$ | C_6H_5	93
$CH_2=CHCOCH_3$	m-$CH_3OC_6H_4CH_2$-MgCl	CuCl (8 mol %,[d] −10°, THF)	erythro- (47), threo- (24) m-$CH_3OC_6H_4(CH_2)_3COCH_3$ (36)	
			C_4H_9-n | CH_3CHCH_2CO	
trans-$CH_3CH=CHCO_2CH_3$	n-C_4H_9MgBr	CuCl (1 mol %)	$CH_3CHCHCO_2CH_3$ (65)	6
trans-$CH_3CH=CHCO_2C_2H_5$	CH_3MgBr	CuCl (1 mol %, −10°)	$(CH_3)_2CHCH_2CO_2C_2H_5$ (54)	97
	n-C_4H_9MgBr	CuCl	$CH_3CHCH_2CO_2C_2H_5$ (70–74) | C_4H_9-n	13
trans-$CH_3CH=CHCO_2C_4H_9$-s	CH_3MgBr	CuCl (1 mol %, 0°)	$CH_3CHCH_2CO_2C_4H_9$-s (I) | R	97, 25
	C_2H_5MgBr	CuCl	I, R = CH_3 (51, 63[f])	25
	n-C_3H_7MgBr	CuCl	I, R = C_2H_5 (56, 79[f])	25, 108
	i-C_3H_7MgBr	CuCl	I, R = n-C_3H_7 (57, 88[f])	14, 25
	n-C_4H_9MgBr	CuCl	I, R = i-C_3H_7 (58, 80[f]) I, R = n-$C_4H_9CH(CH_3)$- $CHCO_2C_4H_9$-s (52)	6
	t-C_4H_9MgCl	CuCl	I, R = n-C_4H_9 (70, 80[f]) I, R = t-C_4H_9 (14, 78[f])	25, 7 25, 108

Note: References 173–204 are on p. 113.

[a] The solvent is diethyl ether unless otherwise noted; stepwise addition means that the copper catalyst was added in several portions.
[b] These are the products formed after hydrolysis of intermediate enolate species.
[c] This yield refers to the reaction run at 0°.
[d] This is the percentage of the number of moles of Grignard reagent used.
[e] Ethyl and alkoxycarbonyl are *trans*.
[f] This is the yield when 1 mol % CuCl was added in several portions during the course of the reaction at −10°.

TABLE I. Copper-Catalyzed Conjugate Addition of Grignard Reagents to α,β-Ethylenic Carbonyl Compounds (*Continued*)

A. Acyclic Carbonyl Substrates

Number of Carbon Atoms	Carbonyl Substrate	Grignard Reagent	Catalyst (Reaction Conditions)[a]	Product(s)[b] and Yield(s)(%)	Refs.
4 (*contd.*)	trans-CH_3CH=$CHCO_2C_4H_9$-s (contd.)	n-$C_8H_{17}MgBr$	CuCl	I, R = n-C_8H_{17} (75)	25
	cis-CH_3CH=$CHCO_2C_4H_9$-s	n-$C_{14}H_{29}MgBr$	CuCl	I, R = n-$C_{14}H_{29}$ (86)	
		n-C_4H_9MgBr	CuCl (stepwise addition)	$CH_3CHCH_2CO_2C_4H_9$-s (51)	7
				$\quad\|$	
				C_4H_9-n	
	trans-CH_3CH=$CHCOO$-menthyl	C_6H_5MgBr	CuCl (stepwise addition, $-15°$)	(−)-$CH_3CHCH_2CO_2H$[g] (I)	90
				$\quad\|$	
				C_6H_5	
				(61; 58, optical yield)	
	(−)-trans-CH_3CH=$CHCOO$-(menthyl cyclohexyl)	C_6H_5MgBr	CuCl (stepwise addition, $-8°$)	I[g] (60; 10, optical yield)	107
	trans-CH_3CH=$CHCOO$-(sugar derivative)	C_6H_5MgBr	CuCl (stepwise addition, $-15°$)	I[g] (42; 5, optical yield)	90
	trans-CH_3CH=$CHCOO$-(sugar derivative)	C_6H_5MgBr	CuCl (stepwise addition, $-15°$)	I[g] (50; 68, optical yield)	90

	Substrate	Reagent	Conditions	Products	Refs.
5	trans-CH$_3$CH=CHCOO-[structure with spiro dioxolane groups]	C$_6$H$_5$MgBr	CuCl (−15°, stepwise addition)	I^g (58; 74, optical yield)	90

	s-C$_4$H$_9$O$_2$CCH=C(CH$_3$)CO$_2$C$_4$H$_9$-s cis-	n-C$_4$H$_9$MgBr	CuCl (1 mol %; 0°)	s-C$_4$H$_9$O$_2$CCHCH(CH$_3$)CO$_2$C$_4$H$_9$-s
				       |
				    C$_4$H$_9$-n (I, 10)
				s-C$_4$H$_9$O$_2$CCH$_2$C(CH$_3$)CO$_2$C$_4$H$_9$-s
				              |
				            C$_4$H$_9$-n (II, 90)
				(total 40)^h (II, 70) (total 21)^h
					8, 7

	trans-CH$_3$CH=CHCOCH$_3$	n-C$_4$H$_9$MgBr	CuCl (1 mol %; 0°)	(I, 30) (II, 70) (total 21)^h (CH$_3$)$_2$CHCH$_2$COCH$_3$, (I, 73)	8, 7
		CH$_3$MgBr	CuCl (5 mol %)	trans-CH$_3$CH=CHC(CH$_3$)$_2$OH (II, 27)	31
			CuCl (10 mol %, excess isoprene)	I (82) II (18)	
			(n-C$_4$H$_9$)$_3$PCuI (1.0 mol %; 25°)	I (95) II (1)	16
			(n-C$_4$H$_9$)$_3$PCuI (25 mol %; isoprene, 25°)	I (95) II (2)	

Note: References 173–204 are on p. 113.

^{a} The solvent is diethyl ether unless otherwise noted; stepwise addition means that the copper catalyst was added in several portions.
^{b} These are the products formed after hydrolysis of intermediate enolate species.
^{g} This is the product after saponification by refluxing in alcoholic potassium hydroxide.
^{h} Di-s-butyl methylsuccinate is the other major product.

TABLE I. COPPER-CATALYZED CONJUGATE ADDITION OF GRIGNARD REAGENTS TO α,β-ETHYLENIC CARBONYL COMPOUNDS *(Continued)*

Number of Carbon Atoms	Carbonyl Substrate	Grignard Reagent	Catalyst (Reaction Conditions)[a]	Product(s)[b] and Yield(s) (%)	Refs.
A. Acyclic Carbonyl Substrates					
5 *(contd.)*	*trans*-$CH_3CH=CHCOCH_3$ *(contd.)*	CH_3MgBr *(contd.)*	$(n\text{-}C_4H_9)_3PCuI$ (1.0 mol %, 25°, 1:1 ether:THF^i)	I (58), II (28)	
			$(n\text{-}C_4H_9)_3PCuI$ (1.0 mol %, 25°, 1:1 ether:DME^j)	I (20), II (77)	
			$CuBr_2$ (1.0 mol %, 25°, 1:1 ether:THF^i)	I (45), II (26)	
			$Cu(OAc)_2$ (1.0 mol %, 25°, 1:1 ether:THF^i)	I (41), II (23)	
			CuI (50 mol %, 0°)	I (87), II (4)	
		$(CH_3)_2Mg$	$(n\text{-}C_4H_9)_3PCuI$ (1.0 mol %, 25°)	I (79), II (17)	
		CH_3Li	$(n\text{-}C_4H_9)_3PCuI$ (1.0 mol %, 25°)	I (8), II (89)	
	trans-$CH_3CH=CHCO_2CH_3$	$n\text{-}C_4H_9MgBr$	CuCl (1 mol %)	CH_3CHCH_2CO 　　| 　$C_4H_9\text{-}n$ $CH_3CHCH_2CO_2CH_3$ (66) 　　| 　$C_4H_9\text{-}n$	6
		C_6H_5MgBr	CuCl (5 mol %)	$CH_3CHCH_2CO_2CH_3$, 　　| 　C_6H_5 (I, 89) *trans*-$CH_3CH=CHC(OH)CH_3$ 　　　　　　　　| 　　　　　　C_6H_5 (II, 11)	31

Substrate	Grignard	Catalyst	Product(s) (% yield)	Refs.
trans-$CH_3CH=C(CH_3)CO_2C_2H_5$	$(C_6H_5)_2Mg$	CuCl (10 mol %)	I (78), II (22)	97
	CH_3MgBr	CuCl	$(CH_3)_2CHCH(CH_3)COCH_3$ (64) $CH_3CHCH(CH_3)COC_4H_9$-n	97
	n-C_4H_9MgBr	CuCl		13
trans-$CH_3CH=C(CH_3)CO_2C_4H_9$-s	CH_3MgBr	CuCl	$(CH_3)_2CHCH(CH_3)COCH_3$ (64)	97
	n-C_4H_9MgBr	CuCl (stepwise addition, $-10°$)	$CH_3CHCH(CH_3)CO_2C_4H_9$-s $\mid$ C_4H_9-n (45)	25
cis-$CH_3CH=C(CH_3)CO_2C_4H_9$-s	n-C_4H_9MgBr	CuCl (stepwise addition)	$CH_3CHCH(CH_3)COC_4H_9$-n (10–17), $\mid$ C_4H_9-n (I, 55)	7
$(CH_3)_2C=CHCO_2C_4H_9$-s	CH_3MgBr	CuCl (1 mol %, $-10°$)	C_4H_9-n (I, 15–20) $(CH_3)_3CCH_2COCH_3$ (33)	97
$(CH_3)_2C=C(CN)CO_2C_2H_5$	n-C_4H_9MgBr	CuCl (1 mol %, 0°, inverse addition[k])	$(CH_3)_2CCH(CN)CO_2C_2H_5$ (77)	8
$(CH_3)_2C=C(CO_2C_2H_5)_2$	CH_3MgI	CuCl (5 mol %, $-20°$)	C_4H_9-n $\mid$ $(CH_3)_3CCH(CO_2C_2H_5)_2$ (87)	100
	n-C_4H_9MgBr	CuCl	C_4H_9-n $\mid$ $(CH_3)_2CCH(CO_2C_2H_5)_2$ (I, 42)	14
	C_6H_5MgBr	CuCl (inverse addition[k])	C_4H_9-n $\mid$ $(CH_3)_2CHCH(CO_2C_2H_5)_2$ (II, 40) I (89), II (3)	14, 8
		CuCl	$(CH_3)_2CCH(CO_2C_2H_5)_2$ (70) $\mid$ C_6H_5	101
$CH_3(CH=CH)_2CO_2C_4H_9$-s	n-C_4H_9MgBr	CuCl (1.4 mol %, stepwise addition)	$CH_3CHCH=CHCH_2CO_2C_4H_9$-$s$ (24) $\mid$ C_4H_9-n $CH_3CHCH_2CH=CHCO_2C_4H_9$-$s$ (74) (total 72) $\mid$ C_4H_9-n	11
trans-$CH_3CH=C(C_2H_5)CO_2C_4H_9$-$s$	n-C_4H_9MgBr	CuCl	$CH_3CHCH(C_2H_5)CO_2C_4H_9$-$s$ (83) $\mid$ C_4H_9-n	14

Note: References 173–204 are on p. 113.

[a] The solvent is diethyl ether unless otherwise noted; stepwise addition means that the copper catalyst was added in several portions.
[b] These are the products formed after hydrolysis of intermediate enolate species.
[i] THF is tetrahydrofuran.
[j] DME is 1,2-dimethoxyethane.
[k] Inverse addition means that the Grignard reagent was added to a mixture of the carbonyl substrate and the copper salt.

TABLE I. COPPER-CATALYZED CONJUGATE ADDITION OF GRIGNARD REAGENTS TO α,β-ETHYLENIC CARBONYL COMPOUNDS (Continued)

A. Acyclic Carbonyl Substrates

Number of Carbon Atoms	Carbonyl Substrate	Grignard Reagent	Catalyst (Reaction Conditions)[a]	Product(s)[b] and Yield(s)(%)	Refs.
7	$C_2H_5C(CH_3)=C(CN)CO_2C_2H_5$	i-C_3H_7MgBr	CuCl (4 mol %, <20°)	$C_2H_5C(CH_3)CH(CN)CO_2C_2H_5$ $\quad\mid$ $\quad C_3H_7$-i (I, 29)	39
			CuCl (inverse addition[k])	I (48)	
			CuBr (3 mol %, <20°, inverse addition[k])	I (46)	
			CuI (2 mol %, <20°, inverse addition[k])	I (56)	
			CuCN (4 mol %, <20°, inverse addition[k])	I (65)	
		n-C_4H_9MgBr	CuCl (2 mol %, −2°)	$C_2H_5C(CH_3)CH(CN)CO_2C_2H_5$ $\quad\mid$ $\quad R$ (II) $R = n$-C_4H_9	102
		i-C_4H_9MgBr	CuCN (4 mol %, <20°, inverse addition[k])	II, $R = i$-C_4H_9 (31)	39
			CuCl (2 mol %, −2°)	II, $R = i$-C_4H_9 (34)	102
		i-C_5H_{11}MgBr	CuCN (4 mol %, <20°, inverse addition[k])	II, $R = i$-C_5H_{11} (81)	39
	trans-$CH_3CH=C(R)CO_2C_4H_9$-s $R = n$-C_3H_7 $R = i$-C_3H_7	n-C_4H_9MgBr	CuCl	$CH_3CHCH(R)CO_2C_4H_9$-s $\quad\mid$ $\quad C_4H_9$-n $R = n$-C_3H_7 (90), $R = i$-C_3H_7 (8), $CH_3CHCH(C_3H_7$-$i)COC_4H_9$-n $\quad\mid$ $\quad C_4H_9$-n (59), $CH_3CHCH(C_3H_7$-$i)CHO$ $\quad\mid$ $\quad C_4H_9$-n (26)	14
	$CH_2=CHCOCH_2CH(CH_3)_2$	$C_6H_5CH_2OCH_2CH_2$-$C\equiv CMgBr$	CuBr	"Unsuccessful"	74

8	CH$_3$(CH=CH)$_3$CO$_2$C$_4$H$_9$-s	n-C$_4$H$_9$MgBr	CuCl (stepwise addition)	CH$_3$CHCH$_2$(CH=CH)$_2$CO$_2$C$_4$H$_9$-s $\quad\mid$ $\quad$C$_4$H$_9$-n (25)	12
	$trans$-CH$_3$CH=C(R)CO$_2$C$_4$H$_9$-s	n-C$_4$H$_9$MgBr	CuCl	CH$_3$CHCH(R)CO$_2$C$_4$H$_9$-s $\quad\mid$ $\quad$C$_4$H$_9$-n R = n-C$_4$H$_9$ (84), R = i-C$_4$H$_9$ (86), R = s-C$_4$H$_9$ (12), CH$_3$CHCH(C$_4$H$_9$-s)COC$_4$H$_9$-n (30), $\quad\mid$ $\quad$C$_4$H$_9$-n CH$_3$CHCH(C$_4$H$_9$-s)CHO (45) $\quad\mid$ $\quad$C$_4$H$_9$-n	14
9	C$_6$H$_5$CH=C(Br)CO$_2$H (I) cis-I $trans$-I	CH$_3$MgBr	CuCl (2 mol %)	cis-C$_6$H$_5$CH=C(Br)COCH$_3$ (29), $trans$-C$_6$H$_5$CH=C(CH$_3$)CO$_2$H (8), I (55) $trans$-C$_6$H$_5$CH=C(Br)COCH$_3$ (41), $trans$-C$_6$H$_5$CH=C(Br)C(CH$_3$)=CH$_2$ (30), I (20)	94
	C$_6$H$_5$CH=C(Br)CO$_2$CH$_3$ (I) cis-I $trans$-I	CH$_3$MgBr	CuCl (2 mol %)	C$_6$H$_5$CH=C(Br)COCH$_3$ (II) I (57), cis-II (31) I (38), $trans$-II (54)	
	$trans$-C$_6$H$_5$CH=CHCO$_2$C$_2$H$_5$	CH$_3$MgBr	CuCl (1 mol %, $-10°$)	C$_6$H$_5$CHCH$_2$COCH$_3$ (77) $\quad\mid$ $\quad$CH$_3$	97
		n-C$_4$H$_9$MgBr	CuCl	C$_6$H$_5$CHCH$_2$CO$_2$C$_2$H$_5$ (15) $\quad\mid$ $\quad$C$_4$H$_9$-n	13
	$trans$-C$_6$H$_5$CH=CHCO$_2$C$_4$H$_9$-s	CH$_3$MgBr	CuCl (1 mol %, $-10°$)	C$_6$H$_5$CHCH$_2$COCH$_3$ (50) $\quad\mid$ $\quad$CH$_3$	97, 25

Note: References 173–204 are on p. 113.

[a] The solvent is diethyl ether unless otherwise noted; stepwise addition means that the copper catalyst was added in several portions.

[b] These are the products formed after hydrolysis of intermediate enolate species.

[k] Inverse addition means that the Grignard reagent was added to a mixture of the carbonyl substrate and the copper salt.

TABLE I. Copper-Catalyzed Conjugate Addition of Grignard Reagents to α,β-Ethylenic Carbonyl Compounds (*Continued*)

A. Acyclic Carbonyl Substrates

Number of Carbon Atoms	Carbonyl Substrate	Grignard Reagent	Catalyst (Reaction Conditions)[a]	Product(s)[b] and Yield(s)(%)	Refs.
9 (*contd.*)	*trans*-$C_6H_5CH=CHCO_2C_4H_9$-*s*	n-C_4H_9MgBr	CuCl (stepwise addition)	$C_6H_5CHCH_2CO_2C_4H_9$-*s* $\|$ C_4H_9-*n* (46–50)	6, 25
	trans-$C_6H_5CH=CHCOO$―[structure with CH$_3$]	$C_6H_5CH_2MgCl$	CuCl (stepwise addition, $-15°$)	$(-)$-$C_6H_5CHCH_2CO_2H$ $\|$ $C_6H_5CH_2$ (I, 56; 17, optical yield)	90
	$(-)$-*trans*-$C_6H_5CH=CHCOO$―[cyclohexyl/isopropyl structure]	$C_6H_5CH_2MgCl$	CuCl (stepwise addition, $-15°$)	I (61; 27, optical yield)	90
	trans-$C_6H_5CH=CHCOO$―[dioxolane structure]	$C_6H_5CH_2MgCl$	CuCl (stepwise addition, $-15°$)	I (40; 22, optical yield)	90

Substrate	Grignard Reagent	Catalyst	Product(s) (% yield)	Ref.
trans-C$_6$H$_5$CH=CHCOO–[spiro dioxolane cyclohexyl]	C$_6$H$_5$CH$_2$MgCl	CuCl (stepwise addition, –15°)	I (58; 17, optical yield)	90
[cyclohexylidene]=C(CN)CO$_2$C$_2$H$_5$	CH$_3$MgI	CuI	[cyclohexyl]–CH(CN)CO$_2$C$_2$H$_5$ (73)	106
n-C$_3$H$_7$C(C$_2$H$_5$)=C(CN)CO$_2$C$_2$H$_5$	n-C$_{12}$H$_{25}$MgBr	CuI (2 mol %, inverse addition^k)	n-C$_3$H$_7$C(C$_2$H$_5$)CH(CN)CO$_2$C$_2$H$_5$ | C$_{12}$H$_{25}$-n (60)	103
trans-CH$_3$CH=C(C$_5$H$_{11}$-i)CO$_2$C$_4$H$_9$-s	n-C$_4$H$_9$MgBr	CuCl	CH$_3$CHCH(C$_5$H$_{11}$-i)CO$_2$C$_4$H$_9$-s | C$_4$H$_9$-n (93)	14
trans-C$_6$H$_5$CH=CHCOCH$_3$	CH$_3$MgBr	CuCl (5 mol %)	C$_6$H$_5$CHCH$_2$COCH$_3$ | CH$_3$ (28)	31
10	C$_6$H$_5$MgBr	CuCl (2.5 mol %)	trans-C$_6$H$_5$CH=CHC(CH$_3$)$_2$OH (72), (C$_6$H$_5$)$_2$CHCH$_2$COCH$_3$ (I, 72), trans-C$_6$H$_5$CH=CHCH(OH)CH$_3$ | C$_6$H$_5$ (II, 28)	
		CuCl (3 mol %, excess isoprene)	I (70), II (30)	

Note: References 173–204 are on p. 113.

^{a} The solvent is diethyl ether unless otherwise noted; stepwise addition means that the copper catalyst was added in several portions.
^{b} These are the products formed after hydrolysis of intermediate enolate species.
^{k} Inverse addition means that the Grignard reagent was added to a mixture of the carbonyl substrate and the copper salt.

TABLE I. Copper-Catalyzed Conjugate Addition of Grignard Reagents to α,β-Ethylenic Carbonyl Compounds
(*Continued*)

Number of Carbon Atoms	Carbonyl Substrate	Grignard Reagent	Catalyst (Reaction Conditions)[a]	Product(s)[b] and Yield(s) (%)	Refs.
A. Acyclic Carbonyl Substrates					
10 (*contd.*)	*trans*-$C_6H_5CH=CHCOCH_3$	$C_6H_5CH(CH_3)CH_2$-MgBr	CuI	$C_6H_5CHCH_2COCH_3$	174
				$|$	
				$C_6H_5CH(CH_3)CH_2$ (—)	
	n-$C_5H_{11}C(CH_3)=C(CN)CO_2C_2H_5$	n-C_5H_{11}MgBr	CuI (1 mol %, 15°)	$(n$-$C_5H_{11})_2C(CH_3)CH(CN)CO_2C_2H_5$ (37)	63
		n-C_7H_{15}MgBr	CuI (1 mol %, 15°)	n-$C_5H_{11}C(CH_3)CH(CN)CO_2C_2H_5$	63
				$|$	
				C_7H_{15}-n (65)	
11	p-$CH_3OC_6H_4C(CH_3)=C(CN)CO_2C_2H_5$	CH_3MgI	CuI (7 mol %)	p-$CH_3OC_6H_4C(CH_3)_2CH(CN)CO_2C_2H_5$ (60)	175
	![cyclohexylidene C(CN)CO2C2H5]	CH_3MgI	CuI	![cyclohexyl CH(CN)CO2C2H5] (65)	106
	n-$C_4H_9C(C_3H_7$-$n)=C(CN)CO_2C_2H_5$	n-$C_{10}H_{21}$MgBr	CuI (2 mol %, inverse addition[k])	n-$C_4H_9C(C_3H_7$-$n)CH(CN)CO_2C_2H_5$	103
				$|$	
				$C_{10}H_{21}$-n (65)	
	n-$C_6H_{13}C(CH_3)=C(CN)CO_2C_2H_5$	n-$C_{10}H_{21}$MgBr	CuI (2 mol %, inverse addition[k])	n-$C_6H_{13}C(CH_3)CH(CN)CO_2C_2H_5$	
				$|$	
				$C_{10}H_{21}$-n (78)	
12	*trans*-m-$CH_3C_6H_4C(CH_3)=C(CN)CO_2C_2H_5$	CH_3MgI	CuCl (9 mol %)	m-$CH_3C_6H_4CH(CH_3)CH(CN)CO_2C_2H_5$ (60)	105
	n-$C_7H_{15}C(CH_3)=C(CN)CO_2C_2H_5$	n-C_7H_{15}MgBr	CuI (1 mol %, 15°)	$(n$-$C_7H_{15})_2C(CH_3)CH(CN)CO_2C_2H_5$ (70)	63
	i-$C_5H_{11}C(C_3H_7$-$n)=C(CN)CO_2C_2H_5$	i-C_5H_{11}MgBr	CuI (1 mol %, 15°)	$(i$-$C_5H_{11})_2C(C_3H_7$-$n)CH(CN)CO_2C_2H_5$ (68)	63
13	$C_6H_5CH=CHCOCH_2CH(CH_3)_2$	CH_3MgI	CuCl	$C_6H_5CHCH_2COCH_2CH(CH_3)_2$	176
				$|$	
				CH_3 (—)	

Substrate	Reagent	Catalyst/Conditions	Product (Yield %)	Ref.
p-CH$_3$OC$_6$H$_4$CH=CHCOCH$_2$CH(CH$_3$)$_2$	CH$_3$MgI	CuCl (10 mol %, stepwise addition)	p-CH$_3$OC$_6$H$_4$CHCH$_2$COCH$_2$CH(CH$_3$)$_2$ with CH$_3$ (70)	177
[Decalin structure with C(CN)CO$_2$C$_2$H$_5$]	CH$_3$MgI	CuI (benzene)	[Decalin with CH(CN)CO$_2$C$_2$H$_5$ and CH$_3$]	106
[AcO-substituted bicyclic structure with =C(CN)CO$_2$C$_2$H$_5$]	CH$_3$MgI	CuI	[AcO-substituted bicyclic with CH(CN)CO$_2$C$_2$H$_5$ and H]	178
n-C$_7$H$_{15}$C(C$_2$H$_5$)=C(CN)CO$_2$C$_2$H$_5$	n-C$_8$H$_{17}$MgBr	CuI (2 mol %, inverse addition)	n-C$_7$H$_{15}$C(C$_2$H$_5$)CH(CN)CO$_2$C$_2$H$_5$ with C$_8$H$_{17}$-n (81)	103
[t-C$_4$H$_9$-cyclohexylidene=CHCOCH$_3$]	CH$_3$MgBr	(n-C$_4$H$_9$)$_3$PCuI (1 mol %, 0°)	[t-C$_4$H$_9$-cyclohexyl with CH$_2$COCH$_3$ and CH$_3$, H] (74)	16
			[t-C$_4$H$_9$-cyclohexyl=CHC(CH$_3$)$_2$OH] (26)	16
[Decalin with C(CN)CO$_2$C$_2$H$_5$] 14	CH$_3$MgI	CuI (benzene)	[Decalin with H, CH(CN)CO$_2$C$_2$H$_5$] (30)	106

Note: References 173–204 are on p. 113.

[a] The solvent is diethyl ether unless otherwise noted; stepwise addition means that the copper catalyst was added in several portions.
[b] These are the products formed after hydrolysis of intermediate enolate species.
[k] Inverse addition means that the Grignard reagent was added to a mixture of the carbonyl substrate and the copper salt.

TABLE I. Copper-Catalyzed Conjugate Addition of Grignard Reagents to α,β-Ethylenic Carbonyl Compounds
(Continued)

A. Acyclic Carbonyl Substrates

Number of Carbon Atoms	Carbonyl Substrate	Grignard Reagent	Catalyst (Reaction Conditions)[a]	Product(s)[b] and Yield(s)(%)	Refs.
	$n\text{-}C_9H_{19}C(CH_3)=C(CN)CO_2C_2H_5$	$n\text{-}C_4H_9MgBr$	CuI (1 mol %, 15°)	$n\text{-}C_9H_{19}C(CH_3)CH(CN)CO_2C_2H_5$ (60) $\quad\quad\quad\quad\quad\quad\quad\|$ $\quad\quad\quad\quad\quad\quad C_4H_9\text{-}n$	63
15	$C_6H_5CH=CHCOC_6H_5$	$p\text{-}CH_3C_6H_4SO_2\text{-}CH_2MgBr$	CuCl (1 mol %, benzene)	$C_6H_5CH=CHC(OH)C_6H_5$ (75) $\quad\quad\quad\quad\quad\quad\quad\|$ $\quad\quad\quad\quad p\text{-}CH_3C_6H_4SO_2CH_2$	92
		CH_3MgBr	CuCl (1 mol %)	$C_6H_5CHCH_2COC_6H_5$ (69), $\quad\quad\|$ $\quad CH_3$ $C_6H_5CH(CH_3)CCOC_6H_5$ (24) $\quad\quad\quad\quad\quad\quad\|\|$ $\quad\quad\quad\quad C_6H_5CCH=CHC_6H_5$	179
	$n\text{-}C_6H_{13}C(C_5H_{11}\text{-}n)=C(CN)CO_2C_2H_5$	$n\text{-}C_7H_{15}MgBr$	CuI (2 mol %, inverse addition)	$n\text{-}C_6H_{13}C(C_5H_{11}\text{-}n)CH(CN)CO_2C_2H_5$ (71) $\quad\quad\quad\quad\|$ $\quad\quad\quad C_7H_{15}\text{-}n$	103
	$n\text{-}C_8H_{17}C(C_7H_{15}\text{-}n)=C(CN)CO_2CH_3$	$n\text{-}C_{10}H_{21}MgBr$	CuI (2 mol %, inverse addition)	$n\text{-}C_8H_{17}C(C_7H_{15}\text{-}n)CH(CN)CO_2CH_3$ (60) $\quad\quad\quad\quad\|$ $\quad\quad\quad C_{10}H_{21}\text{-}n$	103

B. Cyclic Carbonyl Substrates

Number of Carbon Atoms	Carbonyl Substrate	Grignard Reagent	Catalyst (Reaction Conditions)[a]	Product(s)[b] and Yield(s)(%)	Refs.
6	2-Cyclohexenone	$n\text{-}C_3H_7MgBr$	CuI (7 mol %)	3-n-Propylcyclohexanone (86)	55
		$CH_2=C(CH_3)MgBr$	CuI (5 mol %, 0°, THF[c])	3-(1-methylethenyl)cyclohexanone (68)	70

Substrate	Reagent	Catalyst (conditions)	Product(s) (% yield)	Refs.
7	$CH_2=C(CH_3)CH_2MgCl$	CuCl (1 mol %, 0°)	1-(2-methylallyl)-2-cyclohexen-1-ol, HO–C(CH₂C(CH₃)=CH₂)– (95)	62
	$i\text{-}C_4H_9MgCl$	CuCl (1 mol %, 0°)	HO, $C_4H_9\text{-}i$ (35) + $C_4H_9\text{-}i$ ketone (63)	
3-Methyl-2-cyclohexenone	CH_3MgI	CuCl (1 mol %, 2°)	3,3-Dimethylcyclohexanone (60)	116
4-Methyl-2-cyclohexenone	CH_3MgBr	CuCl (10 mol %)	(I) R	
	C_2H_5MgBr	CuCl (10 mol %)	I, R = CH₃ (72:28 trans:cis)	66, 46
	$i\text{-}C_3H_7MgBr$	CuCl (10 mol %)	I, R = C₂H₅ (78:22 trans:cis)	66
	C_6H_5MgBr	CuCl (10 mol %)	I, R = $i\text{-}C_3H_7$ (88:12 trans:cis)	66, 46
			I, R = C₆H₅ (88:12 trans:cis)	
5-Methyl-2-cyclohexenone	CH_3MgI	CuCl (1 mol %)	I (II) (I, 94–96) (II, 4–6) (total 55–56)	87
	CH_3MgBr	CuI (1 mol %)	I (93), II (7) (total > 90)	28
	$(CH_3)_2Mg$	CuI (1 mol %)	I (91), II (9) (total > 90)	
2-Cycloheptenone	C_6H_5MgBr	CuI (5 mol %)	3-Phenylcycloheptanone (33)	110

Note: References 173–204 are on p. 113.

[a] The solvent is diethyl ether unless otherwise noted; stepwise addition means that the copper catalyst was added in several portions.
[b] These are the products formed after hydrolysis of intermediate enolate species.
[i] THF is tetrahydrofuran.

TABLE I. Copper-Catalyzed Conjugate Addition of Grignard Reagents to α,β-Ethylenic Carbonyl Compounds
(Continued)

Number of Carbon Atoms	Carbonyl Substrate	Grignard Reagent	Catalyst (Reaction Conditions)[a]	Product(s)[b] and Yield(s)(%)	Refs.
	B. Cyclic Carbonyl Substrates				
8	COCH₃-cyclohexenone	CH₃MgBr	CuCl (3 mol %, stepwise addition)	COCH₃ + C(CH₃)₂OH cyclohexene (Major) (Minor; total, 40–45)	27
	4-Ethyl-2-cyclohexenone	CH₃MgBr	CuCl (10 mol %)	R, C₂H₅ cyclohexanone (I)	66
		C₆H₅MgBr	CuCl (10 mol %)	R = CH₃ (77:23 *trans:cis*)	
		C₆H₅MgBr	CuI (5 mole %)	I, R = C₆H₅ (89:11 *trans:cis*)	
				3-Phenylcyclooctanone (37)	110
9	2-Cyclooctenone	CH₃MgI	Cu(OAc)₂(THF)	(I) bicyclic I (34)	36
		CH₃MgBr	CuBr	I	180
	trimethylcyclohexenone	CH₃MgBr	CuCl (1 mol %)	(I, 82.5) + (7.0)	4, 181

	CH₃MgI	CuCl (1.4 mol %, <8°)	I (26, large-scale experiment)	182, 183
	CH₃MgBr	CuOAc	(structure with OAc^l, 60)	184
4-i-Propyl-2-cyclohexenone	CH₃MgBr	CuCl (10 mol %)	(I) R = CH₃ (89:11 trans:cis)	66
	C₂H₅MgBr	CuCl (10 mol %)	I, R = C₂H₅ (92:8 trans:cis)	
	CH₃MgBr	CuCl (10 mol %)	OSi(CH₃)₃^m (structure)	185
10 (bicyclic enone structure)	CH₃MgBr	CuBr	(bicyclic ketone structure with CH₃)	112

Note: References 173–204 are on p. 113.

[a] The solvent is diethyl ether unless otherwise noted; stepwise addition means that the copper catalyst was added in several portions.
[b] These are the products formed after hydrolysis of intermediate enolate species.
[i] THF is tetrahydrofuran.
[l] The intermediate enolate was treated with lead(IV) acetate.
[m] The intermediate enolate was treated with trimethylsilyl chloride.

TABLE I. Copper-Catalyzed Conjugate Addition of Grignard Reagents to α,β-Ethylenic Carbonyl Compounds (*Continued*)

B. *Cyclic Carbonyl Substrates*

Number of Carbon Atoms	Carbonyl Substrate	Grignard Reagent	Catalyst (Reaction Conditions)[a]	Products(s)[b] and Yield(s)(%)	Refs.
10 (*contd.*)	(octalenone)	CH_3MgI	$Cu(OAc)_2$ $(THF)^i$	(trans, 10) + (cis, 54)	36
	(octalenone)	CH_3MgBr	CuBr	(I, 60)	186, 5
		CH_3MgI	CuCl CuBr (1 mol %, 0°) $Cu(OAc)_2 \cdot H_2O$ (8 mol %, −10°, 1:1.5 ether:THF)^i	I (80) I (60) I (80)	187 5 84, 36
		$n\text{-}C_4H_9MgBr$	$Cu(OAc)_2 \cdot H_2O$ (6 mol %, THF)^i	($C_4H_9\text{-}n$, 71)	188
		C_6H_5MgBr	CuCl (5 mol %)	(C_6H_5, 25) + (C_6H_5, 66)	85

Substrate	Reagent	Catalyst	Product(s) (Yield %)
(octalone 11)	C₆H₅MgBr	CuCl (10 mol %)	phenyl-OH decalin derivative 125
	m-ClC₆H₄MgBr	Cu(OAc)₂ (8 mol %, room temp, THF^i)	C₆H₅ ketone decalin (60); m-ClC₆H₄ ketone decalin (38) 69
		(n-C₄H₉)₃PCuI	I (40) (I, 23)
(dienone)	CH₃MgI	Cu(OAc)₂·H₂O (10 mol %, −15°, inverse addition, THF^i)	R = CH₃, I (75), II (25; 93 α-R, 7 β-R) 34
	CH₃MgI	Cu(OAc)₂·H₂O (10 mol %, −15°, inverse addition, 1:1 ether:THF^i)	R = CH₃, I (45), II (55; 93 α-R, 7 β-R)
	C₂H₅MgBr	Cu(OAc)₂·H₂O (10 mol %, −15°, inverse addition, THF^i)	R = C₂H₅, I (35–45), II (55–65; 98 α-R, 2 β-R)
	i-C₃H₇MgBr	Cu(OAc)₂·H₂O (10 mol %, −15°, inverse addition, THF^i)	R = i-C₃H₇, I (42), II (58; 100 α-R)

Note: References 173–204 are on p. 113.

[a] The solvent is diethyl ether unless otherwise noted; stepwise addition means that the copper catalyst was added in several portions.
[b] These are the products formed after hydrolysis of intermediate enolate species.
[i] THF is tetrahydrofuran.

TABLE I. COPPER-CATALYZED CONJUGATE ADDITION OF GRIGNARD REAGENTS TO α,β-ETHYLENIC CARBONYL COMPOUNDS (*Continued*)

B. *Cyclic Carbonyl Substrates*

Number of Carbon Atoms	Carbonyl Substrate	Grignard Reagent	Catalyst (Reaction Conditions)[a]	Product(s)[b] and Yield(s)(%)	Refs.
11 (*contd.*)	[structure with NCOCH$_3$]	t-C$_4$H$_9$MgCl	Cu(OAc)$_2\cdot$H$_2$O (10 mol %), $-15°$, inverse addition, THF[i]	R = t-C$_4$H$_9$ I (60), II (40, 100 α-R)	
		C$_6$H$_5$MgBr	Cu(OAc)$_2\cdot$H$_2$O (10 mol %), $-15°$, inverse addition, THF[i]	R = C$_6$H$_5$ I (60)[n]	
	[decalone structure]	C$_6$H$_5$MgBr	CuCl (5 mol %)	No 1,4 addition	85
		CH$_2$=C(CH$_3$)MgBr	CuCl (5 mol %, THF[i])	(50)	71, 189
	[octalone structure]	CH$_3$MgI	Cu(OAc)$_2\cdot$H$_2$O (8 mol %), $-10°$, 1:1.5 ether:THF[i]	(40) (60) CH$_3$ OH	84, 189a
			Cu(OAc)$_2\cdot$H$_2$O (6 mol %), $-10°$, THF[i]	(30)[o] AcO	190

82

Substrate	Reagent	Catalyst/Conditions	Product (yield %)	Ref.
(cyclobutanone with gem-dimethyl)	CH₃MgI	CuCl	(100)	128
(octalone)	i-C₃H₇MgBr	Cu(OAc)₂·H₂O (10 mol %), −15°, THF	Dimer? (30)	34
(dioxolane-octalone)	CH₃MgBr	CuBr (−10°)	(61)	186
(enone decalone)	CH₃MgI	Cu(OAc)₂·H₂O (8 mol %), −50°, THF, inverse addition	R = CH₃, I (82), II (17) (total 91–96)	88
	i-C₃H₇MgBr	Cu(OAc)₂·H₂O (8 mol %), −50°, THF, inverse addition	R = i-C₃H₇, I (52), II (44) (total 84)	
	C₆H₅MgBr	Cu(OAc)₂·H₂O (8 mol %), −50°, THF, inverse addition	R = C₆H₅, I (−), II (97)	
(decalone enone)	CH₃MgBr	Cu(OAc)₂·H₂O (8 mol %), −50°, ether-THF	(High yield of epimeric mixture)	

Note: References 173–204 are on p. 113.

[a] The solvent is diethyl ether unless otherwise noted; stepwise addition means that the copper catalyst was added in several portions.
[b] These are the products formed after hydrolysis of intermediate enolate species.
[i] THF is tetrahydrofuran.
[n] "Nonvolatile ketonic materials" make up the remaining 40 % of product.
[o] The intermediate enolate was treated with acetyl chloride.

TABLE I. Copper-Catalyzed Conjugate Addition of Grignard Reagents to α,β-Ethylenic Carbonyl Compounds
(Continued)

Number of Carbon Atoms	Carbonyl Substrate	Grignard Reagent	Catalyst (Reaction Conditions)[a]	Product(s)[b] and Yield(s)(%)	Refs.
	B. Cyclic Carbonyl Substrates				
12 (contd.)		$CH_2{=}C(CH_3)MgBr$	CuCl (3 mol %, THF[i])	(71–76)	72
13		CH_3MgI	$Cu(OAc)_2 \cdot H_2O$	(I, 76)	38
		C_6H_5MgBr	$Cu(OAc)_2 \cdot H_2O$ (THF[i])	I (56) + (34)	127
		C_6H_5MgBr	CuCl (1 mol %, room temp)	$(C_6H_5)_2CH$ (35)	127
		C_6H_5MgBr	CuCl (1 mol %, ether, reflux)	(45)	191
		C_6H_5MgBr	CuCl	[o]	192

Substrate	Reagent	Catalyst	Product (Yield %)	Ref.
(structure: N-containing bicyclic enone, labeled (I))	CH₃MgI	CuBr (10 mol %, room temp)	(structure: saturated N-bicyclic ketone) (21) + I (54)	141
(structure: octalone with dioxolane)	CH₃MgBr	CuBr	No 1,4 adduct	186
14 (structure: octalone with isopropenyl group, H shown)	CH₃MgI	Cu(OAc)₂·H₂O	(structure: decalone with isopropenyl) (10–25)	122
(structure: octalone with dioxolane and exocyclic methylene, H shown)	C₆H₅CH₂MgCl	Cu(OAc)₂·H₂O (5 mol %)	(I)ᵖ (structure showing CH₂R, AcO, H, dioxolane) I, R = C₆H₅CH₂ (71)	83
	m-CH₃OC₆H₄CH₂-MgCl	Cu(OAc)₂·H₂O (5 mol %)	I, R = m-CH₃OC₆H₄CH₂ (58)	

Note: References 173–204 are on p. 113.

ᵃ The solvent is diethyl ether unless otherwise noted; stepwise addition means that the copper catalyst was added in several portions.
ᵇ These are the products formed after hydrolysis of intermediate enolate species.
ⁱ THF is tetrahydrofuran.
ᵒ The intermediate enolate was treated with acetyl chloride.
ᵖ The intermediate enolate was treated with acetic anhydride.

TABLE I. COPPER-CATALYZED CONJUGATE ADDITION OF GRIGNARD REAGENTS TO α,β-ETHYLENIC CARBONYL COMPOUNDS (*Continued*)

B. Cyclic Carbonyl Substrates

Number of Carbon Atoms	Carbonyl Substrate	Grignard Reagent	Catalyst (Reaction Conditions)[a]	Product(s)[b] and Yield(s) (%)	Refs.
14 (*contd.*)		CH$_3$MgI	Cu(OAc)$_2$·H$_2$O (5 mol %, −10°, 2:1 ether:THF[t])	(16–21)	81
		CH$_3$MgI	Cu(OAc)$_2$·H$_2$O (8 mol %, −10°, THF[t])	(85)	84
15	2-Cyclopentadecenone	CH$_3$MgBr	CuCl	1-Methyl-2-cyclopentadecen-1-ol (major), 3-methylcyclopentadecanone (minor)	109
17		CH$_3$MgI	Cu(OAc)$_2$·H$_2$O	(—)	165
18		C$_6$H$_5$MgBr	(*n*-C$_4$H$_9$)$_3$PCuI (10 mol %, room temp, inverse addition)	(79)	193

Note: References 173–204 are on p. 113.

[a] The solvent is diethyl ether unless otherwise noted; stepwise addition means that the copper catalyst was added in several portions.
[b] These are the products formed after hydrolysis of intermediate enolate species.
[i] THF is tetrahydrofuran.

TABLE I. Copper-Catalyzed Conjugate Addition of Grignard Reagents to α,β-Ethylenic Carbonyl Compounds (*Continued*)

Number of Carbon Atoms	Carbonyl Substrate	Grignard Reagent	Catalyst (Reaction Conditions)[a]	Product(s)[b] and Yield(s)(%)	Refs.
		B. *Cyclic Carbonyl Substrates*			
19		CH_3MgBr	CuCl	(—)	119
20		CH_3MgI	$Cu(OAc)_2$ (THF[f])	(20)	36
		CH_3MgI	CuBr (10 mole % stepwise addition)	(70)	115
	(I)			(II)	133

Substrate	Reagent	Conditions	Product(s) (Yield %)
21: I, R = H / I, R = OH (steroid with COCH₃)	CH₃MgBr	CuCl (5 mol %), 0°, 1:2 ether:THF[i]	II, R = H, 7α-CH₃ (major), 7β-CH₃ (minor) (total 25–30)
	CH₃MgBr	CuCl (5 mol %), −10°, ether:THF[i]	II, R = OH, 7α-CH₃ (minor), 7β-CH₃ (major) (total 40)
	CH₃MgI	CuCl (1 mol %), 0°, ether-toluene	(steroid with COCH₃) (90) 120
	CH₃MgI	CuCl (1 mol %), 10°, 2:3 ether:toluene	(steroid with C(CH₃)OAc) (65)[o]
(steroid with HO and COCH₃)	CH₃MgI	CuCl (5 mol %), 20°, 1:2 ether:THF[i]	(AcO-steroid with C(CH₃)OAc) (50)[p]

Note: References 173–204 are on p. 113.

[a] The solvent is diethyl ether unless otherwise noted; stepwise addition means that the copper catalyst was added in several portions.
[b] These are the products formed after hydrolysis of intermediate enolate species.
[i] THF is tetrahydrofuran.
[o] The intermediate enolate was treated with acetyl chloride.
[p] The intermediate enolate was treated with acetic anhydride.

TABLE I. Copper-Catalyzed Conjugate Addition of Grignard Reagents to α,β-Ethylenic Carbonyl Compounds
(Continued)

Number of Carbon Atoms	Carbonyl Substrate	Grignard Reagent	Catalyst (Reaction Conditions)[a]	Product(s)[b] and Yield(s)(%)	Refs.
21 (contd.)	*B. Cyclic Carbonyl Substrates*				
	(I) R = O I, R = H$_2$ I, R = α-OAc	CH$_3$MgI CH$_3$MgI (6-fold excess) CH$_3$MgI (1.5-fold excess) CH$_3$MgI	CuCl (5 mol %, 20°, 1:2 ether:THF[f]) CuCl (5 mol %, 20°, 1:2 ether:THF[f]) CuCl (5 mol %, 20°, 1:2 ether:THF) CuCl (2 mol %, 20°, 1:1.5 ether:THF[f])	II, R = O, R' = Ac (88) II, R = H$_2$, R' = H (60) II, R = H$_2$; R' = H (93) II, R = α-OAc; R' = Ac (93)	
	(I) I, R = O I, R = H$_2$ I, R = α-OAc	CH$_3$MgI	CuCl (6 mol %, 20°, 1:2 ether:THF[f])	II, R = O (88) II, R = H$_2$ (90–95) II, R = α-OAc (62)	

Substrate	Reagent	Catalyst	Product (yield %)	Ref.
[steroid with HO, CHCH₂OAc, enone]	CH_3MgBr	CuCl (5 mol %), 1:3 ether:THF[i]	[product] (20)	133
[steroid with CH₂, dione]	m-$CH_3OC_6H_4CH_2$-MgCl	$Cu(OAc)_2 \cdot H_2O$ (5 mol %)	[product with $CH_2CH_2C_6H_4OCH_3$-m] (71)	83
22 [steroid with spirolactone, dioxolane, enone]	CH_3MgBr	CuCl (12 mol %), THF[i], inverse addition	[product] (62.5)	134
[steroid with AcO, COCH₃, enone]	CH_3MgI	CuCl	[product with COCH₃]	135

Note: References 173–204 are on p. 113.

[a] The solvent is diethyl ether unless otherwise noted; stepwise addition means that the copper catalyst was added in several portions.
[b] These are the products formed after hydrolysis of intermediate enolate species.
[i] THF is tetrahydrofuran.
[o] The intermediate enolate was treated with acetyl chloride.

TABLE I. Copper-Catalyzed Conjugate Addition of Grignard Reagents to α,β-Ethylenic Carbonyl Compounds (*Continued*)

Number of Carbon Atoms	Carbonyl Substrate	Grignard Reagent	Catalyst (Reaction Conditions)[a]	Product(s)[b] and Yield(s) (%)	Refs.
22 (*contd.*)			*B. Cyclic Carbonyl Substrates*		
		CH$_3$MgI	CuCl (5 mol %, room temp, THF[i])	(75–80)	194
		CH$_3$MgBr	CuBr (~10 mol %, 0°, THF,[i] inverse addition)	(69)	195, 113
		CH$_3$MgBr	CuBr (10 mol %, 0°, THF,[i] inverse addition)	(>35)	195, 113

27		CH₃MgX	CuCl		173
	4-Cholesten-3-one	CH₃MgI	Cu(OAc)₂	5-Methyl-3-coprostanone (6)	196
	5-Cholesten-4-one	CH₃MgI	CuCl	No 1,4 addition	197
	4-Cholesten-6-one	CH₃MgI	Cu(OAc)₂·H₂O	6β-Methylcholestan-4-one (30)	75
		CH₃MgI	Cu(OAc)₂·H₂O	4β-Methylcholestan-6-one (78)	
	3β-Hydroxy-4-cholesten-6-one	CH₃MgI	CuCl (1 mole %, room temp, inverse addition)	3β-Hydroxy-4β-methylcholestan-6-one (I, 35)	126
			Cu(OAc)₂·H₂O (0.2 mol %, 0°, 1:1 ether:THF^c)	I (37)	

Note: References 173–204 are on p. 113.

^{a} The solvent is diethyl ether unless otherwise noted; stepwise addition means that the copper catalyst was added in several portions.
^{b} These are the products formed after hydrolysis of intermediate enolate species.

TABLE II. Copper-Catalyzed Conjugate Addition of Grignard Reagents to α,β-Acetylenic Carbonyl Compounds

Number of Carbon Atoms	Carbonyl Substance	Grignard Reagent	Catalyst (Reaction Conditions)[a]	Product(s)[b] and Yield(s) (%)	Refs.
4	$C_2H_5O_2CC \equiv CCO_2C_2H_5$	n-C_4H_9MgBr	CuCl	Tar	10
	$CH_3C \equiv CCO_2H$	C_6H_5MgBr	CuCl	$CH_3C=CHCO_2H$ (I) $\|$ C_6H_5 I (22 trans, 2 cis)	139
		C_6H_5MgBr	CuCl (5 mol %,[c] room temp)	I	120
	$CH_3C \equiv CCO_2CH_3$	C_6H_5MgBr	CuCl (5 mol %, room temp)	$CH_3C=CHCO_2CH_3$ $\|$ C_6H_5 (19 trans, 47 cis)	120
	$CH_3C \equiv CCO_2C_2H_5$	C_2H_5MgBr	CuCl (3 mol %)	$CH_3C=CHCO_2C_2H_5$ $\|$ C_2H_5 (41 trans, 38 cis)	173
	$CH_3C \equiv CCO_2C_4H_{9\text{-}s}$	n-C_4H_9MgBr	CuCl (1 mol %, $-15°$)	$CH_3C=CHCO_2C_4H_{9\text{-}s}$ (—) $\|$ $C_4H_{9\text{-}n}$	10
7	n-$C_4H_9C \equiv CCO_2H$	C_6H_5MgBr	CuCl	n-$C_4H_9C=CHCO_2H$ (—) $\|$ C_6H_5	139
9	$C_6H_5C \equiv CCO_2H$	CH_3MgBr	CuCl (5 mol %)	$C_6H_5C(CH_3)=CHCO_2H$ (10 trans, 2 cis)	120
		CH_3MgI	CuCl	$C_6H_5C(CH_3)=CHCO_2H$ (—)	139
		n-C_4H_9MgBr	CuCl	$C_6H_5C=CHCO_2H$ (—) $\|$ $C_4H_{9\text{-}n}$	139

$C_6H_5C \equiv CCO_2CH_3$	C_6H_5MgBr	CuCl	$(C_6H_5)_2C=CHCO_2H$	(35)
$C_6H_5C \equiv CCO_2C_2H_5$	CH_3MgBr	CuCl (5 mole %)	cis-$C_6H_5C(CH_3)=CHCO_2CH_3$	(33) 136
$C_6H_5C \equiv CCO_2C_4H_{9}$-$s$	n-C_4H_9MgBr	CuCl (1 mol %, $-10°$)	Tar	10
	n-C_4H_9MgBr	CuCl (1 mol %, $-10°$)	$C_6H_5C=CHCO_2C_4H_{9}$-s $\quad\quad\mid$ $\quad\quad C_4H_{9}$-n	(55)

Note: References 173–204 are on p. 113.

[a] Diethyl ether is the solvent, unless otherwise indicated.
[b] These are the products formed after hydrolysis of intermediate enolate species.
[c] This is the percentage of the number of moles of Grignard reagent used.

TABLE III. CONJUGATE ADDITION OF ORGANOCOPPER REAGENTS TO α,β-ETHYLENIC CARBONYL COMPOUNDS

A. Acyclic Carbonyl Substrates

Number of Carbon Atoms	Carbonyl Substrate	Organocopper Reagent[a]	Conditions[b]	Product(s) and Yield(s) (%)	Refs.
4	CH_2=CHCOCH$_3$	$(CH_2$=CH$)_2$CuLiP$(C_4H_9$-$n)_3$	$-78°$, THF[c]	CH_2=CHCH$_2$CH$_2$COCH$_3$ (70)	73b
5	trans-CH$_3$CH=CHCOCH$_3$	CH_3CuP$(C_4H_9$-$n)_3$	$25°$, LiI present	$(CH_3)_2$CHCH$_2$COCH$_3$ (I, 93)	16
		CH_3Cu[d]	LiI present	I (85)	
		CH_3Cu[e]		I (96)	
6	$(CH_3)_2$C=CHCOCH$_3$	$(CH_2$=CH$)_2$CuLiP$(C_4H_9$-$n)_3$	THF[c]	CH_2=CHC(CH$_3)_2$CH$_2$COCH$_3$ (72)	73b
9	[$(CH_3)_2$C=CH$]_2$CO	$(CH_3)_2$CuLi		$(CH_3)_2$CCH$_2$COCH=C(CH$_3)_2$ (>95)	24b, 65

	n-C_6H_{13}CH=CHCO$_2$C$_2$H$_5$ (95% trans, 5% cis)	$(CH_3)_2$CuLi	—	n-C_6H_{13}CHCHCO$_2$C$_2$H$_5$ (I, 86)
				    $\mid$
				    CH_3
				I (80)
					51

10	n-C_6H_{13}CH=C(CH$_3$)CHO (I)	$(CH_3)_2$CuLi	$0°$, toluene	No reaction	

			$0°$, THF[c]	n-C_6H_{13}CHCH(CH$_3$)CHO (II, 40^f)
				    $\mid$
				    CH_3
					51

	trans-I		$-78°$, 4 hr, toluene	No reaction	
		$(CH_3)_2$CuLi	$-78°$, 4 hr, THF[c]	II (major product)	
		$(CH_3)_2$CuLi	$0°$, 3 min, toluene	II (major product)	
	cis-I		$0°$, 3 min, ether	II (41^f)	
	n-C_6H_{13}CH=C(CH$_3$)CO$_2$C$_2$H$_5$ (80% trans, 20% cis)	$(CH_3)_2$CuLi	$-78°$, 1.5 hr, toluene	n-C_6H_{13}CH=C(CH$_3$)COCH$_3$ (I, 8^f) I (29^g)	
	$(CH_3)_2$C=CHCOCH$_4H_9$-t	CH_2=CH(CH$_2)_3$Cu	$0°$, toluene	CH_2=CH(CH$_2)_3$C(CH$_3)_2$CH$_2$COCH$_2$-C$_4H_9$-t (>80)	65
11	trans-t-C_4H_9CH=CHCOC$_4H_9$-t	$(CH_3)_2$CuLi $(n$-C_4H_9C≡C$)_3$CuLi$_2$	$50°$, dioxane	t-C_4H_9CH(CH$_3$)CH$_2$COC$_4H_9$-t (79) Unidentified product(s)	55
12	t-C_4H_9-⌬-CHCO$_2$H	$(CH_3)_2$CuLi	—	No reaction	16
	t-C_4H_9-⌬-CHCO$_2$C$_2$H$_5$	$(CH_3)_2$CuLi	—	No reaction	16

#	Substrate	Reagent	Conditions	Product (yield %)	Ref.
13	t-C$_4$H$_9$–⟨⟩=CHCOCH$_3$	(CH$_3$)$_2$CuLi	—	t-C$_4$H$_9$–⟨⟩–CH$_2$COCH$_3$	16
15	C$_6$H$_5$CH=CHCOC$_6$H$_5$	C$_6$H$_5$Cu	−5°	(C$_6$H$_5$)$_2$CHCHCOC$_6$H$_5$ (69)	18
		(CH$_2$=CH)$_2$CuLiP(C$_4$H$_9$-n)$_3$	−78°, THF^c	C$_6$H$_5$CHCH$_2$COC$_6$H$_5$ / C$_6$H$_5$CHCH$_2$COC$_6$H$_5$ (89) CH=CH$_2$	73b
	⟨bicyclic⟩–C(CN)CO$_2$C$_2$H$_5$	(CH$_3$)$_2$CuLi	—	⟨bicyclic⟩–CH(CN)CO$_2$C$_2$H$_5$ (80)	104

B. Cyclic Carbonyl Substrates

#	Substrate	Reagent	Conditions	Product (yield %)	Ref.
6	2-Cyclohexenone	(CH$_3$)$_2$CuLi	−78°	3-Methylcyclohexanone (97)	197a
		(CH$_2$=CH)$_2$CuLiP(C$_4$H$_9$-n)$_3$	−78°, THF^c	3-vinylcyclohexanone (65)	73b
		(CH$_2$=CHCH$_2$)$_2$CuLi	−78°	3-allylcyclohexanone CH$_2$CH=CH$_2$ (I, 90–94)	55
		CH$_2$=CHCH$_2$CuP(C$_4$H$_9$-n)$_3$	−78°	I (10–15)	

Note: References 173–204 are on p. 113.

[a] An excess of this reagent was used unless otherwise noted.
[b] A dash in this column indicates that diethyl ether was used as solvent, at 0°.
[c] THF is tetrahydrofuran.
[d] Prepared from dimethylmagnesium and cuprous iodide.
[e] Prepared from methyllithium and cuprous iodide.
[f] This is the yield of tlc-purified material.
[g] The rest of the product was mainly starting material.

TABLE III. Conjugate Addition of Organocopper Reagents to α,β-Ethylenic Carbonyl Compounds (*Continued*)

Number of Carbon Atoms	Carbonyl Substrate	Organocopper Reagent[a]	Conditions[b]	Product(s) and Yield(s) (%)	Refs.
6 (*contd.*)	2-Cyclohexenone (*contd.*)				
		B. Cyclic Carbonyl Substrates			
		cis-(CH$_3$CH=CH)$_2$CuLi	−78°	II, R = CH=CHCH$_3$-*cis*	73
		trans-(CH$_3$CH=CH)$_2$CuLi	−78°	II, R = CH=CHCH$_3$-*trans*	
7	3-Methyl-2-cyclohexenone (I)	(CH$_3$)$_2$CuLi	1 eq LiI	(II, 98) (I, 1)	28
		CH$_3$CuP(C$_4$H$_9$-*n*)$_3$	1 eq LiI	II (97), I (3)	
			—	II (<1), I (81)	
		CH$_3$CuP[(OCH$_3$)$_3$]$_3$	1 eq LiI	II (88), I (9)	
			—	II (<1), I (70)	
	4-Methyl-2-cyclohexenone	CH$_3$Cu[h]	—	2.7 : 1 (total 80)	46
		(CH$_3$)$_2$CuLi	—	10.1 : 1 (total 100)	

	$C_6H_5Cu^j$	—	7	1 (total 75)	
	$C_6H_5Cu^i$	LiBr	8	1 (total 100)	198
		LiI	24	1 (total 100)	29
	$(C_6H_5)_2CuMgBr$		97	3	
	$(C_6H_5)_2CuLi$		Major	Minor	
	$C_6H_5CuP(C_4H_{9\text{-}n})_3$	—	No reaction		
	$p\text{-}FC_6H_4Cu$	—	No reaction		
5-Methyl-2-cyclohexenone	CH_3Cu	1 eq LiI	99 : 1	(total 92–99)	28
	$(CH_3)_2CuLi$	1 eq LiI	98 : 2	(total 97)	
		—	98 : 3	(total 98)	
	$CH_3CuP(C_4H_{9\text{-}n})_3$	1 eq LiI	98 : 2	(total 34)	
	$CH_3CuP(OCH_3)_3$	1 eq LiI	98 : 2	(total 90)	
		1 eq $(n\text{-}C_6H_{13})_4$NI, THF^c	Polymer		
		—	Polymer		

Note: References 173–204 are on p. 113.

[a] An excess of this reagent was used unless otherwise noted.
[b] A dash in this column indicates that diethyl ether was used as solvent, at 0°.
[c] THF is tetrahydrofuran.
[i] This reagent was prepared from methylmagnesium bromide and cuprous chloride.
[h] Phenylmagnesium bromide and cuprous chloride were used to prepare this compound.
[j] Phenyllithium and either cuprous bromide or iodide were used.

TABLE III. CONJUGATE ADDITION OF ORGANOCOPPER REAGENTS TO α,β-ETHYLENIC CARBONYL COMPOUNDS (*Continued*)

Number of Carbon Atoms	Carbonyl Substrate	Organocopper Reagent[a]	Conditions[b]	Product(s) and Yield(s) (%)	Refs.
			B. Cyclic Carbonyl Substrates		
7 (*contd.*)	5-Methyl-2-cyclohexenone (*contd.*)	$(C_6H_5)_2CuLi$	—	(>90) (<10) (total 67)	172
	2-Cycloheptenone	$(CH_3)_2CuLi$	$-78°$	3-Methylcycloheptanone (91)	197a
8	5,5-Dimethyl-2-cyclohexenone	$(CH_3)_2CuLi$	—	(I)	55
		$CH_3CuP(C_4H_9\text{-}n)_3$ $CH_3CuP(OCH_3)_3$ $CH_3Cu(CN)[P(OC_2H_5)_3]_2Li$	1 eq LiI 1 eq LiI 1 eq LiI	I (92) I (93) I (75)	28
		$(n\text{-}C_4H_9C\!\!\equiv\!\!C)_3CuLi_2$	25–30°, dioxane	(II, 61.5)[k] II (major product)	55
		$(C_6H_5C\!\!\equiv\!\!C)_3CuLi_2$	25–30°, dioxane		
	1-Acetylcyclohexene	$(CH_2\!=\!CH)_2CuLiP(C_4H_9\text{-}n)_3$	$-78°$, THF[c]	(70)	73b

Note: References 173–204 are on p. 113.

[a] An excess of this reagent was used unless otherwise noted.
[b] A dash in this column indicates that diethyl ether was used as solvent, at 0°.
[c] THF is tetrahydrofuran.
[k] Spectral and analytical data suggest the major product to be the dimeric ketone shown.

TABLE III. Conjugate Addition of Organocopper Reagents to α,β-Ethylenic Carbonyl Compounds (*Continued*)

Number of Carbon Atoms	Carbonyl Substrate	Organocopper Reagent[a]	Conditions[b]	Product(s) and Yield(s) (%)	Refs.
	B. Cyclic Carbonyl Substrates				
10 (*contd.*)		(CH₃)₂CuLi		(—)	167a
		(CH₃)₂CuLi	—	(60) + (40)	198b
		(CH₃)₂CuLi	—	(1:1, —)	199
11		(CH₃)₂CuLi	—	(<5) + (>95) (total 81)	198a
		(CH₃)₂CuLi	—	(<5) + (>95) (total 84)	198a
		(CH₃)₂CuLi	—	(60) + (40) (total 84)	198a

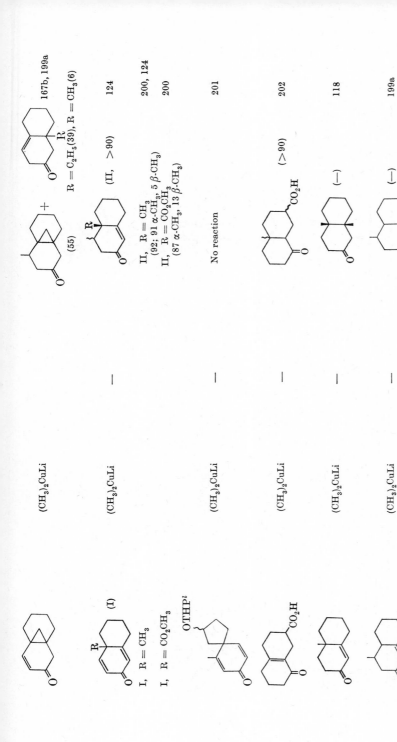

Note: References 173–204 are on p. 113.

[a] An excess of this reagent was used unless otherwise noted.
[b] A dash in this column indicates that diethyl ether was used as solvent, at 0°.
[i] THP is tetrahydropyranyl.

TABLE III. CONJUGATE ADDITION OF ORGANOCOPPER REAGENTS TO α,β-ETHYLENIC CARBONYL COMPOUNDS (*Continued*)

Number of Carbon Atoms	Carbonyl Substrate	Organocopper Reagent[a]	Conditions[b]	Product(s) and Yield(s) (%)	Refs.
		B. *Cyclic Carbonyl Substrates*			
11 (*contd.*)	[structure with OAc]	$(CH_3)_2CuLi$	—	[structure with OAc] (—)	203
	[structure with OTHP[l]]	$(CH_3)_2CuLi$	—	[structure with OTHP[l,m], AcO] (78)	121
	[bicyclic enone]	$(CH_3)_2CuLi$	—	[bicyclic ketone] (84)	166
	[cyclohexanone with =CHSC$_4$H$_9$-n]	$(CH_3)_2CuLi$	—	[cyclohexanone with isopropyl] (>95)	204
12	[bicyclic enone with C$_6$H$_5$]	$(CH_3)_2CuLi$	—	[bicyclic ketone with C$_6$H$_5$] (—)	167a

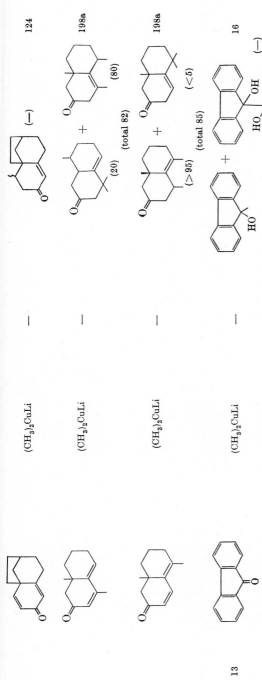

Note: References 173–204 are on p. 113.

[a] An excess of this reagent was used unless otherwise noted.
[b] A dash in this column indicates that diethyl ether was used as solvent, at 0°.
[i] THF is tetrahydropyranyl.
[m] The intermediate enolate was treated with acetyl chloride.

TABLE III. CONJUGATE ADDITION OF ORGANOCOPPER REAGENTS TO α,β-ETHYLENIC CARBONYL COMPOUNDS (*Continued*)

Number of Carbon Atoms	Carbonyl Substrate	Organocopper Reagent[a]	Conditions[b]	Product(s) and Yield(s) (%)	Refs.
	B. Cyclic Carbonyl Substrates				
14		$(CH_3)_2CuLi$	—	(85)	167
		$(CH_3)_2CuLi$	—	(40)	81
		$(CH_3)_2CuLi$	—	(77)	80, 122
		$(CH_3)_2CuLi$	—	No reaction	123
				(I)	

(structure 15: decalone with isopropenyl)	$(CH_3)_2CuLi$ $[CH_2=C(CH_3)CH_2]_2CuLi$ $(C_6H_5CH_2)_2CuLi$	$-10°$ $-78°$ $-25°$	I, R = CH_3 (68) I, R = $CH_2C(CH_3)=CH_2$ (20)[n] I, R = $C_6H_5CH_2$ (45)	82 83
	$(C_6H_5CH_2)_2CuLi$	$-20°$	(structure I: AcO-decalin with dioxolane, CH_2R)	
	$(m\text{-}CH_3OC_6H_4CH_2)_2CuLi$ $(C_6H_5)_2CuLi$	$-20°$ $-20°$	II, R = $C_6H_5CH_2$ (46)[o] II, R = $m\text{-}CH_3OC_6H_4CH_2$ (51)[o] II, R = C_6H_5 (89)[o]	
	$(CH_3)_2CuLi$	—	(structure: decalone with isopropyl) (60)	81
CH_3O_2C (enone, structure)	$(CH_3)_2CuLi$	—	(structure: decalone with isopropenyl, CH_3O_2C, ketone) (>80)	167

Note: References 173–204 are on p. 113.

[a] An excess of this reagent was used unless otherwise noted.
[b] A dash in this column indicates that diethyl ether was used as solvent, at 0°.
[n] The starting material was recovered in 77% yield.
[o] The intermediate enolate was treated with acetic anhydride.

TABLE III. CONJUGATE ADDITION OF ORGANOCOPPER REAGENTS TO α,β-ETHYLENIC CARBONYL COMPOUNDS (*Continued*)

B. Cyclic Carbonyl Substrates

Number of Carbon Atoms	Carbonyl Substrate	Organocopper Reagent[a]	Conditions[b]	Product(s) and Yield(s) (%)		Refs.
16	[C₆H₅-oxazolone with =CHAr] Ar = C₆H₅, o-ClC₆H₄, p-ClC₆H₄, p-CH₃OC₆H₄	Ar′Cu Ar′: C₆H₅, o-CH₃C₆H₄, p-CH₃C₆H₄, p-ClC₆H₄, o-ClC₆H₄, o-CH₃OC₆H₄, p-CH₃OC₆H₄, 1-C₁₀H₇, C₆H₅, p-ClC₆H₄, C₆H₅, C₆H₅, p-CH₃OC₆H₄, p-ClC₆H₄	33% excess Ar′MgBr	[oxazolone with CH(Ar)Ar′ product] 65, 62, 25, 75, 50, 40, 38, 40, 64, 60, 63, 64, 40, 60	+ 1,2-Addition products 2, 3, 30, 1, 2, 20, 40, 19, 3, 0, 0, 2, 10, 0	68
	[tricyclic enone with OC₂H₅ substituent]	(CH₃)₂CuLi	—	No reaction		123

	Substrate	Reagent		Product(s) and Yield(s) (%)	+ 1,2-Addition products
20	![C6H5-oxazoline with CHAr] Ar: 1-C10H7, 2-C10H7	C6H5Cu	33% excess C6H5MgBr	![C6H5-oxazoline with CH(Ar)C6H5] 60 62	68 0 0
21	[decalone with ethylene ketal and CH2CH2C6H4OCH3-m substituent]	(CH3)2CuLi	—	No reaction	123
23	[phenanthrenone with OC2H5 and CH2CH2C6H4OCH3-m]	(CH3)2CuLi	—	No reaction	
27	4-Cholesten-3-one	(CH3)2CuLi	—	5-Methyl-3-coprostanone (50)	79
		(CH3)2CuLi	—	[steroid structure with (C2H5O)2PO-O- group] (55)[p]	

Note: References 173–204 are on p. 113.

[a] An excess of this reagent was used unless otherwise stated.
[b] A dash in this column indicates that diethyl ether was used as solvent, at 0°.
[p] The intermediate enolate was treated with diethyl phosphorochloridate.

TABLE IV. CONJUGATE ADDITION OF ORGANOCOPPER REAGENTS TO α,β-ACETYLENIC CARBONYL COMPOUNDS

Number of Carbon Atoms	Carbonyl Substrate	Organocopper Reagent[a]	Conditions[b]	Product(s) and Yield(s) (%)	Refs.
3	HC≡CCO$_2$CH$_3$	$(n$-C$_4$H$_9)_2$CuLi	$-78°$	$trans$-n-C$_4$H$_9$CH=CHCO$_2$CH$_3$ (81–98)	65
4	CH$_3$C≡CCO$_2$H	t-C$_4$H$_9$CuP(C$_4$H$_9$-n)$_3$	$-78°$	t-C$_4$H$_9$CH=CHCO$_2$CH$_3$ (85 $trans$) (total 75)	67
		C$_6$H$_5$Cu	$-60°$	CH$_3$C=CHCO$_2$H (I)	
				C$_6$H$_5$	
				I (11:77 cis:$trans$)	
				I (10:86 cis:$trans$)	
	CH$_3$C≡CCO$_2$CH$_3$	(C$_6$H$_5)_2$CuLi	$-60°$	CH$_3$CH=CHCO$_2$CH$_3$ (I)	65
		(C$_2$H$_5)_2$CuLi (1.05 eq)	$-78°$, THF[d]	R	
		n-C$_4$H$_9$Cu (1.05 eq)	$-78°$, TMED[e]	I R = C$_2$H$_5$ (97, $trans$) (total 100)	
		n-C$_4$H$_9)_2$CuLi (1.10 eq)	$-78°$, THF[d]	I R = n-C$_4$H$_9$ (85, $trans$) (total 72)	
		$(n$-C$_7$H$_{15})_2$CuLi	$-78°$, THF[d]	I R = n-C$_4$H$_9$ (99, $trans$) (total 100)	
				$trans$-CH$_3$C=C(R)CO$_2$CH$_3$ (II)	64
				C$_7$H$_{15}$-n	
			$-78°$, THF[d], I$_2$,	II R = H (90)	
			$-78°$, THF[d], O$_2$,	II R = I (84)	32
			excess (CH$_3)_2$CuLi	II R = CH$_3$ (10.5)	
		$cyclo$-C$_5$H$_9$CuP(C$_4$H$_9$-n)$_3$	$-78°$	I R = $cyclo$-C$_5$H$_9$ (93.5 $trans$) (total 67–85)	65
		t-C$_4$H$_9$CuP(C$_4$H$_9$-n)$_3$	$-78°$	I R = t-C$_4$H$_9$ (86 $trans$) (total 60–80)	
		$(t$-C$_4$H$_9$CuCH$_3)$Li[f]	$-78°$	I R = t-C$_4$H$_9$ (93 $trans$) (total >90)	
		CH$_2$=CHCH$_2$CuN⌐⌐ (pyrrolidine ring)	$-78°$	I R = CH$_2$CH=CH$_2$ (98 $trans$) (total 55)	
		(CH$_3)_2$C=CHCH$_2$	-80 to $-90°$, excess	CH$_3$C=CHCO$_2$CH$_3$	54
		(n-C$_4$H$_9)_3$PCuCH$_2$	$(n$-C$_4$H$_9)_3$P	(CH$_3)_2$C=CCH$_2$CH$_2$	
				(4:96 cis:$trans$) (total 52)	

Substrate	Reagent	Conditions	Product(s) (% yield)	Yield (%)
$CH_3C\equiv CCON(C_2H_5)_2$	(geranyl-type copper reagent)	$-78°$, THF[d]	(geranyl product with CO_2CH_3) (65)	32
	$CHCH_2CH_2Cu$ / $CH_3C(CH_2)_2CH=C(CH_3)_2$	$-78°$, 3 eq pyrrolidine	$CH_3\diagdown C=C \diagup CON(C_2H_5)_2 \diagup \diagdown H$ (93) with $CHCH_2CH_2$ / $CH_3C(CH_2)_2CH=C(CH_3)_2$	65
$CH_3C\equiv CCON$⟨pyrrolidine⟩	$(C_2H_5)_2CuLi$	$-78°$	$CH_3C=CHCON$⟨ C_2H_5 ⟩ (98.6 *trans*) (total 75)	
$C_2H_5C\equiv CCO_2H$	$(CH_3)_2CuLi$	$-10°$	$CH_3C=CHCO_2H$ (92:8 *cis:trans*) with C_2H_5	
$C_2H_5C\equiv CCO_2CH_3$	CH_3Cu (1.05 eq)	$-78°$, 1.2 eq TMED[e]	$C_2H_5C(CH_3)=CHCO_2CH_3$ (I) (86:0.5 *cis:trans*) (total 90)	54
	$(CH_3)_2CuLi$	$-78°$	I (92:8 *cis:trans*) (total 93)	
		$-100°$	I (97:3 *cis:trans*) (total 95)	
	$t\text{-}C_4H_9CuP(C_4H_9\text{-}n)_3$	$-78°$	$C_2H_5C(CH_3)=CDCO_2CH_3$ (75)[g] $C_2H_5C=CHCO_2CH_3$ (97.4 *trans*) with $C_4H_9\text{-}t$ (total 60–88)	65
$C_2H_5C\equiv CCO_2C_2H_5$	$(C_2H_5)_2C=CH(CH_2)_2Cu$	$-78°$, 2–3 eq pyrrolidine or TMED[e]	$C_2H_5C=CHCO_2C_2H_5$ (>99 *trans*) with $(C_2H_5)_2C=CHCH_2CH_2$ (total >95)	

Note: References 173–204 are on p. 113.

[a] An excess of the reagent was used unless specified otherwise.
[b] Diethyl ether was used as solvent, at 0°, unless otherwise indicated.
[c] Slow addition of ester is essential.
[d] THF is tetrahydrofuran.
[e] TMED is tetramethylethylenediamine.
[f] This mixed lithium diorganocopper reagent was prepared at −30 to −50° from methylcopper and *t*-butyllithium.
[g] Deuterium oxide was used to quench the reaction.

TABLE IV. CONJUGATE ADDITION OF ORGANOCOPPER REAGENTS TO α,β-ACETYLENIC CARBONYL COMPOUNDS (Continued)

Number of Carbon Atoms	Carbonyl Substrate	Organocopper Reagent[a]	Conditions[b]	Product(s) and Yield(s) (%)	Refs.
9	$C_6H_5C\equiv CCO_2H$	CH_3Cu	$25°$	$C_6H_5C(CH_3)=CHCO_2H$ (I) (85 cis, 15 trans)	136
		CH_3Cu	$-60°$	I (82:18 cis:trans) (total 90)	67
		$(CH_3)_2CuLi$	$-60°$	I (7:93 cis:trans)	
	$C_6H_5C\equiv CCO_2CH_3$	CH_3Cu	$25°$	$C_6H_5C(CH_3)=CHCO_2CH_3$ (I) (63 cis, 35 trans)	136
		CH_3Cu	$-60°$	I (36:53 cis:trans)	67
		$(CH_3)_2CuLi$	$-60°$	I (19:35 cis:trans)	
	$C_6H_5C\equiv CCONH_2$	CH_3Cu		No reaction	136
	$C_6H_5C\equiv CCONHCH_3$	CH_3Cu		No reaction	
	$C_6H_5C\equiv CCON(CH_3)_2$	CH_3Cu		No reaction	
	$C_6H_5C\equiv CCO_2^-N^+(C_4H_{9-n})_4$	CH_3Cu	$25°$, THF[d]	trans-$C_6H_5C(CH_3)=CHCO_2^-N^+(C_4H_{9-n})_4$ (80)	
10	n-$C_7H_{15}C\equiv CCO_2CH_3$	$(CH_3)_2CuLi$	$-78°$, THF[d]	n-$C_7H_{15}C(CH_3)=CHCO_2CH_3$ (I) (99.8:0.2 cis:trans) (total 95)	64
			THF[d], ether	I (39:61 cis:trans) (total 90)	
			$-78°$, toluene	I (92:8 cis:trans) (total 50)	
			$-78°$, THF[d] O_2	 $\begin{array}{c} n\text{-}C_7H_{15} \\ \diagdown \\ C=C \\ \diagup \\ CH_3 \end{array} \begin{array}{c} CO_2CH_3 \\ \diagup \\ \diagdown \end{array}_2$ (25)	32
		$(CH_3)_2CuLi$ (large excess)	$-78°$, THF[d] O_2	cis-n-$C_7H_{15}C(CH_3)=C(CH_3)CO_2CH_3$ (32–46)	
		$(CH_3)_2CuLi$	$-78°$, THF[d], I_2,	cis-n-$C_7H_{15}C(CH_3)=C(I)CO_2CH_3$ (II) (95)	
			$-78°$, ether, I_2,	II (68)	
		$(CH_3)_2CuLi$	$-78°$, ether, $(CH_3O)_3P$	n-$C_7H_{15}C(CH_3)=CHCO_2CH_3$ (III) (79:21 cis:trans)	
				III (32:8 cis:trans)	
11	n-$C_7H_{15}C\equiv CCOCH_3$	$(CH_3)_2CuLi$	$-78°$ ether, TMED[e]	n-$C_7H_{15}C(CH_3)=CHCOCH_3$	
			$-78°$, THF[d]	(1:1 cis:trans)	

Note: References 173–204 are on p. 113.

[a] An excess of the reagent was used unless specified otherwise.
[b] Diethyl ether was used as solvent, at 0°, unless otherwise ineadited.
[d] THF is tetrahydrofuran.
[e] TMED is tetramethylethylenediamine.

REFERENCES TO TABLES I–IV

[173] L. Decaux and R. Vessiere, *Compt. Rend.* **267**, 738 (1968).
[174] A. D. Williams and P. J. Flory, *J. Amer. Chem. Soc.*, **91**, 3111 (1969).
[175] A. K. Kundu, N. G. Kundu, and P. C. Dutta, *J. Chem. Soc.*, **1965**, 2749.
[176] K. Subrahmania Ayyar and G. S. Krishna Rao, *Tetrahedron Lett.*, **1967**, 4677.
[177] K. Subrahmania Ayyar and G. S. Krishna Rao, *Can. J. Chem.*, **46**, 1467 (1968).
[178] K. K. Mahalanabis, *Indian J. Chem.*, **5**, 124 (1967).
[179] M. S. Kharasch and D. C. Sayles, *J. Amer. Chem. Soc.*, **64**, 2972 (1942).
[180] H. Christol and R. Vanel, *Bull. Soc. Chim. Fr.*, **1968**, 1398.
[181] O. H. Wheeler and E. Granell de Rodriguez, *J. Org. Chem.*, **29**, 718 (1964).
[182] E. L. Eliel and F. J. Biros, *J. Amer. Chem. Soc.*, **88**, 3334 (1966).
[183] M. Balasubramanian and A. D'Souza, *Indian J. Chem.*, **8**, 233 (1970).
[184] J. W. Ellis, *Chem. Commun.*, **1970**, 406.
[185] G. Stork and P. F. Hudrlik, *J. Amer. Chem. Soc.*, **90**, 4462 (1968).
[186] R. F. Church, R. E. Ireland, and D. R. Shridhar, *J. Org. Chem.*, **27**, 707 (1962).
[187] (a) J. T. Greig and J. D. Roberts, *J. Amer. Chem. Soc.*, **88**, 2791 (1966); (b) W. S. Johnson, A Van der Gen, and J. J. Swoboda, *ibid.*, **89**, 171 (1967); (c) W. Nagata and I. Kikkawa, *Chem. Pharm. Bull. (Tokyo)*, **11**, 289 (1963); (d) W. S. Johnson, P. J. Neustadter, and K. K. Schmiegel, *J. Amer. Chem. Soc.*, **87**, 5148 (1965).
[188] S. Boatman, T. M. Harris, and C. R. Hauser, *J. Amer. Chem. Soc.*, **87**, 82 (1965).
[189] E. Piers, W. de Waal, and R. W. Britton, *Chem. Commun.*, **1968**, 188.
[189a] R. S. Matthews and R. E. Meteyer, *Chem. Commun.*, **1971**, 1576.
[190] J. A. Marshall and A. R. Hochstetler, *J. Amer. Chem. Soc.*, **91**, 648 (1969).
[191] H. E. Zimmerman, *J. Org. Chem.*, **20**, 549 (1955).
[192] P. Angibaud, J-P. Marets, and H. Rivière, *Bull. Soc. Chim. Fr.*, **1967**, 1845.
[193] H. E. Zimmerman and R. L. Morse, *J. Amer. Chem. Soc.*, **90**, 954 (1968).
[194] R. Wiechert, U. Kerb, and K. Kieslich, *Chem. Ber.*, **96**, 2765 (1963).
[195] W. J. Wechter, *J. Org. Chem.*, **29**, 163 (1964).
[196] W. C. Shoppee, T. E. Bellas, and R. Lack, *J. Chem. Soc.*, **1965**, 6450.
[197] O. C. Musgrave, *J. Chem. Soc.*, **1951**, 3121.
[197a] G. H. Posner and J. J. Sterling, unpublished results.
[198] N. T. Luong-Thi and H. Rivière, *Tetrahedron Lett.*, **1971**, 587.
[198a] J. A. Marshall, R. A. Ruden, L. K. Hirsch, and M. Phillippe, *Tetrahedron Lett.*, **1971**, 3795.
[198b] J. A. Marshall and W. F. Huffman, *Synthetic Commun.*, **1**, 221 (1971).
[199] E. Siscovic and A. S. Rao, *Curr. Sci.*, **37**, 286 (1968).
[199a] J. A. Marshall and R. A. Ruden, *Synthetic Commun.*, **1**, 227 (1971).
[200] J. A. Marshall and T. M. Warne, Jr., *J. Org. Chem.*, **36**, 178 (1971).
[201] P. C. Mukharji and P. K. Sen Gupta, *Chem. Ind. (London)*, **1970**, 533.
[202] J. W. Huffman and M. L. Mole, *Tetrahedron Lett.*, **1971**, 501 and *J. Org. Chem.*, **37**, 13 (1972).
[203] J. A. Marshall and G. M. Cohen, *Tetrahedron Lett.*, **1970**, 3865; *J. Org. Chem.*, **36**, 887 (1971).
[204] R. M. Coates and R. L. Sowerby, *J. Amer. Chem. Soc.*, **93**, 1027 (1971).

CHAPTER 2

FORMATION OF CARBON-CARBON BONDS VIA π-ALLYLNICKEL COMPOUNDS

Martin F. Semmelhack

Cornell University, Ithaca, New York

CONTENTS

	PAGE
INTRODUCTION	117
Structure and Bonding	119
CARBON–CARBON BOND FORMATION VIA BIS-(π-ALLYL)NICKEL(0) COMPLEXES	122
As Discrete Reagents	122
Mechanisms	126
Experimental Conditions	128
As Transient Intermediates	128
Mechanisms	130
Scope and Limitations	135
Comparison with other methods	140
Experimental Conditions	143
CARBON–CARBON BOND FORMATION VIA π-ALLYLNICKEL HALIDES	143
As Discrete Reagents	144
The Complex	144
Synthesis via Carbonylation	146
Reaction with Organic Halides	147
Mechanism	148
Scope and Limitations	150
Table I. The Reaction of Organic Halides with π-(2-Methylallyl)nickel Bromide	150
Table II. Coupling Products from π-(2-Methylallyl)nickel Bromide and R–Y–CH$_2$Cl	156
Reaction with Other Functional Groups	158
Comparison with Other Methods	159

Experimental Conditions 160
   Preparation from Nickel Carbonyl 160
   Preparation from Bis-(1,5-Cyclooctadiene)nickel(0) 161
   Reaction with Organic Halides 162
As Transient Intermediates 162
   Coupling of Allylic Halides 162
      Mechanism 163
      Scope and Limitations 165
      Comparison with Other Methods 169
      Experimental Conditions 170
   Carbonylation of Allylic Halides 170
      Mechanism 171
      Scope and Limitations 172
      Comparison with Other Methods 174
      Experimental Conditions 175

EXPERIMENTAL PROCEDURES 175

General Technique 175
Bis-(π-2-methylallyl)nickel (from 2–Methylallylmagnesium Bromide and Nickel Bromide 176
Bis-(π-2-carbethoxyallyl)nickel (from Zinc Reduction of the Corresponding π-Allylnickel Bromide) 176
4,5-Dimethyl-1,4,7-*cis*, *cis*, *trans*-cyclodecatriene 177
π-(2-Methylallyl)nickel Bromide (from 2-Methylallyl Bromide and Nickel Carbonyl) 177
Bis-(1,5-cyclooctadiene)nickel(0) 178
π-(2-Carbethoxyallyl)nickel Bromide [using Bis-(1,5-cyclooctadiene)nickel(0)] 179
4-(2-Methylallyl)cyclohexanol 179
4-Hydroxy-2-methyl-4-phenyl-1-butene 180
cis- and *trans*-Geranylcyclohexane 180
cis-1-(*cis*-Cyclooct-2-en-1-yl)cyclooct-2-ene 181
trans, *trans*-1,5-Cyclooctadecadiene 182
Methyl 2,5-Hexadienoate 182

TABULAR SURVEY 183

Table III. Ligand Displacement Reactions with Bis-(π-allyl)nickel Complexes as Discrete Reagents 184
Table IV. Reactions of the Zerovalent Nickel Catalyst with a 1,3-Diene, a Coreactant, and a Phosphorus Ligand 185
Table V. Reactions of π-Allylnickel Halides as Discrete Reagents . . 188
   A. With Nonallylic Halides 188
   B. With Allylic Halides 190
   C. With Oxygen Functional Groups 191
Table VI. Coupling Reactions of Allylic Compounds with Nickel Carbonyl 192
   A. Intermolecular 192
   B. Intramolecular 194

Table VII. Carbonylation of Allylic Derivatives with Nickel Carbonyl . 196
  A. With Carbon Monoxide in the Absence of Alkynes 196
  B. With Carbon Monoxide and Acetylene 196

REFERENCES TO TABLES III–VII 198

INTRODUCTION

Transition metal complexes containing a planar three-carbon ligand, the π-allyl group, were first described in 1958 with the discovery of (π-1-methylallyl)cobalt tricarbonyl.[1] The synthesis of compounds with π-allyl units bound to other metals followed rapidly;[2] now isolable species have

been reported for all of the members of the first transition series and many other transition metals. The bonding and structure of π-allyl complexes, to be discussed briefly below, is now well understood using the principles of π-bonding[3] which developed from the study of organometallic chemistry after the discovery of ferrocene. Less well known are the effects on the chemical reactivity of an allyl group attached to a metal via a π bond. This chapter deals with the synthetic organic applications of π-allylnickel complexes,[4] a subject which up to now has been systematically investigated primarily in two laboratories, in Germany by G. Wilke and co-workers and in the United States by E. J. Corey and his associates. Other significant but isolated examples of organic preparations involving π-allylnickel complexes are included as well. Unlike most contributions to *Organic Reactions*, this chapter treats a subject still in its infancy.

Only a relatively small number of practical syntheses using π-allylnickel complexes have been reported. Therefore the plan of this chapter is to discuss the mechanism of the reactions in detail unusual for this series in order to stimulate new investigations and applications of these and other π-allyl metal complexes.

Many of the complexes described below are prepared from the most easily available source of zero-valent nickel, nickel tetracarbonyl. This reagent has a deserved reputation as an extremely poisonous and insidious

---

[1] H. B. Jonassen, R. I. Stearns, J. Kenttämaa, D. W. Moore, and A. G. Whittaker, *J. Amer. Chem. Soc.*, **80**, 2586 (1958).

[2] M. L. H. Green and P. L. I. Nagy, *Adv. Organomet. Chem.*, **2**, 325 (1964) provide a review and comprehensive table of π-allylmetal complexes through 1963. See also ref. 17.

[3] For a discussion of π-bonding in organometallic complexes see F. A. Cotton and G. Wilkinson, *Advanced Inorganic Chemistry*, Interscience Publishers, New York, 1962.

[4] A review of synthetic applications of π-allylnickel complexes with somewhat different emphasis has appeared; P. Heimbach, P. W. Jolly, and G. Wilke, *Adv. Organomet. Chem.*, **8**, 29 (1970).

compound, although it is used in large quantities by technical personnel in industrial operations. It must be remembered that nickel carbonyl is poisonous when inhaled, that it is very volatile (bp 43°), that it is more dense than air, that it has a mild odor reminiscent of musty areas, and that no symptoms appear until after a fatal dose may have been inhaled. An experienced research chemist, however, should find the reagent no more forbidding than the many other noxious, volatile compounds that are used routinely. Further details concerning the handling of nickel carbonyl are presented in a later section on experimental procedures.

Two classes of known π-allylnickel complexes can be defined. One class consists of species with formally zero-valent nickel and two uncharged π-allyl ligands [e.g., the simplest member,[5] bis-(π-allyl)nickel(0) (**1**)]; a second class includes compounds of nickel(I) with one π-allyl ligand, an anion (halide, acetylacetonide, etc.), and one or more ligands capable of donating an electron pair (bridging halogen, phosphines, carbon monoxide, etc.). A member of the second class is π-allylnickel(I)-μ-dibromo-π-allyl-nickel(I) (**2**) which will be abbreviated as π-allylnickel bromide in this chapter. The two types of complexes differ in their methods of preparation

and chemical reactivity, and are treated separately. The unifying aspect of π-allylnickel chemistry is the general ability of the allyl ligand to form new carbon–carbon bonds. Important examples are coupling of the allyl halides, reaction with unsaturated hydrocarbons, and reaction with carbon–halogen bonds. The reactions, illustrated in the three accompanying equations, have been used as key steps in the synthesis of humulene,[6]

---

[5] An alternative name for **2** is bis-(trihaptoallyl)nickel(0); F. A. Cotton, *J. Amer. Chem. Soc.*, **90**, 6230 (1968).

[6] E. J. Corey and E. Hamanaka, *J. Amer. Chem. Soc.* **89**, 2758 (1967).

santalenes,[7] and farnesene,[8] as well as in simple and efficient preparations of medium-sized ring olefins.[9, 10]

Structure and Bonding

According to McClellan[11] and Nordlander and Roberts,[12] it is possible to distinguish three types of allylmetal compounds on the basis of their proton nmr spectra. Static σ-allyl species produce a pattern of absorption very similar to that of allyl bromide, a doublet for the $-CH_2-$ and the characteristic multiplet of $-CH=CH_2$. An example in the transition metal series is σ-allylmanganese pentacarbonyl,[13] which is characterized also by a C=C stretching band at 1620 cm^{-1} in its infrared spectrum, typical of an unperturbed vinyl group. Dynamic allyl ligands are most often observed for the nontransition metals such as allylmagnesium halides.[12] The nmr spectrum consists of an AX_4 pattern indicating that the two CH_2 groups are indistinguishable, owing to rapid equilibration between ends of the allyl systems. The π-allyl ligand is indicated by an nmr spectrum consisting of three signals with intensity ratio 1:2:2; one of the

signals, corresponding to H_3 in 3, is a multiplet of approximately seven peaks, and H_1 and H_2 appear as doublets.

Most common for the solid state of transition metal-allyl complexes is the planar π-allyl arrangement,[3] with a p orbital from each of three carbon atoms forming delocalized molecular orbitals which interact with the metal via π bonding. This arrangement has been referred to as a "half sandwich" to emphasize the similarity in bonding compared to ferrocene, a "sandwich" complex.[14] In solution with noncoordinating solvents such as benzene and chloroform, π-allylnickel bromide **(2)**[8, 15] and bis-(π-allyl)-nickel **(1)**[9] have been shown by nmr spectroscopy to contain the

---

[7] E. J. Corey and M. F. Semmelhack, *J. Amer. Chem. Soc.*, **89**, 2755 (1967).
[8] M. F. Semmelhack, Ph.D. Thesis, Harvard University, 1967.
[9] G. Wilke, *Angew. Chem. Int. Ed. Engl.*, **5**, 151 (1966).
[10] E. J. Corey and E. Wat, *J. Amer. Chem. Soc.*, **89**, 2757 (1967).
[11] W. R. McClellan, H. H. Hoehn, H. N. Cripps, E. L. Muetterties, and B. L. Howk, *J. Amer. Chem. Soc.*, **83**, 1601 (1961).
[12] J. E. Nordlander and J. D. Roberts, *J. Amer. Chem. Soc.*, **81**, 1769 (1959).
[13] H. D. Kaese, R. B. King, and F. G. A. Stone, *Z. Naturforsch.*, **15b**, 682 (1960).
[14] F. A. Cotton and G. Wilkinson, *Advanced Inorganic Chemistry*, Interscience Publishers, New York, 1962, p. 654.
[15] G. Burger, Thesis, University of Munich, 1962.

planar delocalized π-allyl ligand. X-ray diffraction studies have confirmed the π-allyl ligand structure as indicated for π-(2-carbethoxyallyl)-nickel bromide in Fig. 1.[16] An interesting detail of this structure is the angle of approximately 110° between the plane of the allyl ligand and the plane containing the metal and halogen atoms.[16a] The tilt is helpful in understanding the nmr spectra of π-allyl ligands.

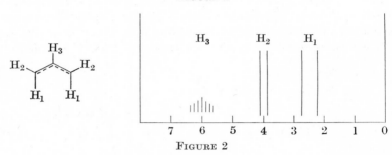

FIGURE 1

FIGURE 2

A typical nmr spectrum of a π-allyl ligand is presented in Fig. 2. The coupling constant J_{12} is approximately zero, similar to nonequivalent hydrogens in terminal methylene groups, while J_{13} is 13.3 Hz and J_{23} is 7.0 Hz. The difference in chemical shifts of H_1 and H_2 is due to the fact that H_1 is closer to the metal atom and more highly shielded.[17] Thus substituents on the terminal carbons of the allyl ligand have two possible orientations, termed *anti* (H_1 in Fig. 2) and *syn* (H_2). Theoretical treatments of the bonding in π-allyl complexes exist in quantitative terms[18] and in a brief qualitative description.[3] It is immaterial whether the π-bonded allylic system is regarded as a radical, an anion, or a cation. According to molecular orbital theory, the bonding orbital in an allyl radical would be doubly occupied and the nonbonding orbital singly occupied; in an allyl anion both orbitals would be full, while in an allyl cation only the bonding orbital would be full. Hence, depending upon the viewpoint, the π-allylic

---

[16] (a) M. Churchill and T. O'Brien, Harvard University, private communication; (b) A. Sirigu, *Chem. Commun.*, **1969**, 256; C. H. Dietrich and R. Uttech, *Z. Kristallogr.*, **122**, 60 (1965).

[17] E. O. Fischer and H. Werner, *Metal π-Complexes*, Elsevier, 1966, p. 177.

[18] (a) S.F.A. Kettle and R. Mason, *J. Organomet. Chem.*, **5**, 573 (1966); (b) W. W. Fohleman, L. C. Cusacks, and H. B. Jonassen, *Chem. Phys. Letters*, **3**, 52 (1969).

system can supply four, three, or two π electrons for coordination to the metal. However, this ambiguity is of no decisive importance, since the number of electrons supplied by the allyl group to the molecule can be formally compensated for by a change in the valency of the metal atom. In fact, the π-allyl ligand on nickel shows chemical properties consistent with an electron density slightly greater than that of the allyl radical, as indicated by additions to reactive carbonyl groups[7] and slow reaction with water to produce propylene[19]; the protonation reaction is accelerated by mineral acid.

It is important to keep in mind the number of electrons being donated to the metal atom from all of the ligands. Many of the reactions discussed below involve the simultaneous coordination of several organic moieties; the number of ligands that can be accommodated by one nickel atom is a factor in determining the selectivity of the reactions. The upper limit on the number of ligands is suggested by the inert gas rule,[3] which for nickel means a maximum of 18 valence electrons. For zero-valent nickel (*e.g.*, in **1**), the metal atom contributes 10 electrons, and thus all the ligands combined must provide at most 8 more electrons. However, in complex **1**, each allyl ligand (counted as allyl radicals to maintain overall zero charge on the complex) contributes three π electrons, and the nickel atom does *not* reach the inert gas (18-electron) configuration. This is a general rule for the pseudo-square planar coordination geometry that is common for π-allyl-nickel complexes—the favorable electronic configuration involves a total of 16 electrons. Ligand field theory provides a direct explanation of this observation in terms of the relative energies of the metal d orbitals in the presence of a square planar ligand field.[3] It turns out that one d orbital, usually designated $d_{x^2-y^2}$, is directed toward the ligands; therefore it is of rather high energy and is usually left unfilled.

It is not possible to attribute with confidence the reactions of π-allyl-nickel complexes to the π-bonded form of the allyl ligand. In solution in polar coordinating solvents such as dimethylformamide or in the presence of a molar equivalent of a good electron donor such as triphenylphosphine, the nmr spectrum of the allylnickel complex changes to the AX_4 pattern characteristic of a rapidly equilibrating σ-bonded allyl unit.[20] Consistent with the general instability of σ-bonded carbon ligands on nickel, few σ-allyl complexes of nickel have yet been isolated, although evidence for their existence is available from spectroscopy in solution.[9, 20] Because of the uncertainty as to which intermediate allylnickel species is responsible for carbon-carbon bond formation, the mechanisms presented

[19] (a) E. J. Corey and M. F. Semmelhack, unpublished results; (b) A. V. Volkov, O. P. Parenago, V. M. Frolov, and B. A. Dolgoplosk, *Dokl. Akad. Nauk SSSR*, **187**, 574 (1969) [*C.A.* **71**, 91662 (1970)].

[20] (a) D. Walter, Dissertation, Technische Hochschule, Aachen, 1965; (b) D. Walter and G. Wilke, *Angew. Chem. Int. Ed. Engl.*, **5**, 897 (1966).

below are usually hypothetical. They are rationalizations of the product distributions obtained under various conditions and are presented as guide lines to be used in planning further applications of the reactions.

CARBON-CARBON BOND FORMATION VIA BIS-(π-ALLYL)NICKEL COMPLEXES

As Discrete Reagents

Soon after the elucidation of structure of the earliest π-allyl metal complexes, Wilke and co-workers isolated bis-(π-allyl)nickel[21] (**1**) and a few substituted bis-(π-allyl)nickel complexes during their very successful investigation of π-allyl complexes of all the transition metals.[9] At first, interest in the complexes arose from their possible intermediacy in the oligomerization of 1,3-butadiene,[20] but study of their reactivity revealed several other useful applications in synthesis.

Bis-(π-allyl)nickel is an orange-yellow substance with mp 1° and high volatility. It can be sublimed easily at low temperature and even codistils with ether; it is extremely sensitive to oxygen, usually catching fire spontaneously in air. Two methods of synthesis are now available, one from nickel(II) salts and the other from nickel(0) reagents. The first and most direct route is an example of the general method of synthesis of π-allylmetal complexes,[20] the reaction of allylmagnesium halide with nickel dibromide in ether at 0°. No yield has been reported for this reaction.[21] An alternative method, involving two steps, is the preparation

$$2 \diagup\!\!\!\diagdown\!\!\!\diagup MgBr + NiBr_2 \xrightarrow[0°]{Et_2O} \text{[bis-(π-allyl)Ni] } \mathbf{1} + 2MgBr_2$$

of π-allylnickel bromide followed by disproportionation to bis-(π-allyl)-nickel and nickel dibromide. The disproportionation is induced by strongly coordinating solvents such as dimethylformamide,[22] hexamethylphosphoramide,[22] liquid ammonia,[23] and water.[24] From solutions of π-allyl-

[structure: dimeric π-allylnickel bromide with CO$_2$Et groups] ⇌ (Polar solvent) [structure: bis-(π-allyl)nickel with CO$_2$Et groups] + NiBr$_2$

---

[21] (a) G. Wilke and B. Bogdanovic, *Angew. Chem.*, **73**, 756 (1961); (b) G. Wilke, *Angew. Chem. Int. Ed. Engl.*, **2**, 110 (1963).

[22] E. J. Corey, L. F. Hegedus, and M. F. Semmelhack, *J. Amer. Chem. Soc.*, **90**, 2417 (1968).

[23] U. Birkenstock, Dissertation Technische Hochschule, Aachen, 1966; Belg. Pat. 702682 (1968).

[24] Studiengesellschaft Kohle m.b.H., Fr. Pat. 1,543,303, 1968 [*C.A.* **72**, 43881s (1969)].

nickel bromide complexes in dimethylformamide and hexamethylphosphoramide, volatile bis-(π-allyl)nickel species are distilled in yields up to 60%.[22] Similarly, using ammonia, bis-(π-allyl)nickel is obtained in 100% yield by sublimation from the hexaammoniate of nickel dibromide.[23] Alternatively, washing an ammonia or water solution of π-allylnickel bromide with pentane-ether affords the bis-(π-allyl)nickel complex in high yield.[24] Direct reduction of π-allylnickel bromide with zinc-copper couple in dimethylformamide also leads to complete disproportionation, giving the bis-(π-allyl)nickel in more than 90% yield.[22]

The disproportionation reaction is more convenient than the route via allylic Grignard reagents and also allows the synthesis of complexes containing functional groups that would interfere with the Grignard synthesis.

The stability and reactions of bis-(π-allyl)nickel complexes are critically dependent upon the structure of the ligand. Bis-(π-2-methylallyl)nickel is more stable than bis-(π-allyl)nickel and only slowly decomposes on storage at 25° under inert gas.[9] In contrast, bis-(π-cyclohexenyl)nickel and bis-(π-cycloheptenyl)nickel decompose rapidly at less than 25° and cannot be stored easily.[9] The eight-carbon analog, bis-(π-cyclooctenyl)nickel, is more stable, depositing nickel from crystals in a sealed tube only after several days at 25°.[9] Bis-(π-cyclopentenyl)nickel has not been characterized owing to facile loss of hydrogen, which affords the very stable π-

cyclopentadienyl-π-cyclopentenyl nickel system.[9, 25] A special method of synthesis[9] leads to bis-(π-cyclooctatrienyl)nickel (4), according to Scheme 1 (p. 124. top).

Perhaps because of their thermal instability, bis-(π-allyl)nickel complexes have not found wide use as synthetic reagents. Two general reactions, however, have been uncovered: coupling of the allyl ligands to form 1,5-hexadiene derivatives, and ketone formation by insertion of carbon monoxide during coupling.

The direct coupling reaction is very general, occurring whenever a bis-(π-allyl)nickel complex is mixed with a donor ligand such as a phosphine, carbon monoxide, or even 1,3-butadiene.[9] Reaction of certain bis-(π-allyl)nickel complexes (5, 6, and 7) with iodine also leads to coupling of the allyl ligands.[21] (See p. 124, bottom).

[25] E. O. Fischer and H. Werner, *Chem. Ber.*, **92**, 1423 (1959).

SCHEME 1

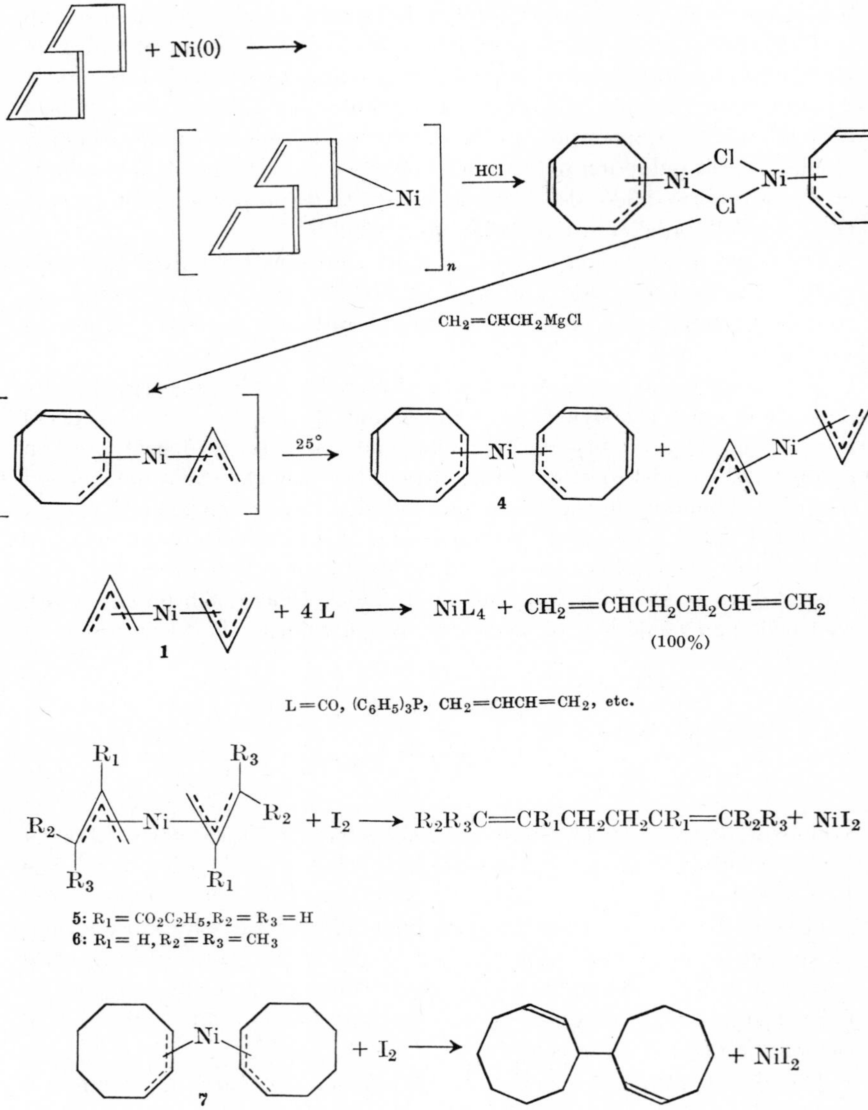
+ Ni(0)
Ni
n
HCl
Cl
Ni
Ni
Cl
CH2=CHCH2MgCl
Ni
25°
Ni
4
+
Ni
Ni
1
+ 4 L
NiL4 + CH2=CHCH2CH2CH=CH2
(100%)
L=CO, (C6H5)3P, CH2=CHCH=CH2, etc.
R1
R3
R2
Ni
R2
+ I2
R2R3C=CR1CH2CH2CR1=CR2R3+ NiI2
R3
R1
5: R1= CO2C2H5, R2= R3 = H
6: R1= H, R2= R3 = CH3
Ni
7
+ I2
+ NiI2

The use of carbon monoxide as the donor ligand produces symmetrical ketones from certain bis-(π-allyl)nickel complexes. Bis-(π-2-methylallyl)-nickel reacts with one molar equivalent of carbon monoxide at $-80°$ to form a 1:1 adduct, with excess carbon monoxide at higher temperature to produce mainly 2,5-dimethyl-1,5-hexadiene accompanied by di-(2-methyl-prop-2-enyl) ketone as a minor product.[9] Better yields of carbonyl insertion

$$\text{(bis-}\pi\text{-methallyl)Ni} \xrightarrow{CO} CH_2{=}C(CH_3)CH_2CH_2C(CH_3){=}CH_2 +$$

$$CH_2{=}C(CH_3)CH_2\overset{O}{\underset{\|}{C}}CH_2C(CH_3){=}CH_2$$

are obtained from bis-(π-cyclooctenyl)nickel[9] [92% of di-(2-cyclooctenyl) ketone]; bis-(π-2-carbethoxyallyl)nickel and bis-(π-1,1-dimethylallyl)-nickel also afford the corresponding symmetrical ketones with carbon monoxide.[22] The contrasting fact that bis-(π-allyl)nickel, bis-(π-crotyl)-nickel,[9] and others give only hydrocarbon coupling products when treated with carbon monoxide is still unexplained. No general rule can be presented at this time to allow prediction of the extent of coupling vs. carbon monoxide insertion when a bis-(π-allyl)nickel complex interacts with carbon monoxide.

The propensity of ligands of bis-(π-allyl)nickel complexes to couple with each other makes less likely the possibility of bringing the π-allyl group into reaction with organic substrates. Reactions with alkyl and aryl iodides are slower and less efficient than the corresponding reaction of a π-allylnickel bromide complex (see p. 147), but are otherwise not different. Rapid coupling does occur with allylic halides, but a mixture of products is obtained exactly as is produced from a π-allylnickel halide complex (see p. 155).[22] One example of reaction of bis-(π-allyl)nickel with an olefin

$$\mathbf{1} + 2\,CH_2{=}C(CH_3)CH_2Br \longrightarrow \begin{cases} CH_2{=}CHCH_2CH_2CH{=}CH_2 \\ + \\ CH_2{=}C(CH_3)CH_2CH_2CH{=}CH \\ + \\ CH_2{=}C(CH_3)CH_2CH_2C(CH_3){=}CH_2 \end{cases}$$

$$\mathbf{1} + CF_2{=}CF_2 \longrightarrow \mathbf{8}$$

has been reported: the reaction of tetrafluoroethylene to afford the unique complex 8.[26] With 1,3-butadiene, the π-allyl ligands couple quantitatively

[26] J. Browning, D. J. Cook, C. S. Cundy, M. Green, and F. G. A. Stone, *Chem. Commun.*, **1968**, 929.

and the butadiene itself is cyclotrimerized to 1,5,9-cyclododecatriene; the allyl complex is simply a source of active nickel.

Mechanisms

The reaction of triethylphosphine with bis-(π-2-methylallyl)nickel affords a 1:1 adduct which is shown to contain a σ-allyl ligand by the appearance of infrared absorption at 1605 cm^{-1}.[9] Additional triethylphosphine at higher temperature promotes coupling and formation of tetrakis-(triethylphosphine)nickel. Thus a plausible mechanism for the coupling reaction is stepwise addition of the donor ligand leading to σ-bonded allyl

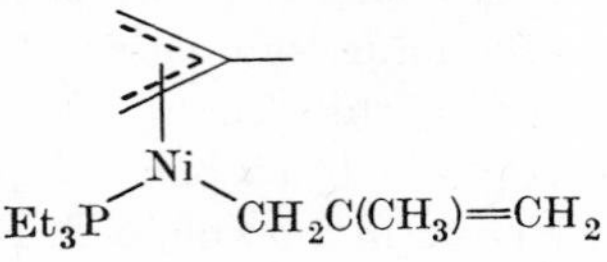

$$Et_3P-Ni-CH_2C(CH_3)=CH_2$$

ligands which then undergo radical-like coupling (Scheme 2).

SCHEME 2

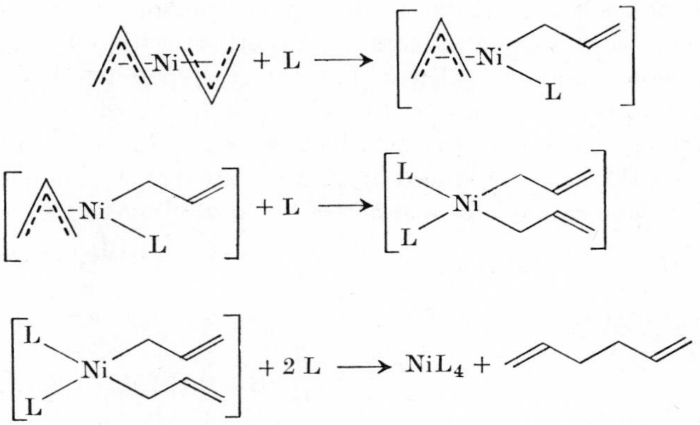

The σ-bonded intermediate is not appropriate, however, for reaction of bis-(π-crotyl)nickel with carbon monoxide because of the high stereospecificity of the reaction.[9] Bis-(π-crotyl)nickel can exist as six geometrical isomers, depending on the relative position of the methyl substituents

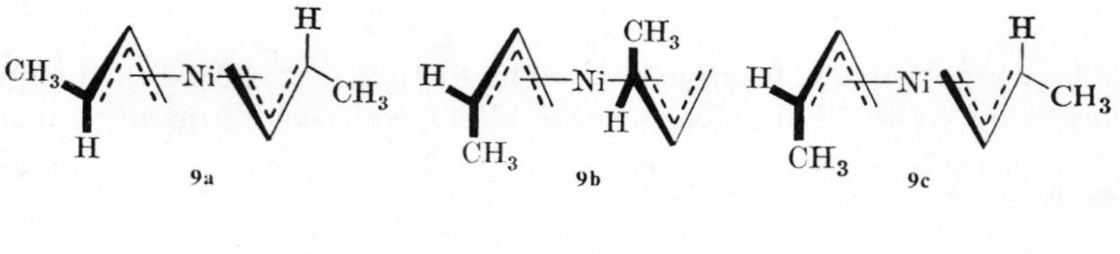

9a 9b 9c

(three are shown above, **9a–c**). Crystalline bis-(π-crotyl)nickel from crotylmagnesium bromide and nickel dibromide reacts with carbon monoxide at $-40°$ to give 98% *trans, trans*-octa-2,6-diene and 2% 3-methylhepta-1,5-diene in a combined yield of 96–98%. This result is attributed to stereoselective coupling from *syn, syn*-bis-(π-crotyl)nickel, **9d**,[9] and is inconsistent with *cis,trans* equilibrium of the double bonds via intermediate σ-crotyl ligands. At higher reaction temperature, the yield of

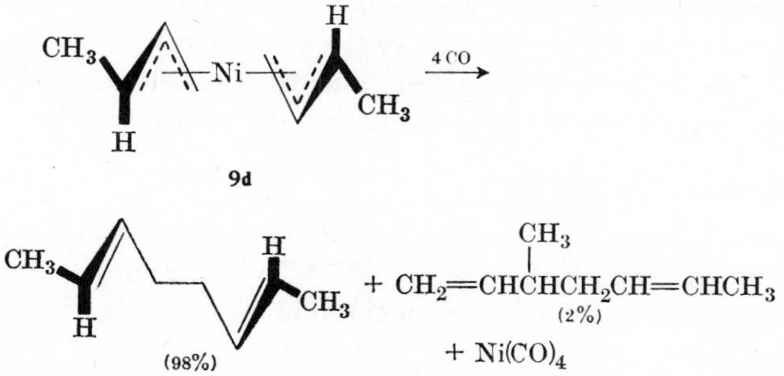

trans, trans product drops to 58% with corresponding increase in the yield of 3-methylhepta-1,5-diene and another isomer.

Insertion of carbon monoxide during coupling of π-allyl ligands has not been studied in detail, but a plausible mechanism, based on the generally accepted mechanism for carbonylation of π-allyl complexes,[27] is outlined in

SCHEME 3

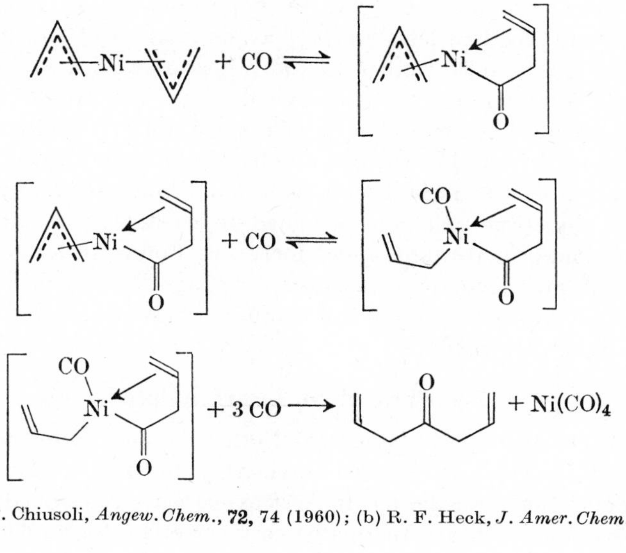

---

[27] (a) G. P. Chiusoli, *Angew. Chem.*, **72**, 74 (1960); (b) R. F. Heck, *J. Amer. Chem. Soc.*, **85**, 2013 (1963).

Scheme 3(p. 127). No information is available concerning the stereoselectivity of the insertion; no insertion products have been observed for reaction of carbon monoxide with bis-(π-allyl)nickel complexes such as bis-(π-crotyl)nickel where the geometry of the products would provide a measure of the stereoselectivity of the insertion.

A more convenient technique for the synthesis of 1,5-hexadiene derivatives from π-allylnickel complexes is discussed on p. 162, together with a comparison with other methods of allylic coupling. Unique to bis-(π-allyl)nickel complexes is the highly stereoselective coupling observed so far in one case, bis-(π-crotyl)nickel, and the efficient carbon monoxide insertion in certain π-allyl ligands such as π-cyclooctenyl.

Experimental Conditions

The preparation of bis-(π-allyl)nickel complexes via the Grignard route has been described in detail only once.[21a] The reaction is carried out by adding a solution of allylmagnesium chloride in ether to a suspension of an anhydrous nickel salt, for example nickel dibromide, in ether at low temperature. At $-10°$ the solution becomes gold-orange and the product is isolated from the ether. For volatile complexes like bis-(π-allyl)nickel and bis-(π-2-methylallyl)nickel, the ether must be separated by careful fractionation below 25°. The orange product can be sublimed and crystallized. It is pyrophoric and decomposes on storage at 25° or even lower. Solutions in water are moderately stable, and the complex easily dissolves in organic solvents without change.[21a]

The alternative method using disproportionation of a π-allylnickel halide is only slightly less direct. The preparation of the π-allylnickel halide is operationally similar to the preparation of an allyl Grignard reagent. The advantage to this method is its compatibility with most functional groups in contrast to the Grignard route. A limited number of examples are known, but there is no obvious limit to the applicability of the method. In the most generally useful technique, the π-allylnickel halide in dimethylformamide is added to a suspension of zinc-copper couple granules in pentane under inert gas. Complete formation of the zero-valent nickel complex is signaled by a color change from red to yellow and the bis-(π-allyl)nickel complex is isolated from the pentane layer in moderate yield.

As Transient Intermediates

The reactions of 1,3-butadiene and other simple 1,3-dienes with zero-valent nickel has been studied intensively by Wilke, Heimbach, and co-workers. They find the results best explained by intermediate π-allylnickel species which are thermally very sensitive, incompletely characterized, and never available as discrete, on-the-shelf reagents. The reactions, which are truly catalytic in nickel, have produced a variety of cyclic and linear polyolefins. They constitute a beautiful example of two important effects of metal coordination: the template effect where coordination of an intermediate with a metal atom gives orientation and activation for further reaction; and the ligand control effect, where a molecule can be added which will block a coordinating position on the metal and force the metal-catalyzed reaction to take a new path.

The most general and therefore the most important synthetic applications of transient bis-(π-allyl)nickel intermediates develop from the reaction of bis-(acetylacetonato)nickel(II) with an alkylaluminum in the presence of 1,3-butadiene. In the absence of other alkenes, alkynes, or donor ligands like triphenylphosphine, 1,3-butadiene is cyclotrimerized to 1,5,9-cyclododecatriene in high yield and high catalytic efficiency.[21b, 9] In the presence of carefully chosen phosphine ligands in catalytic amounts the 1,3-butadiene is cyclodimerized to any of three products: 1,2-divinyl-

cyclobutene,[28] 4-vinylcyclohexene,[9] or 1,5-cyclooctadiene.[9] The product distribution is dependent on contact time with the nickel catalyst (1,2-

divinylcyclobutene catalytically rearranges to 1,5-cyclooctadiene[28]) and on the steric and electronic properties of the phosphine ligand. Any one of the three cyclodimers can be obtained in high selectivity with proper choice of reaction conditions.

If a mixture of 1,3-butadiene and an alkene or alkyne is passed into the zero-valent nickel catalyst, the alkene or alkyne is incorporated in the product. For example, ethylene and 1,3-butadiene form *cis,trans*-cyclo-

[28] W. Brenner, P. Heimbach, H. Hey, E. W. Muller, and G. Wilke, *Ann.*, **727**, 161 (1969).

deca-1,5-diene in high yield.[29] Many interesting products have been observed in these reactions when 1,3-dienes, alkenes, and alkynes bearing substituents are included. Even the simplest substituents lead to much more complicated product mixtures, however, and often the effect of a new substituent on the course of the reaction cannot be predicted beforehand.

Mechanisms

Several questions can be asked about this reaction: What is the nature of the nickel catalyst? What are the forces which overcome the usual transannular nonbonded interactions during medium ring synthesis? How does the insertion of alkenes and alkynes proceed, and how does the phosphine ligand direct the cyclodimerization?

A key discovery was that of Wilke and his co-workers, who found that the reaction of bis-(acetylacetonato)nickel(II) with alkylaluminums leads to a very reactive form of nickel here referred to as Ni(0) which is probably an alkylnickel species that coordinates rapidly with even weakly bonding ligands such as simple olefins.[21b] In the absence of coordinating groups the alkylnickel complex rapidly decomposes to nickel metal. This soluble, zero-valent alkylnickel has been termed "naked nickel" to emphasize its ability to associate with weakly bonding ligands.[4]

The reactive nickel catalyst coordinates with two molecules of 1,3-butadiene, and the ends of the diene units couple to produce a C_8 bis-allyl ligand which associates with the nickel as the bis-(π-allyl)nickel complex **10**.[20b] It is likely that a 1,3-butadiene molecule reversibly coordinates as indicated in **10** to give nickel the 18-electron inert-gas configuration. Complex **10** has not been isolated, but in the presence of tri-(2-biphenylyl) phosphite a crystalline species believed to be **11** is isolable at low tem-

$$3\ CH_2{=}CHCH{=}CH_2 + Ni(0) \longrightarrow \quad \mathbf{10}$$

perature. Brief warming to 20° or additional phosphite leads to rapid formation of 1,5-cyclooctadiene.[21b, 28]

$$(RO)_3P \longrightarrow \quad \mathbf{11}$$

In the absence of other olefins or phosphines the 1,3-butadiene in complex **10** couples with the bis-π-allyl moiety to form a linear C_{12} ligand

[29] P. Heimbach and G. Wilke, *Ann.*, **727**, 183 (1969).

(in complex **12**), an 8-electron donor.[21b, 30] Then another diene unit takes up

a ligand position on nickel causing ring closure to 1,5,9-cyclododecatriene, a 6-electron donor (in complex **13**). The first catalytic cycle is completed by the addition of another 1,3-butadiene unit, which displaces the C_{12} triene and regenerates **10**. Complex **12**, although thermally very unstable, has been implicated by spectral and chemical characterization. For instance, hydrogenation of a solution expected to contain **12** affords n-dodecane in high yield.[30] The species with a cyclododecadiene ligand (**13**) is somewhat easier to handle and has been studied in detail.[31] Of the four possible geometrical isomers of cyclododecatriene, three are formed—the all-*cis* does not appear under any conditions. The distribution of the other three isomers is dependent on reaction temperature and reactant concentration and can be directed to mainly the all-*trans*-cyclododecatriene in the presence of, for instance, pyridine.[30] It has been suggested that the configuration of the allyl ligands in **10** and **12** is affected by the pyridine and thus determines the geometry of the olefin bonds of the product.[32] Evidence from nmr spectra indicates that two configurations of **12** are important: **12a** and **12b**.

[30] B. Bogdanovic, P. Heimbach, H. Kroner, G. Wilke, E. G. Hoffman, and J. Brandt, *Ann.*, **727**, 143 (1969).
[31] K. Jonas, P. Heimbach, and G. Wilke, *Angew. Chem. Int. Ed. Engl.*, **7**, 949 (1968).
[32] V. Birkenstock, Dissertation, Technische Hochschule, Aachen, 1966; cited in ref. 4.

When 1,3-butadiene is passed into a solution containing zero-valent nickel and one mole of a phosphine or phosphite per mole of nickel, the first intermediate is **11**, in which a phosphorus ligand is present instead of 1,3-butadiene as in **10**.[21b] The tightly bound phosphine blocks one coordination site so that no additional 1,3-butadiene can coordinate unless the bis-π-allyl ligand reacts to reduce the number of occupied coordination sites. Thus the 8-carbon ligand is converted to 1,5-cyclooctadiene which is subsequently displaced from the nickel by 1,3-butadiene to regenerate **11** and free 1,5-cyclooctadiene.[28]

$$2 \diagup\!\!\!\diagdown \xrightarrow[R_3P]{Ni(0)} R_3P\!\!-\!\!Ni \xrightarrow{2 \diagup\!\!\!\diagdown} 11 + \bigcirc$$

11

Other C_8H_{12} isomers also appear in the reaction. At low percentage of conversion, 1,2-divinylcyclobutene is present in significant amounts;[28] with certain phosphine ligands (*e.g.*, triphenylphosphine), 4-vinylcyclohexene is obtained in up to 27% yield.[21b] Proper choice of ligand, especially with tri-(2-biphenylyl) phosphite, allows conversion of 1,3-butadiene to a mixture of 96.5% 1,5-cyclooctadiene and 3% 4-vinylcyclohexene.[28] The reactions outlined in Scheme 4 have been proposed to account for the product distribution.[4]

The concept of electron transfer from metal to ligand via π bonding[33] is important in understanding the control exerted by phosphines and phosphites in Scheme 4. Phosphorus ligands attach to metals by a dative σ bond and π backbonding. The degree of electron transfer from metal to phosphorus depends on the other ligands on the metal and the substituents on the phosphorus. A σ-bonded allyl ligand contributes less electron density to the nickel atom than does the π-allyl ligand, and the σ form is favored when the other ligands are strong donors (*e.g.*, phosphines). Conversely, the π-allyl arrangement, providing more electron density to nickel, is favored when phosphites (good electron acceptors) are present on the nickel.

In scheme 4, σ-bound allyl groups are involved in the route to 4-vinylcyclohexene only. With L = triphenylphosphine (a good donor ligand) the π-allyl ligand can shift to a σ-allyl, and larger amounts of 4-vinylcyclohexene are formed. With L = phosphite (weaker donor, good acceptor) the bis-π-allyl structures (**11a** and **11b**) are favored (maximum electron transfer from allyl to nickel), and nearly exclusive formation of 1,5-cyclooctadiene is observed. As indicated by Scheme 4, at least part of the 1,5-cyclooctadiene is formed via 1,2-divinylcyclobutene.

[33] F. A. Cotton and G. Wilkinson, *Advanced Inorganic Chemistry*, Interscience Publishers, New York, 1962, p. 611.

SCHEME 4

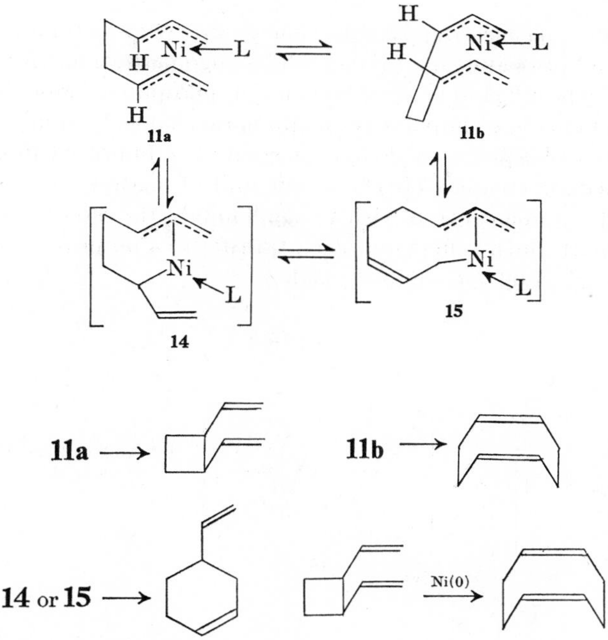

The mechanism of insertion of unsaturated hydrocarbons into the 8-carbon ligand is illustrated using ethylene as an example. The intermediates **16** and **17** have not been characterized and are short-lived at best, but *cis,trans*-cyclododecadiene is isolated in 80% yield.[34] Similarly, 2-butyne is co-oligomerized with two molecules of 1,3-butadiene to afford 1,2-dimethyl-*cis,cis,trans*-1,4,7-cyclodecatriene.[35] In this case an additional ligand, triphenylphosphine, gives the best yield (87.5%), a fact

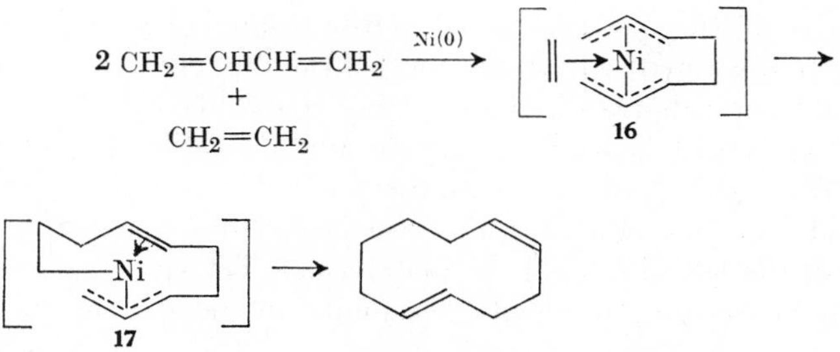

---

[34] W. Brenner, P. Heimbach, K. Ploner, and F. Thömel, *Angew. Chem.*, **81**, 744 (1969).
[35] P. Heimbach, P. W. Jolly, and G. Wilke, *Adv. Organomet. Chem.*, **8**, 76 (1969).

which is not accounted for in the mechanism and which has not been explained.

Olefins bearing electron-withdrawing groups (*e.g.*, acrylic esters) also require an additional ligand for smooth co-oligomerization with two units of diene; the best ligand is tri-(2-biphenylyl) phosphite.[36] More interesting from a mechanistic standpoint is the dramatic change brought about by a less hindered phosphite, triphenyl phosphite. As indicated in Scheme 5, the intermediate complex (18) from one unit of methyl acrylate and two units of 1,3-butadiene reacts with a second unit of the ester to give 19 (not characterized) which suffers hydrogen transfer to a mixture of open-chain triene diesters in 92% combined yield.

SCHEME 5

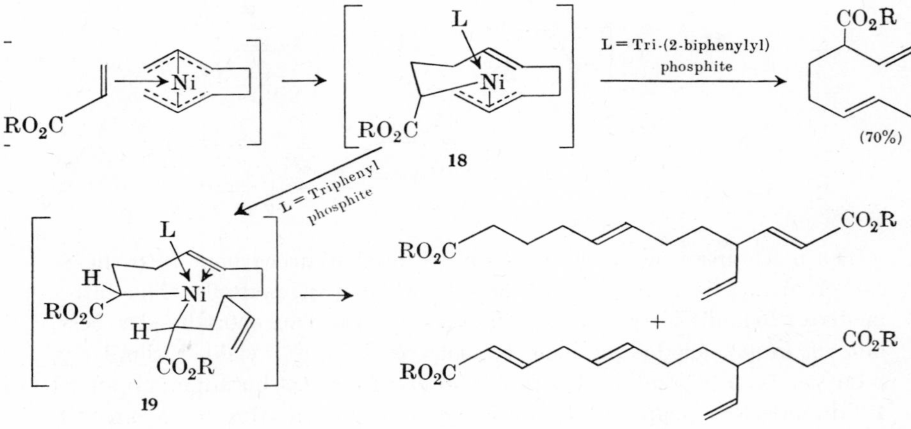

The mechanistic questions posed at the beginning of this section are only partly answered. All of the mechanisms involve one or more hypothetical intermediates whose individual credibility rests on analogy to complex 11, which is itself a delicate and incompletely characterized species. The usual coordination requirements of nickel dictate the size of the ligand (*i.e.*, the number of diene or other units per ligand) and the position of the ligand around the metal atom. The combination of these two features overrides the usual transannular nonbonded interactions and provides a pathway to medium-sized rings.

[36] P. Heimbach, G. Schomburg, and G. Wilke, unpublished results cited in ref. 4, p. 69.

Scope and Limitations

From the foregoing discussion it should be clear that the reaction of Wilke's zero-valent nickel catalyst with 1,3-butadiene can be controlled to give cyclization of two units of diene (1,2-divinylcyclobutane, 4-vinylcyclohexene, and 1,5-cyclooctadiene), cyclization of three units of diene (1,5,9-cyclododecatriene), cyclization of one unit of alkene with two units of diene, and cyclization of one unit of alkyne with two units of diene. Wilke's group has also considered extension of these reactions to substituted 1,3-dienes, substituted alkenes and alkynes, nickel-catalyzed additions of R_2NH to 1,3-dienes, and to the synthesis of open-chain polyenes.

Adding substituents to 1,3-butadiene often leads to a drastic change in the product distribution. For example, the catalyst system [zero-valent nickel, tri-(2-biphenylyl) phosphite] which converts 1,3-butadiene to 1,5-cyclooctadiene and 4-vinylcyclohexene in a 97:2 ratio[37] also cyclodimerizes 2,3-dimethyl-1,3-butadiene, but the ratio of eight-membered ring products to six-membered ring products is 6:86. [38]The rate of reaction is lower by a factor of 10–100 compared to that of 1,3-butadiene, but the yields remain high. Isoprene gives cyclodienes with an eight-membered ring to six-membered ring ratio of 55:35, while piperylene affords cyclodienes in the ratio 91:5 for eight-:six-membered rings. Four isomers of dimethyl-*cis,cis*-1,5-cyclooctadiene have been characterized from *cis*-piperylene,[4, 38] **20** and **21** being the principal products.

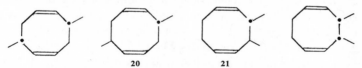

                 20   **21**

Cyclocodimerization of a substituted 1,3-diene with 1,3-butadiene also proceeds readily, although a deficiency of 1,3-butadiene is usually employed to avoid preferential formation of 1,5-cyclooctadiene.[36] With piperylene, isoprene, or 2,3-dimethyl-1,3-butadiene, the main product is a 1:1 cyclocodimer in 84–92 % yield based on the substituted diene. Analysis of the isomer mixtures obtained has not yet been reported.

$$\begin{array}{c} CH_2{=}CHCH{=}CH_2 \\ + \\ CH_3CH{=}CHCH{=}CH_2 \end{array} \xrightarrow{Ni(0)}$$

---

[37] (a) P. Heimbach and W. Brenner, *Angew. Chem. Int. Ed. Engl.*, **5**, 961 (1966); (b) P. Heimbach, *ibid.*, **5**, 961 (1966); (c) P. Heimbach and R. Schimpf, *ibid.*, **7**, 727 (1968); (d) W. Brenner, P. Heimbach, and G. Wilke, *Ann.*, **727**, 194 (1969).

[38] H. Hey, Thesis, Bochum University, 1969.

Dienes bearing electron-withdrawing substituents also enter into codimerization with 1,3-butadiene, using triphenylphosphine as the added ligand. Sorbic ester (ethyl penta-2,4-dieneoate) affords a cyclooctadiene derivative in 45% yield and a vinylcyclohexene derivative in 5% yield.[4] The sorbic ester has been considered a better electron-donor ligand than alkyl-1,3-dienes, and so a better electron-accepting ligand, triphenylphosphine, provides the best results.[35]

Divinylcyclobutanes appear to be the initial products from a 1,3-diene and the zero-valent-nickel/phosphorus-ligand catalyst.[28] In the cyclodimerization of 1,3-butadiene itself, 1,2-divinylcyclobutane is present in 38% yield at 85% conversion, but disappears at 100% conversion via a nickel-catalyzed Cope rearrangement to 1,5-cyclooctadiene. The formation of methyl-substituted divinylcyclobutanes from piperylene gives a mixture of products and is unlikely to be synthetically useful. The fact that cis- and trans-piperylene give different divinylcyclobutane isomers has been helpful in understanding the mechanism of the cycloaddition.[35,38]

Co-oligomerization of ethylene with 1,3-butadiene is six times faster than cyclotrimerization of the diene alone, and thus provides an efficient route to 10-membered rings.[29] A side product is the open-chain triene **22** from hydrogen transfer in intermediate **23**. Low temperature (0–20°), no additional phosphine, and a moderate excess of ethylene give the best yield of cis,trans-1,5-cyclodecadiene, about 80%, and only small amounts of butadiene dimers and trimers. The open-chain triene is produced in 10–20% yield and is difficult to separate from the cyclodecadiene. A more

23   22

easily purified sample of 1,5-cyclodecadiene is obtained with one mole of triphenyl phosphite per mole of nickel, but the yield goes down to 30–40%.[29] Co-oligomerization of 1,3-butadiene with alkyl-substituted alkenes has not been studied systematically, but reaction of styrene and 1,3-butadiene in the presence of zero-valent nickel and a phosphite ligand gives 1-phenyl-*cis,trans*-3,7-cyclodecadiene.[4] The yield is only moderate owing to substantial hydrogen transfer to give a mixture of 1-phenyl-decatriene isomers (see below).

$$C_6H_5CH=CH_2 + CH_2=CHCH=CH_2 \xrightarrow{Ni(0)} \left[\text{complex} \right] \longrightarrow \text{1-phenyl-cyclodecadiene}$$

Alkynes also co-oligomerize with 1,3-butadiene to provide a rather general route to 1,2-disubstituted cyclodecatrienes.[34, 37] The reaction has been exploited beautifully by Heimbach and co-workers in the synthesis of a variety of medium and large ring systems. The simplest example, using 2-butyne with excess 1,3-butadiene and a zero-valent nickel-triphenyl-phosphine catalyst, affords 4,5-dimethyl-*cis,cis,trans*-1,4,7-cyclodecatriene in 95% yield based on 2-butyne. The cyclodecatriene is free of other isomers and is accompanied by only 13% of other hydrocarbons from oligomerization of 1,3-butadiene.[37d] Equally good yields of ten-membered ring compounds can be obtained if alkyl-substituted alkynes, or alkynes in which the functional groups are separated from the triple bond by at

$$CH_3C{\equiv}CCH_3 + 2CH_2=CHCH=CH_2 \xrightarrow[Ph_3P]{Ni(0)} \text{product} \quad (95\%)$$

least two methylene groups, are used.[34] Larger rings are obtained from the reaction of two units of alkyne with two units of 1,3-butadiene if the alkyne is present in higher concentration.[4, 34] For example, 1,4-dimethoxy-2-butyne leads to 1,2,3,4-(tetramethoxymethyl)-*cis,cis,trans,trans*-1,3,6,10-cyclododecatetraene in 38% yield. The big rings have been used as intermediates for the synthesis of substituted aromatics via Cope re-

$$CH_3OCH_2C{\equiv}CCH_2OCH_3 + CH_2=CHCH=CH_2 \xrightarrow[Ph_3P]{Ni(0)} \text{cyclododecatetraene product}$$

arrangement followed by base-catalyzed isomerization,[37c] for the preparation of long-chain diketones via hydrogenation and ozonolysis,[37b] and for the preparation of even larger rings using cyclododecyne and cyclotetradecadiyne.[37a]

In most of these cyclizations open-chain oligomers of 1,3-butadiene or co-oligomers of the diene with alkenes and alkynes are obtained as by-products, usually in minor amounts. However, in certain cases the nickel-catalyzed reactions can provide efficient routes to long-chain polyenes. The general pathway to open-chain alkenes was illustrated for the reaction of ethylene with two units of 1,3-butadiene to afford 1,trans-4,9-decatriene, **22** (p. 136). Ligand control is relatively ineffective in promoting hydrogen transfer, but raising the reaction temperature to 80° gives an 80% yield of C_{10} products of which 69% is the linear triene **22**. At 0° the triene comprises only 4% of the C_{10} products, the remainder being *cis,trans*-1,5-cyclodecadiene.

The cyclo-cooligomerization of sorbic ester with 1,3-butadiene can be directed to an open-chain product by addition of ethylene to the reaction mixture.[4] The undecatrienoate **26** is formed in over 80% yield, probably via ethylene insertion into intermediate **24** followed by hydrogen transfer in **25**.

It was pointed out above that styrene affords phenylcyclodecadiene with 1,3-butadiene and zero-valent nickel-phosphite catalyst. Without the phosphite ligand and at elevated temperature, a mixture of 1-phenyldecatriene isomers comprises 96% of the C_{16} products (40% of the total product). The mixture of open-chain products can be simplified to geometric isomers of 1-phenyl-1,4,8-decatriene by isomerization over a π-allylnickel halide-aluminum trihalide catalyst system.[39]

In the presence of N–H or O–H bonds the ubiquitous bis-(π-allyl)nickel intermediate **11** from 1,3-butadiene and zero-valent nickel undergoes protonation at temperatures above 60° followed by loss of a proton and formation of an open-chain octatriene.[40] (Equations on p. 140.) A side reaction that has interesting synthetic possibilities is aminoalkylation, which becomes the major pathway at lower temperatures (75% yield of **27**).[40]

$$2\ CH_2=CHCH=CH_2 + R_2NH \xrightarrow[\text{Phosphite}]{\text{Ni(0), 20°}}$$

$$R_2NCH_2CH=CHCH_2CH_2CH=CHCH_3$$
27 (75%)

[39] B. Bogdanovic and G. Wilke, *Brennst.-Chem.*, **49**, 323 (1968).
[40] P. Heimbach, *Angew. Chem. Int. Ed. Engl.*, **7**, 882 (1968).

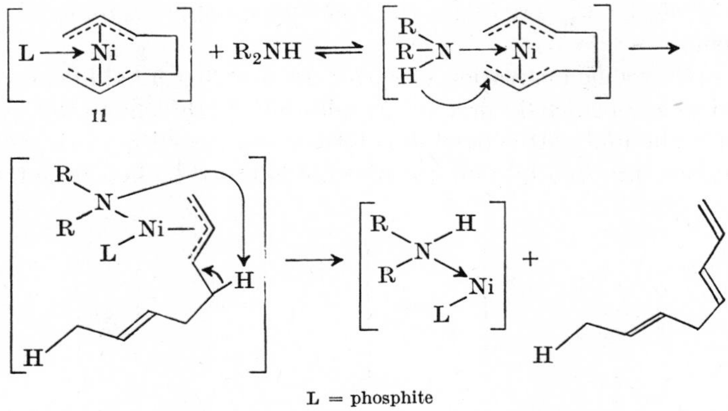

L = phosphite

The reaction requires a phosphite ligand for efficient formation of the octatrienes; naked nickel affords the amino triene from incorporation of three units of 1,3-butadiene.

$$3\ CH_2=CHCH=CH_2 + R_2NH \xrightarrow[\text{phosphite}]{\text{Ni(0)}\atop\text{No}}$$
$$R_2NCH_2CH=CHCH_2CH_2CH=CHCH_2CH_2CH=CHCH_3$$

Clearly the use of bis-(π-allyl)nickel intermediates for oligomerization of 1,3-dienes gives efficient catalytic formation of medium-sized carbocycles that would be difficult to prepare by other routes. But the present state of knowledge provides limited predictive power, and one cannot immediately choose conditions guaranteed to work smoothly with an untried substituted 1,3-diene. The same limitation exists for the reaction of olefins and acetylenes in cycloco-oligomerization with 1,3-butadiene: each new reactant will require careful choice of conditions, including the phosphorus ligand which might be necessary. The proportion of cyclic *vs.* open-chain isomers depends on a delicate balance of factors, including the substituents on the reactants.

Comparison with Other Methods

Oligomerization of 1,3-butadiene has been carried out on a variety of catalyst systems. A catalyst prepared from titanium tetrachloride and diethylaluminum chloride rapidly converts 1,3-butadiene to the cyclotrimer, *trans,trans,cis*-1,5,9-cyclododecatriene in yields over 80%.[41] Other catalysts are more or less efficient and may produce different double-bond isomers. For example, chromyl chloride and triethylaluminum in benzene afford a 60:40 mixture of the all-*trans* and *trans,trans,cis* isomers.[21b] The

---

[41] (a) G. Wilke, *Angew. Chem.*, **69**, 397 (1957); (b) H. W. B. Reed, *J. Chem. Soc.*, **1954**, 1931.

catalysts and intermediates in these examples are less well understood than in the zero-valent nickel system, and no example of ligand control has been reported with either the titanium- or chromium-based species.

Bis-(cyclooctatetraene)iron oligomerizes 1,3-butadiene to produce the linear trimer, 1,3,6,10-dodecatetraene (**28**), as the main product, accompanied by 1,5-cyclooctadiene.[42] A likely mechanism parallels that suggested above for nickel, but ring closure around the iron atom is not rapid, hydrogen transfer predominates, and the product is the open-chain tetraene. Codimerization also is observed but without cyclization; 1,3-butadiene and ethylene lead to *cis*-1,4-hexadiene.[42]

$$Fe(C_8H_8)_2 + 2\ CH_2{=}CHCH{=}CH_2 \longrightarrow$$

28

Bis-(π-allyl)nickel performs as a zero-valent catalyst, giving all the reactions discussed above with 1,3-butadiene. However, bis-(π-allyl)palladium, a more stable complex, reacts only slowly with 1,3-butadiene and produces no cyclic products. Linear polyenes such as dodecatetraene and higher oligomers are the major products.[9] In the explanation of Wilke, "it may be very tentatively concluded from this result that the chain ends of the intermediate complex (*e.g.*, complex **29**) are prevented from coming together because of the greater atomic volume of palladium."[9] Similarly, bis-(π-allyl)platinum is very stable and does not react with 1,3-butadiene.[9] Tris-(π-allyl)cobalt gives either polymerizations of 1,3-butadiene or selective dimerization to 5-methylhepta-1,3,6-triene in 90% yield.[9]

29

A palladium(II) species has been used to give dimerization of 1,3-butadiene with concomitant addition of phenol, a sequence formally

$$C_6H_5OH + 2\ CH_2{=}CHCH{=}CH_2 \xrightarrow{PdCl_2}$$

OC_6H_5

+

C_6H_5O

(*trans:cis* = 18:1)

---

[42] A. Carbonaro, A. Greco, and G. Dall'Asta, *Tetrahedron Lett.*, **1967**, 2037.

parallel to addition of amines to 1,3-butadiene via nickel catalysts.[43] An example of ligand control is apparent in this case, as triphenylphosphine and low reaction temperature lead to 1,3,7-octatriene in 98% purity and 85% yield.[43]

Another nickel catalyst system produces 1,4-hexadienes from 1,3-butadiene and α olefins.[44] The catalyst is derived from reduction of bis-(tri-n-butyl)phosphinenickel(II) chloride with diisobutylaluminum chloride and is thought to contain a nickel-hydride bond. The mechanism is based

$$X = CH_3, H, \text{ or } Cl$$
$$L = (n\text{-}C_4H_9)_3P$$

on analogy, with no direct evidence. Ethylene and various dienes react in low conversion (generally about 30%) and moderate yield (60–70%); with propylene and 1,3-butadiene, the overall yield of 2-methyl-1,4-hexadiene is 8%, but is 25% on the basis on the 1,3-butadiene that reacts.[44]

The only method of ring synthesis that approaches the nickel-catalyzed cyclizations of 1,3-butadiene in efficiency of medium-sized ring formation without the use of high dilution is the acyloin condensation.[45] Each method has important advantages and limitations. Generally, the nickel-catalyzed route suffers from the severe constraint of working well with only one 1,3-diene, 1,3-butadiene itself, and a limited number of co-reactants such as alkenes and alkynes. An additional limitation is that present examples include even-membered rings only; extrapolation to odd-membered rings is not obvious. The acyloin reaction is conceptually much more general, but is in fact limited by the severe reaction conditions (alkali metal) which cause changes, usually undesirable, in almost every common organic functional group.

[43] E. J. Smutny, *J. Amer. Chem. Soc.*, **89**, 6793 (1967).
[44] R. G. Miller, T. J. Kealy, and A. L. Barney, *J. Amer. Chem. Soc.*, **89**, 3756 (1967).
[45] (a) V. Prelog, *Helv. Chim. Acta*, **30**, 1741 (1947); (b) S. M. McElvain, *Org. Reactions*, **4**, 256 (1948).

Experimental Conditions

The general experimental conditions for cyclo-oligomerization of 1,3-dienes are described by the developers of the nickel-catalyzed reactions in a recent review[4] and in a series of detailed papers.[28–30, 37a] The catalyst is most conveniently prepared in the presence of 1,3-butadiene by reaction of an organoaluminum compound (*e.g.*, diethylaluminum ethoxide or triisobutylaluminum) with nickel acetylacetonate and a suitable ligand if necessary.[46] Mechanistic studies where a discrete catalyst is required can be carried out with bis-(1,5-cyclooctadiene)nickel or 1,5,9-cyclododecatrienenickel and one equivalent of the appropriate phosphorus ligand. If no additional ligands are used, then passing 1,3-butadiene into the solution at 25° or lower leads to 1,5,9-cyclododecatriene. A gram of nickel in the catalyst converts several hundred grams of 1,3-butadiene per hour. The catalysts are very sensitive to molecules that can coordinate strongly to nickel; carbon monoxide, for example, must be rigorously excluded. Careful control of temperature is critical to selective reaction. The cyclodimerization of 1,3-butadiene to 1,5-cyclooctadiene requires a catalyst from one mole of nickel acetylacetonate, one mole of tri-(2-biphenylyl) phosphite, and one mole of an alkylaluminum.[21b] The reaction temperature is ideally 80°, which leads to 800–900 g of 1,5-cyclooctadiene per hour per gram of catalyst. The catalyst's life is nearly unlimited, provided impurities are absolutely excluded.[21b] The apparatus and detailed appearance of these reactions have not been described.

CARBON-CARBON BOND FORMATION VIA π-ALLYLNICKEL(II) HALIDES

The π-allylnickel(I) halide dimers[4, 7] (*e.g.*, **2**) find synthetic application in two ways. First, they can be prepared as discrete reagents in any of several ways differing in convenience and severity of conditions. They can be isolated, crystallized, stored for long periods, and simply weighed out for reaction like any other well-defined thermally stable, moderately air-sensitive compound. Routine operation with these reagents is consequently much simpler and more accurate than with other common organometallic reagents such as the organo-magnesium, -lithium, -zinc, or -cadmium species. Second, the π-allylnickel(I) halides are transient intermediates in the formation of 1,5-hexadiene derivatives from allylic halides and nickel carbonyl. This simple conversion has been applied in some sophisticated syntheses, and clearly has great potential as a mild and selective method of carbon-carbon bond formation.

---

[46] G. Wilke, E. W. Muller, and M. Kröner, *Angew. Chem.*, **73**, 33 (1961).

As Discrete Reagents

Three types of synthetically useful reactions of π-allylnickel(I) halides have been reported. Carbonylation occurs to form β,γ-unsaturated esters;[47] more complex examples involving simultaneous addition to acetylenes and

$$\text{[Ni}_2\text{Br}_2(\text{allyl})_2\text{]} + 8\text{ CO} \xrightarrow{\text{ROH}} 2\text{ CH}_2\text{=CHCH}_2\text{CO}_2\text{R} + 2\text{ Ni(CO)}_4 + 2\text{ HBr}$$

carbonylation are also reported.[47, 48] Carbon-halogen bonds are replaced by carbon-carbon bonds.[7] Finally, the allyl ligands, behaving as mild

$$\text{[Ni}_2\text{Br}_2(\text{allyl})_2\text{]} + 2\text{ RBr} \longrightarrow 2\text{ CH}_2\text{=CHCH}_2\text{R} + 2\text{ NiBr}_2$$

nucleophiles, slowly attack reactive carbonyl groups[7] and undergo Michael addition to α,β-unsaturated esters.[49]

$$\text{[Ni}_2\text{Br}_2(\text{allyl})_2\text{]} + 2\text{ RR'CO} \longrightarrow 2\text{ CH}_2\text{=CHCH}_2\text{C(OH)RR'} + 2\text{ NiBr}_2$$

The Complex

Before discussing the chemistry of these complexes further, it may be useful to take a short digression and indicate how they are prepared and handled. Several routes to the π-allylnickel halides have been uncovered, but recent improvements in the methods used by Fisher and Burger to prepare the first allyl complexes of nickel(I) make their route still the most generally useful.[50] After allyl halide and excess nickel carbonyl are heated in benzene, and the solvent and excess nickel carbonyl are removed, the residual red solid can be sublimed at 90–100° to yield the π-allyl complexes (e.g., 2) in 6–10% yield. The same conditions produce 75–90% yields (optimum for X = Br) if the product is isolated by crystallization to avoid thermal decomposition.[7]

[47] G. P. Chiusoli and L. Cassar, *Angew. Chem. Int. Ed. Engl.*, **1**, 124 (1967).
[48] (a) L. Cassar, G. P. Chiusoli, and M. Foa, *Tetrahedron Lett.*, **1967**, 285; (b) R. F. Heck, *J. Amer. Chem. Soc.*, **85**, 2013 (1963).
[49] M. Dubini, F. Montino, and G. P. Chiusoli, *Chim. Ind.* (Milan), **47**, 839 (1965).
[50] E. O. Fischer and G. Bürger, *Z. Naturforsch.*, **161**, 77 (1961).

Wilke and co-workers prepared a variety of π-allylnickel halides in quantitative yield by reaction of allyl halides with bis-(1,5-cyclooctadiene)nickel(0).[9] Although the nickel reagent is less easily obtained than

$$\text{Ni(COD)}_2 + CH_2=CHCH_2Br \xrightarrow{100\%} \text{[π-allyl-Ni-Br-Ni-π-allyl]}_2$$

the commercially available nickel carbonyl, the procedure can be carried out at $-10°$, and is useful for the preparation of thermally sensitive π-allylnickel complexes. The reaction of bis-(π-allyl)nickel with hydrogen halides leads to π-allylnickel(I) halides in high yield, but gives at most 50%

$$2 \text{ [π-allyl-Ni-π-allyl]}_1 + 2 HBr \longrightarrow \text{[π-allyl-Ni-Br-Ni-π-allyl]}_2 + 2 CH_3CH=CH_2$$

utilization of the allyl ligand.[9] In addition, bis-(π-allyl)nickel complexes are very sensitive species to use as routine synthetic starting materials. Allylmagnesium halides react with nickel(II) halides in ether to produce bis-(π-allyl)nickel(0).[21] π-Allylnickel(I) halides are likely intermediates in this reaction, but as yet the method has not been used as a practical route to the nickel(I) species.

The π-allylnickel(I) halides themselves serve as useful starting materials for preparation of other complexes that have the same general type of bonding as the dimers, and may show unique properties of reactivity for organic synthesis. For instance, "bridge splitting" occurs when a strong electron-pair donor is added to a solution of π-allylnickel(I) halide; triphenylphosphine gives triphenylphosphine-π-allylnickel(I) halide.[4, 51] Both of the halogens on nickel in the dimeric species can be replaced to

$$2(C_6H_5)_3P + \text{[π-allyl-Ni-Br-Ni-π-allyl]}_2 \longrightarrow 2 \text{[π-allyl-Ni-P(C}_6\text{H}_5)_3\text{-Br]}$$

give new, halide-free monomeric complexes. Sodium cyclopentadienide and thallium acetylacetonate react with π-allylnickel(I) bromide to give π-allyl-π-cyclopentadienylnickel **(30)**[52] and acetylacetonato-π-allylnickel

---

[51] R. F. Heck, J. W. C. Chien, and D. S. Breslow, *Chem. Ind.* (London), **1961**, 986.

(31),[8, 53] respectively. The acetylacetonato complex 31 differs significantly from the π-allylnickel bromide only in its color (pale yellow) and its volatility (it sublimes rapidly at 50°/0.01 torr); its reactions and thermal stability parallel closely those of the bridged dimers such as 2.

Crystallization of π-allylnickel bromide from ether at low temperature (0° to −78°) produces red needles that are soluble in polar organic solvents, slightly soluble in water, and slowly destroyed by alcohols, water, and acid due to protonation of the allyl ligands.[19] They are stable indefinitely at 25° under inert gas. Solutions of the complex, and to a lesser extent the crystals, are very sensitive to oxygen; the red color in solution is lost immediately upon introduction of air.

Synthesis Via Carbonylation

Carbonylation of π-allylnickel halides has been studied briefly, with only a few important synthetic applications. The reaction parallels the carbonylation of the much more easily available and stable π-allylpalladium chloride dimers.[54] Similar products can be obtained by direct carbonylation of allylic halides with nickel carbonyl without isolation of the intermediate nickel complex.[47] Both of these facts help to explain the lack of synthetic interest in carbonylation of the π-allylnickel(I) halides.

Two important general reactions are known. π-Allylnickel halides react rapidly with carbon monoxide to afford a 3-butenoyl halide and nickel carbonyl.[48b] In the presence of water or alcohols the corresponding

---

[52] E. O. Fischer and G. Bürger, *Chem. Ber.*, **94**, 2408 (1961).

[53] S. D. Robinson and B. L. Shaw, *J. Chem. Soc.*, **1963**, 4806.

[54] (a) W. Dent, R. Long, and G. Whitfield, *J. Chem. Soc.*, **1964**, 1588; (b) J. Tsuji, S. Imamura, and J. Kiji, *J. Amer. Chem. Soc.*, **86**, 4350 (1964); J. Tsuji, J. Kiji, and M. Morikawa, *Tetrahedron Lett.*, **1963**, 1811.

3-butenoic acid or esters are formed. In the presence of acetylene the same reactants incorporate a molecule of acetylene to give a *cis*-hexa-2,5-dienoyl halide[48b] which is usually carried further to the carboxylic acid or ester.

$$\text{2} + 10\ CO + 2\ HC\equiv CH \longrightarrow$$

$$2 \ \text{(cis-hexa-2,5-dienoyl bromide)} + 2\ Ni(CO)_4$$

Phenylacetylene furnishes derivatives of 2-cyclopentenone and 2-cyclohexenone.[55] Practical syntheses of unsaturated carboxylic acid derivatives by this route have not been reported; this pathway probably occurs in the catalytic carboxylation of allylic halides with nickel carbonyl and is discussed later under the subject of reactions involving transient π-allylnickel halides (see p. 162).

Reaction with Organic Halides

In 1967 it was reported that π-allylnickel halides react with organic halides in polar aprotic media to produce allyl-substituted molecules and nickel(II) dihalides.[7] The potential utility of such a procedure in synthesis

$$\text{2} + 2\ RBr \xrightarrow{\text{Polar solvent}} 2\ CH_2{=}CHCH_2R + 2\ NiBr_2$$

is clear. In principle, any molecule containing the $-C{=}C-\overset{|}{\underset{|}{C}}-\overset{|}{\underset{|}{C}}-$ grouping could be prepared by appropriate choice of organic halide and π-allyl ligand. Important advantages are the very weak basicity and high selectivity of the π-allyl reagent; π-allylnickel halides are relatively unreactive toward common functional groups other than halides. A dramatic example of the efficiency of the π-allylnickel halide reagents is the synthesis of α-santalene (32).[7] Corey and co-workers had earlier prepared this sesquiterpene in 20% yield from Grignard reagent 33 and γ,γ-dimethylallyl bromide.[56] By comparison, conversion of γ,γ-dimethylallyl bromide to π-(1,1-dimethylallyl)-nickel bromide (34) followed by reaction

---

[55] G. P. Chiusoli, G. Bottaccio, and C. Venturello, *Tetrahedron Lett.*, **1965**, 2875.
[56] E. J. Corey, S. W. Chow, and R. A. Scherrer, *J. Amer. Chem. Soc.*, **79**, 5773 (1957).

with the tricyclic iodide **35** produced α-santalene in 95% yield for the second step (80% overall).

$$(CH_3)_2C=CHCH_2Br \longrightarrow \mathbf{34} \longrightarrow \mathbf{32}$$

Mechanism. Very little detailed mechanistic work has been done on this reaction but, from relative reactivity and selectivity information and from the general mechanism of oxidative-addition reactions of transition metals,[57] a mechanism can be formulated as illustrated in Eqs. 1–6. This example involves the reaction of π-allylnickel bromide and an organic bromide (RBr) in a polar coordinating solvent like dimethylformamide.

$$\mathbf{2} \underset{-S}{\overset{+S}{\rightleftharpoons}} \mathbf{36} \quad \text{S=Solvent} \qquad (\text{Eq. 1})$$

$$\mathbf{36} \underset{-S}{\overset{+S}{\rightleftharpoons}} \mathbf{37} \qquad (\text{Eq. 2})$$

$$\mathbf{37} \rightleftharpoons \qquad (\text{Eq. 3})$$

[57] J. P. Collman, *Accts. Chem. Res.*, **1**, 136 (1968).

$$\left[\text{allyl-Ni} \begin{array}{c} \text{Br} \\ -\text{S} \\ \text{S} \end{array} \right] + \text{R}-\text{Br} \underset{+2\text{S}}{\overset{-2\text{S}}{\rightleftharpoons}} \left[\text{allyl-Ni} \begin{array}{c} \text{R} \\ -\text{Br} \\ \text{Br} \end{array} \right] \quad \text{(Eq. 4)}$$

$$\mathbf{37} \qquad\qquad\qquad \mathbf{38}$$

$$\left[\text{allyl-Ni} \begin{array}{c} \text{R} \\ -\text{Br} \\ \text{Br} \end{array} \right] \longrightarrow \text{allyl-R} + \text{NiBr}_2 \quad \text{(Eq. 5)}$$

$$\mathbf{38}$$

$$\left[\text{allyl-Ni} \begin{array}{c} \text{R} \\ -\text{Br} \\ \text{Br} \end{array} \right] \rightleftharpoons \text{allyl-Br} + \left[\begin{array}{c} \text{S} \quad \text{R} \\ \text{Ni} \\ \text{S} \quad \text{Br} \end{array} \right] \quad \text{(Eq. 6)}$$

$$\mathbf{38} \qquad\qquad\qquad \mathbf{39}$$

The reaction is clearly not a nucleophilic displacement by the allyl ligand, because halides at sp^2 centers (aryl, vinyl) are approximately as reactive toward the π-allylnickel bromide as is methyl bromide, and generally more reactive than simple primary bromides (see below). The solvent plays a crucial role as a ligand in the proposed steps; this is consistent with the fact that the reaction is very slow in ether solvents compared to polar coordinating solvents such as dimethylformamide and hexamethylphosphoramide. In Eq. 1 the solvent breaks up the dimer structure in a step that is likely to occur but is not crucial to the overall mechanism. Complexes like **36** have been isolated only when S is the strongly coordinating triphenylphosphine,[51] and then the complex is not reactive toward organic halides. Equation 2 represents the π-allyl $\rightarrow \sigma$-allyl equilibrium that is known to occur in dimethyl sulfoxide and dimethylformamide, and is likely to occur in similar solvents. The complex **37** has not been isolated and is expected to be much less stable than **36**. Equation 3 represents a degenerate 1,3-sigmatropic shift in this example, but leads to two different complexes when there are unsymmetrically substituted allyl ligands. Then oxidative addition of the alkyl halide to a π-allyl intermediate produces a nickel(II) halide with two σ-bonded ligands, **38** (Eq. 4), which is expected to be very unstable with respect to irreversible radical-like coupling of the carbon ligands to form a 1-alkene and nickel(II) dibromide (Eq. 5). All of the intermediates are hypothetical at this time.

Equations 4 and 6 taken together amount to a special case of halogen-metal exchange and become important only for those organic halides in which R contains a coordinating atom or functional group (R = allyl,

R′SCH$_2$—, R′COCH$_2$—, *etc.*) to stabilize complex **39**. This is discussed in more detail below.

Scope and Limitations. An indication of the scope of the reaction can be found from the reactivity of π-(2-methylallyl)nickel bromide **(40)** with organic halides. Simple alkyl halides are inert to complex **40** in hydrocarbon or ether solvents.[7] In dimethylformamide, dimethyl sulfoxide, hexamethylphosphoramide, or N-methylpyrrolidone, a smooth reaction

$$\text{40} + 2\,RX \xrightarrow{\text{DMF}} \text{methallyl-R}$$
$$X = \text{Br or I}$$

occurs with a wide variety of organic halides; iodides are more reactive than bromides which, in turn, are more reactive than chlorides of similar structure. Table I displays a series of halides arranged in order of decreasing reactivity toward complex **40** as indicated by the time necessary for the intense color of the π-allyl complex to be replaced by the emerald green of the nickel(II) halide product.[7]

Several cases require special mention. Cyclohexyl bromide and cyclohexyl *p*-toluenesulfonate do not react with complex **40**.[8] Attempts to force the reaction by raising the temperature led to thermal decomposition of the nickel complex and, at best, low yields of methallylcyclohexane. Cyclopropyl bromide does not yield methallylcyclopropane but instead gives products from ring opening to the allyl system. This particular

TABLE I. The Reaction of Organic Halides with π-(2-Methylallyl)-nickel Bromide

Halide	Product	Time (temp, °)	Yield, %
Iodobenzene	Methallylbenzene	1 hr (22)	98
Cyclohexyl iodide	Methallylcyclohexane	3 hr (22)	91
Methyl iodide	2-Methyl-1-butene	10 hr (22)	90[a]
Vinyl bromide	2-Methyl-1,4-pentadiene	15 hr (22)	70[a]
Methyl bromide	2-Methyl-1-butene	19 hr (22)	90[a]
Benzyl bromide	2-Methyl-4-phenyl-1-butene	2 hr (22), then 4 hr (60)	91
3-Phenylpropyl bromide	2-Methyl-6-phenyl-1-hexene	46 hr (65)	92

[a] The yield was obtained by quantitative glpc analysis.

distribution of products is characteristic of an allyl halide coupling with

π-(2-methylallyl)nickel bromide. Similarly, cyclopropylcarbinyl bromide reacts with complex **40** during 63 hours at 60° to produce one volatile product, 2-methyl-1,6-heptadiene. Facile ring opening to the allyl carbinyl system is a characteristic of cyclopropylcarbinyl radical[58] and anion[59]; the

corresponding cation usually gives cyclobutyl products in addition to allylcarbinyl products.[60] A likely mechanism for this ring opening involves complex **41** which could undergo fast ring opening (as do cyclopropylcarbinyl magnesium halides[61]) to give complex **42**, followed by coupling to give the observed product, 2-methyl-6-heptadiene.

Dihalides undergo disubstitution with the appropriate quantity of π-(2-methylallyl)nickel bromide, as shown by the conversion of 1,6-diiodohexane to 2,11-dimethyl-1,11-dodecadiene in 95% yield and of 1,4-dibromobenzene to 1,4-dimethallylbenzene in 97% yield.[7] Elimination of bromine as nickel bromide occurs rapidly when 1,2-dibromoethane is treated with π-(2-methylallyl)nickel bromide **(40)** at 22° in dimethylformamide, and 2,5-dimethyl-1,5-hexadiene (dimer of the allyl ligands) is produced quantitatively.[8]

[58] R. Breslow in *Molecular Rearrangements*, P. de Mayo, Ed., Vol. I., Interscience Publishers, New York, 1963, p. 293.
[59] R. Breslow in *Molecular Rearrangements*, P. de Mayo, Ed. Vol. I, Interscience Publishers, New York, 1963, p. 281.
[60] J. D. Roberts and R. H. Mazur, *J. Amer. Chem. Soc.*, **73**, 2509 (1951).
[61] J. P. Dinshaw, C. L. Hamilton, and J. D. Roberts, *J. Amer. Chem. Soc.*, **87**, 5144 (1965).

$ICH_2(CH_2)_4CH_2I$ + [complex **40**] ⟶ $CH_2=C(CH_3)CH_2-CH_2(CH_2)_4CH_2-CH_2C(CH_3)=CH_2$ (95%)

$Br-C_6H_4-Br$ + [complex **40**] ⟶ 1,4-bis(2-methylallyl)benzene (97%)

Several functional groups are less reactive than the halogen of aryl and alkyl iodides and bromides toward the π-allyl ligand and do not interfere with replacement of halogen with an allyl group. Conversion of *trans*-4-iodocyclohexanol to 4-(2-methylallyl)cyclohexanol using complex **40** in dimethylformamide is complete within 12 hours at 23° to give a mixture of epimers (38:62) in 89% yield, showing that the hydroxyl group had no significant effect on the reaction rate or direction.[7] The π-allyl complex from ethyl 2-(bromomethyl)acrylate converts 1-iodo-3-chloropropane to

[trans-4-iodocyclohexanol] + [complex **40**] ⟶ 4-(2-methylallyl)cyclohexanol (89%)

ethyl 6-chloro-2-methylenehexanoate in 96% yield. The ester group and chlorine are not affected.

[π-allyl Ni complex with $CO_2C_2H_5$ groups] + $I(CH_2)_3Cl$ ⟶ $CH_2=C(CO_2C_2H_5)(CH_2)_3Cl$ (96%)

Three kinds of isomerization might be encountered in couplings of a halide with π-allyl ligands, and at least one example casting light on each of them has been reported. The alkyl halide might undergo inversion or retention of configuration or racemization during reaction with the nickel species; a pure *trans*-allylic halide can be converted to the π-allylnickel complex and coupled with an alkyl halide to give a *trans* or *cis* alkene or both; the organic halide can end up attached at either end of the allyl system.

The mixture of epimers obtained from pure *trans*-4-iodocyclohexanol indicates that the reaction proceeds with partial epimerization. A possible explanation for the mixture of products is inversion of a σ-bonded cyclohexylnickel intermediate such as **43** before coupling. No information about inversion of a carbon σ-bonded to nickel is available to support this suggestion.

When a *trans*-allylic halide such as *trans*-geranyl bromide (**44**) is converted to its π-allylnickel bromide complex and the complex is allowed to react with cyclohexyl iodide in dimethylformamide, the resulting coupling leads to both the *cis* and the *trans* products; the double bond which participated in the π-bonding to the nickel is isomerized during reaction. In the example cited here, the mixture is 40% *cis* and 60% *trans* in an overall yield of 60%.[7] The facile π → σ equilibrium of allyl ligands in polar media provides an explanation of the double-bond isomerization.[7, 20] The π complex **45** from geranyl bromide can form a σ-allylnickel species in two ways, giving structures **46** and **47**. The stereointegrity of the double bond is lost if the σ-complex **47** participates in the π → σ equilibrium.

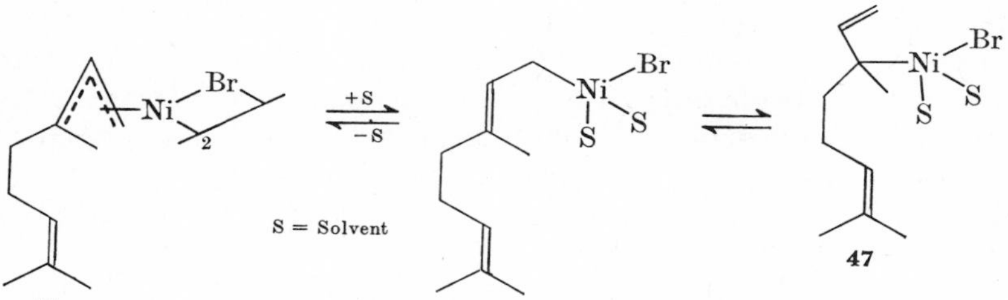

S = Solvent

45   **46**   **47**

Finally, there is the possibility of reaction at either end of an allyl ligand. High selectivity is shown here, as evidenced by the fact that the reaction of complex **45** with cyclohexyl iodide gives only the product from coupling at the less substituted end of the allyl system. Another example involves the reaction of π-(1,1-dimethylallyl)nickel(I) bromide with methyl iodide in dimethylformamide. The only C_6 product is 2-methyl-2-pentene, obtained in 90% yield.[8] No other isomer could be detected by vapor-phase chromatography.

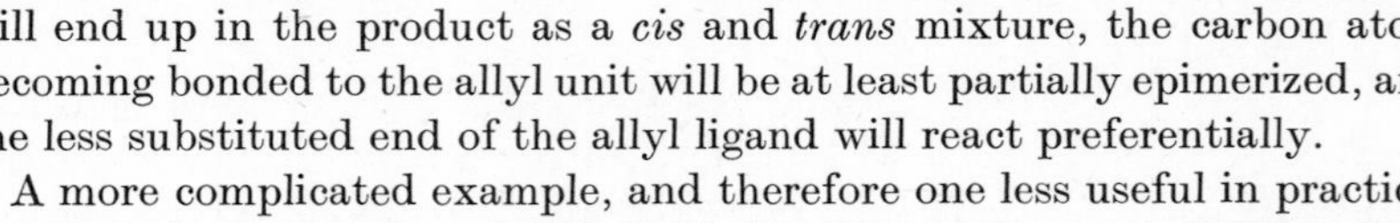

$$+ CH_3I \longrightarrow (CH_3)_2C=CHCH_2CH_3$$

These examples provide some guide lines for the selectivity to be expected: it is likely that the double bond participating in the π-allyl ligand will end up in the product as a *cis* and *trans* mixture, the carbon atom becoming bonded to the allyl unit will be at least partially epimerized, and the less substituted end of the allyl ligand will react preferentially.

A more complicated example, and therefore one less useful in practical synthesis, is the reaction of π-allylnickel halides with organic halides of the general structure R–Y–CH$_2$X, where X is halogen and Y is a functional group which forms a strong donor bond to nickel. Allyl bromide is a member of this group, where X = Br and Y = carbon π bond. Other examples that have been studied are α-halosulfides (Y = sulfur),[8] α-haloketones (Y = carbonyl),[8] and α-haloethers (Y = oxygen).[8] The reactions are complicated because halogen-metal exchange between allylnickel species and the organic halide (expressed in general terms in Eqs. 4 and 6 and repeated here for R–Y–CH$_2$–X) now becomes favorable. That is, complex **48** is stabilized by the presence of the group Y; when Y is a carbon π bond, the overall equilibrium constant for the reaction must be approximately 1.0.

The effect of halogen-metal exchange is to make available in solution the initial π-allylnickel complex (allyl$_1$-nickel bromide), the allylic halide (allyl$_2$-bromide), and the products from exchange: allyl$_1$-bromide and allyl$_2$-nickel bromide. Then either allylic bromide can couple with either π-allylnickel complex to give any of three products.[62] In fact, the reaction

$$\left.\begin{array}{c}\pi\text{-Allyl}_1\text{-Ni-Br}\\ \text{Allyl}_2\text{-Br}\\ \updownarrow\\ \pi\text{-Allyl}_2\text{-Ni-Br}\\ \text{Allyl}_1\text{-Br}\end{array}\right\}\xrightarrow{\text{Coupling}} \text{Allyl}_1\text{-allyl}_2 + \text{allyl}_1\text{-allyl}_1 + \text{allyl}_2\text{-allyl}_2$$

of π-(2-methylallyl)nickel bromide with allyl bromide affords all three coupling products (C$_6$, C$_7$, and C$_8$) in 95% yield and, roughly, the statistical distribution.[8, 62] The distribution probably does not reflect attainment of equilibrium in the exchange of allyl groups before coupling, because

[62] E. J. Corey, M. F. Semmelhack, and L. S. Hegedus, *J. Amer. Chem. Soc.*, **90**, 2416 (1968).

reaction of π-allylnickel bromide with 2-methylallyl bromide produces a different ratio: C_6 35%, C_7 24%, C_8 41%. In all reported cases of an allylic halide reacting with a π-allylnickel halide, the cross-coupling product (allyl$_1$-allyl$_2$) is formed in less than 55% yield, and sometimes as low as 20%.[8] Table II presents the results for other Y groups.[8]

TABLE II. Coupling Products from π-(2-Methylallyl)-nickel Bromide and R—Y—CH$_2$Cl

Halide (1 mol)	Product (mol)
$C_6H_5SCH_2Cl$	C_8 (0.20)
	$C_6H_5SCH_2CH_2C(CH_3)$=CH_2 (0.24)
	$C_6H_5SCH_2CH_2SC_6H_5$ (0.33)
$C_6H_5OCH_2Cl$	C_8 (not determined)
	$C_6H_5OCH_2CH_2C(CH_3)$=CH_2 (0.50)
	$C_6H_5OCH_2CH_2OC_6H_5$ (0.10)
CH_3COCH_2Cl	C_8 (0.24)
	$CH_3COCH_2CH_2C(CH_3)$=CH_2 (0.46)
	$CH_3COCH_2CH_2COCH_3$ (not determined)

Considerable effort has been focused on understanding this reaction in the hope of controlling it for synthetic purposes. It represents a direct method for preparation of unsymmetrically substituted 1,5-hexadiene derivatives such as the abundant natural products in the terpene class. For example, a simple synthesis of the naturally occurring sesquiterpene, β-farnesene (49), was attempted by this route. The usual mixture of stereoisomers (cis-β-farnesene:trans-β-farnesene = 55:45) was obtained in low (20%) yield. The major products are from symmetrical coupling: 3,6-dimethylene-1,7-octadiene and three C_{20} isomers (e.g., 50) resulting from coupling of geranyl units.

π-Geranylnickel bromide + **45** → **49a** + **49b**

+ **50** trans,trans-Bigeranyl

An extensive series of attempts has been made to control the coupling of allylic halides with π-allylnickel halides and produce the cross-coupling product selectively.[8] A simple case, the reaction of allyl bromide with π-(2-methylallyl)nickel bromide, was studied under varied conditions and monitored by vapor-phase chromatographic analysis for C_6, C_7 and C_8.[8] Solvents in which the reaction occurred at a reasonable rate (dimethylformamide, tetrahydrofuran, 1,2-dimethoxyethane) all give the same mixture of products ($C_6:C_7:C_8 = 1:2:1$).

Changes in the ligands on the nickel complex [iodide, chloride, or acetylacetonate in place of bromide; triphenylphosphine-(π-2-methylallyl)nickel bromide] do not change the mixture of products significantly. Allyl chloride, allyl iodide, allyl p-toluenesulfonate, and allyl N,N-dimethylsulfamate (51) all react rapidly with π-(2-methylallyl)nickel bromide and afford the same product mixture as does allyl bromide. Other leaving groups (acetate, thiolacetate, dithiobenzoate, and O-methyl-

xanthate)[8] were also tested in place of bromide, but they failed to react rapidly with the nickel complex at any temperature below the decomposition point of the complex (about 70° in dimethylformamide). However, allyl pyrrolidinedithiocarbamate (52) reacts completely with π-(2-methylallyl)nickel bromide within 21 hours at 25° in dimethylformamide and produces the cross-coupling product (C_7) in 65% yield, accompanied by only 5% of the symmetrical coupling products (C_6 and C_8).[8] Application of this procedure in the farnesene synthesis produced β-farnesene (cis: trans = 40:60) in 46% yield, accompanied by only 3-5% of the C_{20} isomers and no 3,6-dimethylene-1,7-octadiene. It appears that the dithiocarbamate leaving group affords a simpler mixture of products but in only moderate yield.

In summary, halogen-metal exchange becomes important for reactions of π-allylnickel halides with molecules of the type R—Y—CH$_2$—X, where Y is an atom or group which can participate in bonding to nickel. Allyl halides belong to this class and lead to a mixture of coupling products. The method is more selective when the leaving group is pyrrolidinedithiocarbamate, but the yields are not more than 65%. As usual, the double bond in the allyl ligand ends up in both cis (ca. 40%) and trans (ca. 60%) geometry.

Reaction with Other Functional Groups

In reaction with alkyl and aryl halides, the π-allyl ligand does not show characteristics of a nucleophilic reagent.[7] Under more vigorous conditions the π-allylnickel halides behave like other organometallic species in attacking reactive carbonyl groups and epoxide rings.[7,8] For example, benzaldehyde and cyclopentanone undergo alkylation in moderate to high yield in strongly polar media at 50–60°. Under similar conditions, styrene

$$\text{40} + \underset{R}{\overset{R'}{>}}C=O \longrightarrow RR'CCH_2C(CH_3)=CH_2 \atop |\atop OH$$

oxide gives the product from S_N1-like attack of π-(2-methylallyl)nickel

$$C_6H_5\overset{O}{\triangle} + \text{40} \longrightarrow \underset{CH_2C(CH_3)=CH_2}{\overset{C_6H_5}{\underset{|}{CHCH_2OH}}}$$

bromide. Apparently nickel is assisting in the opening of the epoxide ring, because simple nucleophilic attack would be expected to occur at the 2 position in styrene oxide.[63] Benzophenone and methyl benzoate fail to react with the nickel complex after many hours at 65° in dimethylformamide.[8] Acrolein undergoes 1,2 addition by the allyl ligand in dimethyl-

$$CH_2=CHCHO + \text{40} \longrightarrow CH_2=CHCHOHCH_2C(CH_3)=CH_2$$

formamide,[8] while methyl acrylate affords the 1,4 addition product and a product in which there was replacement of hydrogen by allyl after

$$CH_2=CHCO_2CH_3 + \text{2} \longrightarrow \begin{array}{c} CH_2=CHCH_2CH_2CO_2CH_3 \\ + \\ CH_2=CHCH_2CH=CHCO_2CH_3 \end{array}$$

reaction in benzene.[49] This is the only reported example of the use of benzene as a solvent for reactions of π-allylnickel halides.

These additions of the allyl ligand to carbonyl and epoxide groups may find certain special synthetic applications, but generally offer no overall

[63] R. R. Russell and C. A. Vanderwerf, *J. Amer. Chem. Soc.*, **69**, 11 (1947).

advantage compared to magnesium, lithium, or zinc reagents. The important conclusion to be drawn from these reports is that oxygen functional groups are relatively unreactive toward π-allylnickel halides; halo esters, halo epoxides, halo ketones, especially iodides, should react selectively at the halogen.

Comparison with Other Methods

The efficient formation of carbon-carbon bonds by reaction of an allyl-metal species with an organic halide is virtually unique to the π-allylnickel halides. The same overall conversion results, of course, when an organomagnesium derivative is allowed to react with an allylic halide.[64] This is a well-known synthetic method, but suffers in contrast to the π-allylnickel route because of the low selectivity of Grignard reagents. Organomagnesium (and organolithium) compounds react with many functional groups other than allylic halides, and often at a higher rate. Thus the reaction of a Grignard reagent with an allylic halide is limited to reactants containing no carbon-oxygen or carbon-nitrogen double bonds, no acidic hydrogens, and no epoxide rings, for example. None of these functional groups is likely to interfere with the formation of the π-allylnickel complexes or their subsequent coupling reaction. Allylmagnesium and allyl-lithium reagents are usually preferred to the nickel complex for the synthesis of alcohols by addition to carbonyl groups, because they are much more reactive.

An important advantage of the π-allylnickel halides is the ease of purification and the convenience of measuring a precise amount of dry solid. The direct formation of allylmagnesium or allyl-lithium reagents from the metal and the allylic halide is usually not quantitative because the allylmetal species couples with the allylic halide. However, an indirect method is now available that allows preparation of allylmagnesium halides in high purity using (tetra-allyl)tin.[65]

Allylzinc halides and diallylzinc appear to be intermediate in reactivity between π-allylnickel halides and allylmagnesium halides.[66] The zinc derivatives react with esters and aryl ketones, in contrast to the π-allylnickel halides. Certain activated halides and alkoxides can be displaced by the allylzinc halides, but simple alkyl or aryl halides have not been reported to react.[67] An interesting selectivity of the allylzinc[66] and allylmagnesium halides[68] appears in the reaction of substituted allyl species with carbonyl

---

[64] M. S. Kharasch and O. Reinmuth, *Grignard Reactions of Nonmetallic Substances*, Prentice-Hall, Englewood Cliffs, N.J., 1954, p. 1078.
[65] D. Seyferth and M. A. Weiner, *J. Org. Chem.*, **26**, 4797 (1961).
[66] D. Abenhaim, E. Henry-Basch, and P. Freon, *Bull. Chim. Soc. Fr.*, **9**, 4038 (1969).
[67] G. Courtois and L. Miginiac, *Bull. Soc. Chim. Fr.*, **9**, 3330 (1969).
[68] W. G. Young and J. D. Roberts, *J. Amer. Chem. Soc.*, **68**, 649 (1946).

compounds. For example, crotylzinc bromide reacts preferentially at the more substituted end of the allylic system. No comparable test of selectivity

$$CH_2=CHCH_2ZnBr \xrightarrow{\begin{array}{c}n\text{-}C_4H_9OCH_2N(C_2H_5)_2\\ \\ ClCH_2OCH_3\end{array}} \begin{array}{c}CH_2=CHCH_2CH_2N(C_2H_5)_2\\ \\ CH_2=CHCH_2CH_2OCH_3\end{array}$$

in attack on carbonyl groups has appeared for π-crotylnickel bromide but,

$$CH_3CH=CHCH_2ZnBr + R_2C=O \rightarrow R_2C(OH)CH(CH_3)CH=CH_2$$

in all reactions with alkyl or aryl halides, substituted π-allylnickel reagents react preferentially at the less substituted end of the allyl ligand.

Experimental Conditions

The preparation and isolation of π-allylnickel halides requires rigorous exclusion of oxygen and temperatures below 60–80° while the complex is in solution, but the crystals can be handled for short periods of time in air. Nuclear magnetic resonance data are very useful in characterizing the complexes as solutions in chloroform, carbon tetrachloride, or benzene. Trace quantities of oxygen cause line broadening, probably via formation of paramagnetic or insoluble nickel species. By working in an inert atmosphere and filtering the sample solution, spectra with high resolution are obtained. The solubility of the complexes depends on the halogen bridge, in the order chloride < bromide < iodide for solvents like chloroform or benzene. Additional alkyl substituents on the allyl ligand increase solubility in hydrocarbon solvents and lower the melting point; π-geranylnickel bromide is very soluble in pentane and separates as a liquid at $-78°$.[8]

Preparation from Nickel Carbonyl. *The hazardous nature of nickel carbonyl cannot be overemphasized. It is poisonous and very volatile (bp 43°), with a mild but characteristic odor often described as musty or the smell of new-mown hay. Any concentrations that can be detected by the average person are at the danger level. A nonluminous flame from a Bunsen burner or alcohol lamp will become tinged with yellow in the presence of small concentrations of nickel carbonyl and can serve as a detection device where the flame does not produce other hazards.*[69] *The toxic effects due to inhalation of nickel carbonyl may not be apparent for some hours, when the early symptoms will include headache, giddiness, vomiting or nausea, and breathlessness. After a delay period of 12–18 hours, the following late symptoms may appear: severe breathlessness increasing on slight exertion, pain in chest, dry hard cough, and*

---

[69] "Nickel Carbonyl Is Dangerous," Technical Bulletin from the Matheson Company.

a blue coloration of face and ears. Initial treatment should involve keeping the patient lying down and warm, administering oxygen as soon as possible, and general treatment for shock.[69] *Nickel carbonyl is spontaneously flammable in air, but an induction period is usually necessary before the liquid inflames. All operations with nickel carbonyl should be carried out in a good hood.*

In the author's experience, no difficulty is encountered if the nickel carbonyl is drawn into an Erlenmeyer flask from an inverted lecture cylinder, immediately withdrawn into a syringe, and injected into the reaction vessel which has been previously evacuated and filled with inert gas several times to provide an inert atmosphere. Larger-scale preparations, where the volume of nickel carbonyl is inconvenient to handle by syringe techniques, are best carried out by inverting the lecture cylinder of nickel carbonyl over an addition funnel with a slow flow of argon exiting from the addition funnel. After the requisite quantity of nickel carbonyl is in the funnel, the system is closed and addition of nickel carbonyl to the reaction mixture can proceed dropwise.

The time and temperature for complete formation of the π-allylnickel halide depends on the allyl halide used. Overreaction is to be avoided because, at the temperature necessary to form the complex (ca. 60–70°), slow decomposition occurs, especially with π-allylnickel chlorides. The reaction is conveniently monitored by the rate of evolution of carbon monoxide. The best procedure for preparation of a new π-allylnickel halide is to mix the allylic halide, nickel carbonyl (3- to 6-fold excess), and benzene under inert gas and warm the mixture rapidly to 65°. After a short induction period the gas evolution will become vigorous and a deep red color will appear in solution. After ½ to 1 hour, the gas evolution becomes very slow or stops. The reaction mixture is cooled and the excess nickel carbonyl and benzene are removed by applying aspirator vacuum. The red residue, usually solid, is dissolved in anhydrous air-free ether at 25°, filtered under inert gas, concentrated at 25° until crystals appear, and finally cooled for several hours at −78° to complete crystallization. The supernatant liquid is decanted to reveal red crystals amounting to 50–85% of the theoretical yield.

Preparation from Bis-(1,5-cyclooctadiene)nickel. Because of the vigorous reaction conditions necessary to get complete formation of the π-allylnickel halides using nickel carbonyl, the yield of product is seldom greater than 85%, and can be significantly lower for thermally sensitive complexes. Wilke and co-workers report that bis-(1,5-cyclooctadiene)nickel is a much more active source of zero-valent nickel, and formation of π-allylnickel halides from allylic halides proceeds rapidly and quantitatively below 0°.[20b] The method is less convenient than that using nickel carbonyl because the nickel reagent is extremely sensitive to oxygen in solution and is not commercially available. However, bis-(1,5-cyclooctadiene)nickel is easily prepared by reduction of nickel acetylacetonate with triethylaluminum in the presence of 1,5-cyclooctadiene using inert-atmosphere techniques.[70] It is clearly the preferred reagent for the preparation of thermally sensitive π-allylnickel halides.

Reactions with Organic Halides. The very simple general procedure for the reaction of organic halides and carbonyl compounds with π-allylnickel halides is to dissolve the complex in a polar aprotic solvent, add one equivalent of the organic halide at 25°, and stir until the initial deep red changes to emerald green. If necessary, the temperature can be raised, but decomposition of the π-allylnickel halide becomes rapid above 70° in polar solvents. The mixture of nickel dihalide and organic product is partitioned between water and an organic solvent such as petroleum ether, and the product is isolated from the petroleum ether. Common solvents for this reaction are dimethylformamide, hexamethylphosphoramide, dimethyl sulfoxide, and N-methylpyrrolidone. Only very reactive organic halides like allylic halides and α-halosulfides couple with π-allylnickel halides in less polar solvents.

As Transient Intermediates

Coupling of Allylic Halides

It was a very significant observation of Webb and Borcherdt that allylic halides react with nickel carbonyl in methyl alcohol to produce carbon monoxide, nickel(II) halide, and 1,5-hexadienes.[71] This first reported

$$2 \text{ CH}_2\text{=C(CH}_3\text{)CH}_2\text{Cl} + \text{Ni(CO)}_4 \longrightarrow \text{CH}_2\text{=C(CH}_3\text{)CH}_2\text{CH}_2\text{C(CH}_3\text{)=CH}_2 + \text{NiCl}_2 + 4\text{ CO}$$

suggestion of the unique reactions of allylic halides and nickel species eventually led to the π-allylnickel halides discussed in the preceding section. The overall procedure, coupling of allylic halides with zero-valent metals to produce 1,5-hexadiene derivatives, is a well-known reaction; magnesium metal and allyl bromide produce 1,5-hexadiene, often as an undesirable side reaction during preparation of allylmagnesium bromide.[72]

[70] B. Bogdanovic, M. Kröner, and G. Wilke, *Ann*, **699**, 1 (1966).
[71] I. D. Webb and G. T. Borcherdt, *J. Amer. Chem. Soc.*, **73**, 2654 (1951).
[72] M. S. Kharasch and O. Reinmuth, *Grignard Reactions of Nonmetallic Substances*, Prentice-Hall, Englewood Cliffs, N.J., 1954, p. 1078.

Unique to the nickel carbonyl reaction is the nonbasic and essentially non-nucleophilic character of the intermediate allylnickel species. This means that the coupling can be carried out on allyl halides with moderately acidic

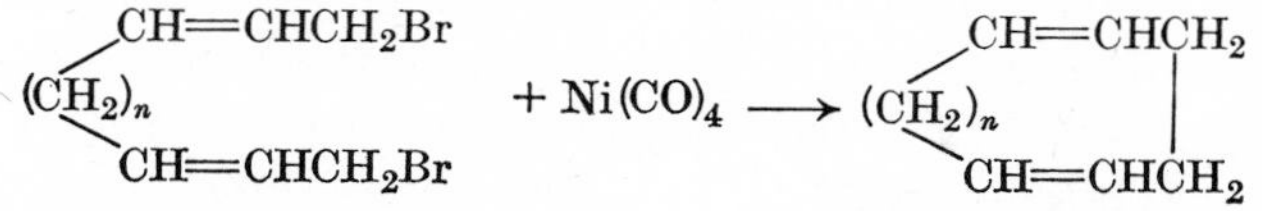

$$(CH_2)_n \underset{CH=CHCH_2Br}{\overset{CH=CHCH_2Br}{\diagup}} + Ni(CO)_4 \longrightarrow (CH_2)_n \underset{CH=CHCH_2}{\overset{CH=CHCH_2}{\diagup}}$$

hydrogen atoms or functional groups sensitive to nucleophiles. Also, in analogy with the cyclo-oligomerizations discussed on p. 131, the nickel atom can behave as a template upon which an α,ω-diallylic dihalide undergoes coupling to produce rings.

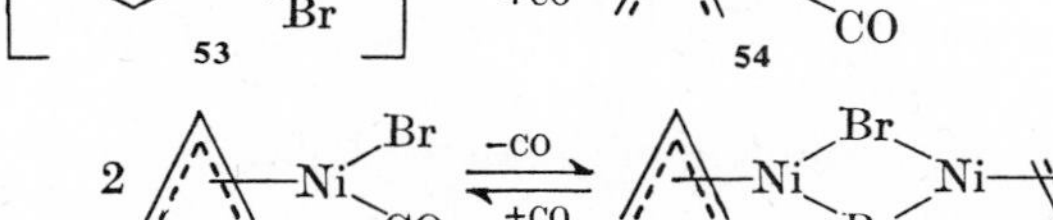

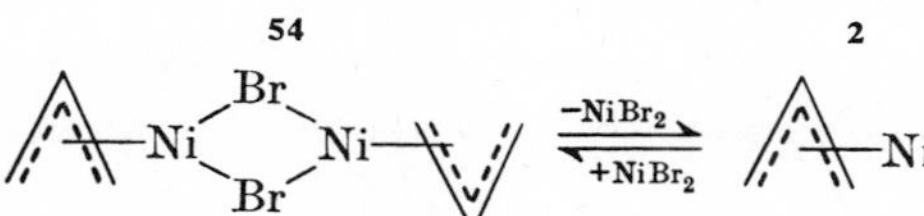

Mechanism. When allyl bromide is added to nickel carbonyl in nonpolar solvents, no reaction occurs at 25°; at elevated temperature (65°) the product is π-allylnickel bromide. In ether solvents (tetrahydrofuran, tetraethyleneglycol dimethyl ether), and (much faster) in very polar solvents (dimethylformamide, hexamethylphosphoramide, dimethyl sulfoxide), reaction of allyl bromide and nickel carbonyl proceeds at 25° with disappearance of the halide and appearance of a deep-red color.[62] Subsequently, 1,5-hexadiene appears in the solution, and the end of the reaction is signaled by a color change from red to green (in the polar solvents) or from red to a pale-yellow supernatant liquid and an orange precipitate (in the ether solvents). Thus it appears that the mechanism involves rapid formation of an allylnickel intermediate followed by decomposition to 1,5-hexadiene and nickel(II) halide. The nature of the allylnickel intermediate is not known in detail, but one or more of the intermediates in the following reaction sequence (Eqs. 7–10) is likely to be important, especially σ-allyldicarbonylnickel bromide (**53**) and π-allylcarbonylnickel bromide (**54**). Equation 7 is an oxidative-addition step common for zero-valent metal complexes [57] and probably is reversible in the presence of a strong donor ligand like carbon monoxide. Complex **53** has not been observed and is likely to be relatively unstable, as are most nickel complexes with σ-bonded ligands. The second step (Eq. 8) is a π-allyl ⇌ σ-allyl equilibrium which is promoted by carbon monoxide or a coordinating solvent. Complex **54** has not been fully characterized but has been identified as a pale-orange species during carbonylation of π-allylnickel halides,[51,73] the reverse of Eq. 9. When complex **54** loses carbon monoxide, the well-known thermally stable π-allylnickel bromide results.

A crucial part of these steps is the role of carbon monoxide (from nickel carbonyl) in taking up coordination positions of the nickel, a role which may also be played by the polar solvent. Corey and co-workers[8,22,62] have investigated the mechanistic possibilities in some detail, but the exact intermediate complex which leads to the 1,5-hexadiene is not known. Some possibilities can be eliminated, however. π-Allylnickel bromide by itself in polar solvents does not produce 1,5-hexadiene unless the solution is heated to 65° for an extended period of time. Bis-(π-allyl)nickel is known to be in equilibrium with π-allylnickel bromide in very polar media (dimethylformamide, water), and also must be considered a precursor of 1,5-hexadiene.[22,24] Carbon monoxide converts both π-allylnickel bromide

$$[\text{allyl-Ni}(\mu\text{-Br})_2\text{Ni-allyl}] + 4\ CO \longrightarrow$$

$$CH_2{=}CHCH_2CH_2CH{=}CH_2 + NiBr_2 + Ni(CO)_4$$

$$[\text{allyl-Ni-allyl}] + 4\ CO \longrightarrow CH_2{=}CHCH_2CH_2CH{=}CH_2 + Ni(CO)_4$$

and bis-(π-allyl)nickel to 1,5-hexadiene very rapidly at 25°. The latter conversion does not represent the final stages of 1,5-hexadiene formation from allyl bromide because it occurs rapidly even in nonpolar solvents and, in at least one case, coupling of an allylic halide (cyclooctenyl bromide) with nickel carbonyl leads to a product different from that from carboynlation of the corresponding bis-(π-allyl)nickel complex (p. 165).[22]

On the other hand, the formation of 1,5-hexadiene from π-allylnickel bromide and carbon monoxide is complete and quantitative within

---

[73] G. P. Chiusoli, *Chim. Ind.* (Milan), **41**, 504 (1959).

1 minute at 25° in dimethylformamide; no efficient formation of 1,5-hexadiene occurs in nonpolar media (toluene), but the color of π-allylnickel

bromide disappears and allyl bromide is formed in low yield (ca. 16%) in toluene[22, 74] (thus supporting the complete reversibility of the reaction sequence 7–10).

In summary, carbon monoxide is important to the mechanism, probably in activating the allyl ligand and preventing the buildup of π-allylnickel bromide. The reaction does not proceed by reaction of unchanged allyl bromide with a π-allylnickel species because the 1,5-hexadiene forms much more slowly than allyl bromide disappears. The exact mode of carbon-carbon bond formation is not known in detail but is probably radical-like coupling; however, no experiments designed to trap free radicals in the coupling have been reported.

Scope and Limitations. In addition to allylic bromides used as examples above, the corresponding chlorides,[70] iodides,[62] tosylates[62] and acetates[75] couple efficiently with nickel carbonyl in polar media. The rates of coupling are in the order I > Br > Cl and OTs > OAc; bromides are usually the most convenient starting materials because of their ease of formation without rearrangement from the corresponding alcohols[76] and their rapid rate of reaction.

The geometry of the product from coupling of allylic halides that can exist as *cis* and *trans* forms is generally independent of the geometry of the starting material. Substituted allyl halides also offer the possibility of

cis,cis or *trans,trans*   *trans,trans*

[74] L. S. Hegedus, Ph.D. Thesis, Harvard University, 1969.
[75] N. Bauld, *Tetrahedron Lett.*, **1962**, 859.
[76] J. M. Osbond, *J. Chem. Soc.*, **1961**, 5270.

coupling at either end of the allyl system. In all reported examples, with the exception of certain cyclizations, the coupling is very selective in forming the carbon-carbon bond at the less substituted ends. The 8-carbon diallylic halide 55 affords a six-membered ring product preferentially by coupling at the secondary end of the allyl units.[10]

$$\begin{array}{c} CH=CHCH_2Br \\ | \\ CH_2 \\ | \\ CH_2 \\ | \\ CH=CHCH_2Br \\ \mathbf{55} \\ \text{(Mixture of isomers)} \end{array} + Ni(CO)_4 \longrightarrow \quad + \quad$$

(42%)   (5%)

Intramolecular coupling was found by Corey and co-workers to be very effective as a mild method of synthesis of medium-sized carbocyclic rings.[6, 10, 77] A series of long methylene chains with allylic bromide groups on each end was prepared and treated with nickel carbonyl under various conditions to determine the scope of the cyclization. The yields of 12-, 14-, and 18-membered rings ($n = 6, 8, 12$) are in the range 60–80% when high dilution is used to inhibit intermolecular coupling. As indicated above, 8-

$$\begin{array}{c} CH=CHCH_2Br \\ | \\ (CH_2)_n \\ | \\ CH=CHCH_2Br \end{array} + Ni(CO)_4 \rightarrow \begin{array}{c} CH=CHCH_2 \\ | \\ (CH_2)_n \\ | \\ CH=CHCH_2 \end{array}$$

$n = 6$, 59% yield (*trans,trans*)
$n = 8$, 70% yield (*trans,trans*)
$n = 12$, 84% yield (*trans,trans*)

and 10-membered rings cannot be made efficiently by this procedure.

A beautifully conceived application of the intramolecular coupling reaction is the synthesis of the unique natural sesquiterpene, humulene (**56**), from the dibromotriene **57**, whose internal double bond appears particularly well-suited for assisting in wrapping the triene around a nickel atom, thus favoring intramolecular cyclization.[6, 78] The cyclization is not very specific, and a mixture of three monomeric products is obtained in about a 40% combined yield; humulene comprises 12% of the mixture, corresponding to a 5% overall yield. A more efficient humulene synthesis is provided by the reaction of **58**, the *trans,cis,trans* isomer of dibromotriene **57**, with nickel carbonyl, which affords four cyclic products of unspecified structure in *ca*. 40% yield. Isomerization of the *cis*-disubstituted double bond with light and diphenyl disulfide produces humulene in 10% overall

[77] E. J. Corey and E. Hamanaka, *J. Amer. Chem. Soc.*, **86**, 1641 (1964).
[78] E. Hamanaka, Ph.D. Thesis, Harvard University, 1967.

yield. That the internal double bond is effective in promoting intramolecular cyclization is not clear from the humulene synthesis, but a simpler system was examined[77] which does demonstrate the importance of the internal double bond. The C_{12}-triene dibromide **59** reacts with nickel carbonyl without using high dilution techniques to produce two isomeric C_{12} rings in 68% combined yield. In contrast, the C_{12}-diene dibromide **60** cyclizes in only 4% yield under identical conditions. The most likely explanation of these results is participation of the internal double bond in coordination with nickel after reaction of one of the allylic halide ends, thus reducing the number of available conformations of the chain and favoring intramolecular reaction (p. 168).

Cyclo-oligomerization of a 4-carbon unit, superficially parallel with 1,3-butadiene and nickel(0) catalysts, occurs when 1,1-bis-(chloromethyl)-ethylene is allowed to react with nickel carbonyl in tetrahydrofuran at 50°

for 51 hours.[79] The major product is a 9-membered cyclic triene, 1,4,7-trimethylenecyclononane, along with low yields of the cyclic dimer, 1,4-dimethylenecyclohexane, and a carbonyl insertion product, 3,6-dimethylenecycloheptanone. Attempts to find oligomers of moderate molecular weight ($>C_{30}$) were unsuccessful, and the remaining products appear to be polymers. With magnesium, 1,1-bis-(chloromethyl)ethylene is converted to methylenecyclopropane in 30% yield;[80] with nickel carbonyl, methylenecyclopropane was not detected (<2.5% yield). The most straightforward mechanism for this reaction is stepwise coupling of C_4-dichloride units until the chain is long enough for cyclization around a nickel atom. The intermediates (C_8-dichloride **61** and C_{12}-dichloride **62**) could not be detected during the reaction, but both have been synthesized and are efficient

---

[79] E. J. Corey and M. F. Semmelhack, *Tetrahedron Lett.*, **1966**, 6237.
[80] B. C. Anderson, *J. Org. Chem.*, **27**, 2720 (1962).

precursors of 1,4,7-trimethylenecyclononane.[79] The C_{12}-dichloride **62** affords 1,4,7-trimethylenecyclononane in 64% yield eight times faster than 1,1-bis-(chloromethyl)ethylene under identical conditions of reaction with nickel carbonyl. Control of these cyclo-oligomerizations by changing the ligand on nickel has an effect very much like that observed in the cyclo-oligomerizations of 1,3-butadiene with nickel(0) catalysts.[21a] Triphenylphosphinenickel tricarbonyl reacts with 1,1-bis-(chloromethyl)ethylene to give the same low yield of 1,4-dimethylenecyclohexane (10%) as nickel carbonyl, but the formation of 1,4,7-trimethylenecyclononane is completely inhibited.

On the other hand, the C_8-dichloride **61** cyclizes efficiently (62%) with triphenylphosphinenickel tricarbonyl; with nickel carbonyl, polymer is the major product.

No report of other cyclo-oligomerizations of small allylic dihalides has yet appeared. This reaction is a unique and limited application of the coupling of allylic halides with nickel carbonyl, but has all the potential of the versatile cyclo-oligomerizations of 1,3-butadiene.

Comparison with Other Methods. The formation of 1,5-hexadiene derivatives with nickel carbonyl is one of the most selective methods known. Other metals perform the same overall conversion, but generally the intermediates are reactive toward acidic hydrogen atoms and other functional groups. Most of the alternatives are very well known and will not be discussed in detail. Magnesium[72] and lithium[81] produce 1,5-hexadiene from allyl bromide, but the reactions are not compatible with hydroxyl, carbonyl, and many other substituents. Relatively few examples have been studied with nickel carbonyl, but it is clear that most organic functional groups will not interact with the intermediate allylnickel complexes under the conditions of the coupling reaction. Alkyl, aryl, or benzyl halides should not interfere, as they do not react with nickel

[81] W. Kawai and S. Tsutsumi, *J. Chem. Soc.* (Japan), **81**, 109 (1960).

carbonyl at 25° and react relatively slowly with π-allylnickel complexes in tetrahydrofuran, the usual solvent for allylic coupling with nickel carbonyl. More examples must be tested before careful definition of the scope of the reaction is possible.

Experimental Conditions. The coupling of an allylic halide with nickel carbonyl is usually carried out by preparing a mixture of the allylic halide and appropriate solvent in a flask bearing a three-way stopcock, reflux condenser, and rubber septum. The system is alternately evacuated to aspirator pressure and filled with inert gas (nitrogen or argon) three or more times, and nickel carbonyl (3- to 5-fold molar excess) is added by syringe to the solution at 25°. An excess of nickel carbonyl is necessary because a portion of the reagent is carried off with the carbon monoxide formed during the reaction. In ether solvents the reaction is generally slow at 25°, and the temperature is raised to 50° until the initial deep red color disappears, leaving a yellow suspension. Above 65°, solutions of nickel carbonyl begin to decompose to carbon monoxide and nickel metal unless a positive pressure of carbon monoxide is maintained. In dimethylformamide and the other very polar aprotic solvents, the reaction is vigorous at 25°; nickel carbonyl should be added at a rate controlled to avoid foaming as the carbon monoxide is evolved. A deep red color appears rapidly and changes to green [solvated nickel(II) halide] when the reaction is complete. Usually a few hours at 25° is sufficient. The carbon monoxide is allowed to escape into the hood through a mercury bubbler tube in order to monitor the rate of gas evolution.

After the reaction is judged complete by observation of very slow or zero gas evolution and by the color changes indicated above, the excess nickel carbonyl is removed by applying an aspirator vacuum or by adding a volume of anhydrous ether equal to the volume of the reaction mixture and distilling the ether at atmospheric pressure. The nickel carbonyl will codistill and can be destroyed by adding the ether distillate cautiously to a dilute solution of bromine. *Before removing nickel carbonyl directly into the aspirator flow, it must be ascertained that the hood plumbing does not connect with an open drain in the room where the vapors of nickel carbonyl could reappear.* The residue after removal of the nickel carbonyl is poured into water, and washed with an organic solvent appropriate for the expected product, and the product is isolated from the organic layer. In some cases, difficult emulsions occur, but a small amount of hydrochloric acid is very effective in bringing the nickel salts into aqueous solution.

Carbonylation of Allylic Halides

The coupling of allylic halides with nickel carbonyl depends on the use of a polar solvent, usually aprotic, and no more than atmospheric pressure of carbon monoxide. Under essentially identical conditions but at 2 or 3 atmospheres of carbon monoxide, the reaction takes a different course and leads to 3-butenoyl halides (in inert solvents) or, more generally, to 3-butenoic acid (in water) and methyl 3-butenoate (in methyl alcohol).[47]

$$CH_2=CHCH_2Br + CO \xrightarrow[CH_3OH]{Ni(CO)_4} CH_2=CHCH_2CO_2CH_3 + HBr$$

The yields generally do not exceed 50%. Probably because of stabilization of the nickel carbonyl, the reaction is suppressed at higher pressures of carbon monoxide unless higher temperatures are also applied.

More intriguing from a synthetic standpoint is the incorporation of acetylene during carbonylation.[47] The overall reaction involves an allylic halide, acetylene, carbon monoxide, and a nickel catalyst. In contrast to

$$\text{\Large\textbackslash}Br + CO + HC\equiv CH \xrightarrow[CH_3OH]{Ni(0)} \text{\Large\textbackslash}COCH_3 + HBr$$

the simple carbonylation of allylic halides, this reaction proceeds at atmospheric pressure of carbon monoxide and produces the cis-2,5-hexadienoic acid derivatives in yields up to 80%. Nickel carbonyl has been used as the catalyst, often in stoichiometric amounts; however, a simple variation has been reported in which the nickel catalyst is formed *in situ* by reduction of nickel chloride with a manganese-iron alloy in the presence of thiourea.[82] The allyl halide is added to the methanolic suspension, and a gaseous mixture of acetylene and carbon monoxide is bubbled through to give high conversion and high yields of methyl cis-2,5-hexadienoate.

Mechanism. It has been demonstrated that π-allylnickel complexes are probably intermediates in the carbonylation of allylic halides with nickel carbonyl.[48b] The presently accepted mechanism of formation of 2-butenoyl halides is outlined in Eqs. 11–14.[4,47] Clearly, π-allylnickel

$$\text{\Large\textbackslash}Br + Ni(CO)_4 \rightleftharpoons \left[\text{\Large\textbackslash}Ni\begin{smallmatrix}Br\\-CO\\CO\end{smallmatrix}\right] + 2\,CO \qquad \text{(Eq. 11)}$$

$$\left[\text{\Large\textbackslash}Ni\begin{smallmatrix}Br\\-CO\\CO\end{smallmatrix}\right] \rightleftharpoons \triangle\!-\!Ni\begin{smallmatrix}Br\\CO\end{smallmatrix} + CO \qquad \text{(Eq. 12)}$$

[82] G. P. Chiusoli, M. Dubini, M. Ferraris, F. Guerrieri, S. Merzoni, and G. Mondelli, *J. Chem. Soc., C*, **1968**, 2890.

$$\left[\text{CH}_2=\text{CHCH}_2\text{Ni}(\text{CO})_2\text{Br} \right] + \text{CO} \rightleftharpoons \left[\text{CH}_2=\text{CHCH}_2\text{C(O)Ni(CO)}_2\text{Br} \right] \quad \text{(Eq. 13)}$$

$$\left[\text{CH}_2=\text{CHCH}_2\text{C(O)Ni(CO)}_2\text{Br} \right] + 2\,\text{CO} \longrightarrow \text{CH}_2=\text{CHCH}_2\text{C(O)Br} + \text{Ni(CO)}_4 \quad \text{(Eq. 14)}$$

complexes need not be invoked to explain the overall conversion. But the equilibria represented by Equations 12 and 13 are known to occur starting from π-allylnickel bromide and carbon monoxide, and are likely to be important in the carbonylation system.[47]

The acetylene insertion is thought to occur by initial coordination of acetylene and then insertion between the allyl ligand and nickel to give a σ-vinylnickel complex.[4,47] Carbonyl insertion followed by displacement of the acyl ligand by carbon monoxide produces *cis*-2,5-hexadienoyl halide (Eqs. 15–18).

$$\left[\text{allyl-Ni(CO)}_2\text{Br} \right] + \text{HC}{\equiv}\text{CH} \rightleftharpoons \left[\text{allyl-Ni(CO)(HC{\equiv}CH)Br} \right] \quad \text{(Eq. 15)}$$

$$\left[\text{allyl-Ni(CO)(HC{\equiv}CH)Br} \right] + \text{CO} \rightleftharpoons \left[\text{hexadienyl-Ni(CO)}_2\text{Br} \right] \quad \text{(Eq. 16)}$$

$$\left[\text{hexadienyl-Ni(CO)}_2\text{Br} \right] + \text{CO} \rightleftharpoons \left[\text{hexadienyl-C(O)Ni(CO)}_2\text{Br} \right] \quad \text{(Eq. 17)}$$

$$\left[\text{hexadienyl-C(O)Ni(CO)}_2\text{Br} \right] + 2\,\text{CO} \longrightarrow \text{hexadienyl-C(O)Br} + \text{Ni(CO)}_4 \quad \text{(Eq. 18)}$$

Scope and Limitations. A feature of the carbonylation of allylic halides which can be a limitation and at other times an advantage is that carbonylation occurs selectively at the less substituted end of the allylic system, independent of the initial position of the halogen. For example, crotyl chloride and 3-chloro-1-butene give the same product, methyl 2-pentenoate (in methyl alcohol).[73]

$$\left.\begin{array}{c}CH_3CH{=}CHCH_2Cl\\ \text{or}\\ CH_3CHClCH{=}CH_2\end{array}\right\} \xrightarrow[CH_3OH]{Ni(CO)_4,\ CO} CH_3CH{=}CHCH_2CO_2CH_3$$

Allylic halides bearing electron-attracting substituents such as ester, nitrile, or amide groups on the carbons of the allyl unit do not give insertion of carbon monoxide if a small amount of water is present in the alcoholic medium. Instead, a reductive reaction occurs, nickel carbonyl replacing the allylic halogen with hydrogen.[83] Interestingly, the nonconjugated tautomer is usually formed selectively.

$$ClCH_2CH{=}CHCN + Ni(CO)_4 \xrightarrow{H_2O} CH_2{=}CHCH_2CN + Ni(OH)Cl$$

The insertion of an acetylene unit is not limited to acetylene itself, but also occurs with monoalkyl- and monoaryl-alkynes.[84] However, all of the reported examples of incorporation of substituted acetylenes give products from further reaction of the initially formed cis-2,5-hexadienoyl halide with the nickel catalyst. Acyl halides generally react with nickel carbonyl,[85] and unsaturated acyl halides often give cyclization to mixtures of lactones and cyclopentenones, depending on the solvent and the exact structure of the acyl halide.[86] These reactions of acyl halides do not appear to involve π-allylnickel intermediates and are not discussed in detail here.

It is not necessary to start with an allylic halide, because allylic alcohols, ethers, and esters will also undergo carbonylation or carbonylation/acetylene insertion if hydrogen chloride is added to the reaction mixture.[87] In this case the allylic halide is probably formed *in situ*.

$$\diagup\!\!\!\diagdown\!\!\!\diagup\!\!\!\diagdown^{OR} \xrightarrow[CH_3OH]{Ni(CO)_4,\ HCl} \diagup\!\!\!\diagdown\!\!\!\diagup\!\!\!\diagdown^{CO_2CH_3}$$

R = H, alkyl, acyl

The 2,5-hexadienoic acid derivatives which are available by this route are useful synthetic intermediates. Isomerization to the 3,5-diene and 2,4-diene isomers is rapid under basic conditions.[88] Sulfide radicals cause

[83] G. P. Chiusoli, G. Bottaccio, and A. Cameroni, *Chim. Ind.* (Milan), **44**, 131 (1962).
[84] G. P. Chiusoli and G. Bottaccio, *Chim. Ind.* (Milan), **47**, 165 (1965).
[85] L. Cassar and G. P. Chiusoli, *Tetrahedron Lett.*, **1966**, 2805.
[86] L. Cassar, G. P. Chiusoli, and M. Foa, *Tetrahedron Lett.*, **1967**, 285.
[87] (a) G. Chiusoli and S. Merzoni, *Chim. Ind.* (Milan), **45**, 6 (1963); (b) J. B. Mettalia, Jr. and E. H. Specht, *J. Org. Chem.*, **37**, 3941 (1967).
[88] G. P. Chiusoli and S. Merzoni, *Chim. Ind.* (Milan), **43**, 259 (1961).

isomerization about the *cis* double bond, providing a route to the corresponding 2,5-*trans*-hexadienoic acids.[89] Heating the *cis*- or *trans*-hexadienoic acid in base, acid, or acetic anhydride/zinc chloride affords phenols (or phenol acetates) in good yield.[90] The combination of carbonylation/acetylene insertion followed by isomerization was employed in a

$$\text{[cyclohexadiene-C(O)-OH]} \xrightarrow[\text{ZnCl}_2]{(CH_3CO)_2O} \text{[phenyl-OCOCH}_3\text{]}$$

synthesis of *d,l*-methyl *trans*-chrysanthemate.[91]

$$\text{[isobutenyl-Br]} + CO + HC\equiv CH \xrightarrow[\text{CH}_3\text{OH}]{\text{Ni(CO)}_4} \text{[diene-CO}_2\text{CH}_3\text{]}$$

$$\text{[diene-CO}_2\text{CH}_3\text{]} \xrightarrow{(C_6H_5)_2\overset{+}{S}-\bar{C}(CH_3)_2} \text{[cyclopropane product-CO}_2\text{CH}_3\text{]}$$

Comparison with Other Methods. The conversion of an allylic halide to a 3-butenoic acid derivative without metal catalysts is often accompanied by migration of the double bond to give a 2-butenoic acid or ester. The nickel-catalyzed reactions do not suffer this complication, and an exactly parallel procedure using catalytic quantities of palladium dichloride[92] is also very specific. The yields are low to moderate (50%) with a few exceptions, and the procedure requires high-pressure apparatus

$$CH_2=CHCH_2Br \xrightarrow[\text{CO (100 atm)}]{\text{PdCl}_2, \text{CH}_3\text{CH}_2\text{OH}} CH_2=CHCH_2CO_2CH_2CH_3$$

and elevated temperatures (120°). An advantage of the palladium catalyst is that many different leaving groups are reactive. Allyl alcohol and its *p*-toluenesulfonate, acetate, phenyl ether, and ethyl ether are all converted to ethyl 3-butenoate under these conditions. The carbonylation appears to involve intermediate π-allylpalladium complexes.[93] Allyl alcohol is also carbonylated by carbon monoxide and catalytic amounts of tris[tris(*p*-fluorophenyl)phosphine]platinum.[94]

The direct formation of *cis*-2,5-hexadienoate derivatives from acetylene, carbon monoxide, and allylic halides using nickel carbonyl has no parallel in chemistry. Simple α,β-unsaturated esters are formed by carbonylation

---

[89] G. P. Chiusoli, G. Agnes, C. A. Ceselli, and S. Merzoni, *Chim. Ind.* (Milan), **46**, 21 (1964).
[90] G. P. Chiusoli and G. Agnes, *Chim. Ind.* (Milan), **46**, 25 (1964).
[91] E. J. Corey and M. Jautelat, *J. Amer. Chem. Soc.*, **89**, 3192 (1967).
[92] J. Tsuji, J. Kiji, S. Imamura, and M. Morikawa, *J. Amer. Chem. Soc.*, **86**, 4350 (1964).
[93] J. Tsuji, J. Kiji, and M. Morikawa, *Tetrahedron Lett.*, **1963**, 1811.
[94] G. W. Parshall, *Z. Naturforsch.*, **18b**, 772 (1963).

of monosubstituted alkynes using carbon monoxide and nickel carbonyl,[95] and the mechanism is suggested[48b] to be analogous to that outlined in Eqs. 15–18.

Experimental Conditions.[47] The carbonylation of allylic halides with nickel catalysts requires a moderate pressure (3–4 atm) of carbon monoxide in order to suppress direct coupling to 1,5-hexadiene derivatives. The reaction is generally carried out at 25° with the exclusion of oxygen. Aromatic hydrocarbon and ether solvents are suitable for synthesis of 3-butenoyl halides; methyl alcohol is the most commonly used solvent, leading directly to methyl 3-butenoate derivatives. Aqueous media afford 3-butenoic acid, but certain allylic halides, especially those with electron-withdrawing groups on the allyl unit, undergo replacement of halogen by hydrogen in the presence of water. Coupling of 1,5-hexadiene derivatives is also an important side reaction of allylic halides with electron-withdrawing groups; efficient carbonylation is achieved only at high pressures of carbon monoxide at 100° with rigorous exclusion of water. Although the stoichiometry requires only a catalytic amount of nickel carbonyl, side reactions often destroy the catalyst. In practice, nickel carbonyl is generally used in small molar excess.

The insertion of acetylene during carbonylation of allylic halides is a simple reaction experimentally. The solvent, methyl alcohol, is saturated at atmospheric pressure with a gas mixture containing equal amounts of carbon monoxide and acetylene; air is excluded. Then the allylic halide, usually the chloride, and nickel carbonyl are slowly added simultaneously. The gas mixture is introduced into the solution at a rate controlled to provide a slow flow of unreacted gases from the reaction mixture. After addition of the reactants is complete, the gas introduction is continued for about an hour, or until the gases are no longer absorbed by the reaction mixture (as monitored by flow meters before and after the reaction vessel). Improved yields are often obtained through the addition of an inorganic base (*e.g.*, magnesium oxide or magnesium carbonate) which neutralizes the hydrogen halide formed. The product is isolated by distillation of the reaction mixture, with appropriate caution in collecting the more volatile fractions containing nickel carbonyl. The solid catalyst is used in place of nickel carbonyl by adding the solid mixture (powdered iron alloy, nickel chloride, and thiourea) portionwise during the reaction.[82]

EXPERIMENTAL PROCEDURES

General Technique. The hazardous nature of nickel carbonyl has been pointed out earlier on p. 160. The techniques used in handling

[95] G. N. Schrauzer, *Adv. Organomet. Chem.*, **2**, 1 (1964).

organometallic complexes (weighing, recrystallizing, filtering) may be unfamiliar to some organic chemists who wish to use the synthetic methods outlined in this chapter. An excellent complete explanation of the necessary laboratory technique is available[96] as well as a shorter summary.[97] These references will be very useful to anyone wishing to use the π-allylnickel reagents in organic synthesis.

Bis-(π-2-methylallyl)nickel (from 2-Methylallylmagnesium Bromide and Nickel Bromide).[74] After the general procedure of Wilke,[21a] nickel bromide (anhydrous, 18.7 g, 0.086 mol) is suspended in 100 ml of anhydrous diethyl ether in a 1-l three-necked flask fitted with a 500-ml dropping funnel, a mechanical stirrer, and a Claisen head containing a low-temperature thermometer and a three-way stopcock. The system is alternately evacuated to aspirator vacuum and filled with inert gas (nitrogen or argon) several times, and cooled to $-10°$. 2-Methylallylmagnesium bromide (0.17 mol) in 300 ml of ether is added dropwise to the vigorously stirred solution during 2 hours. The mixture is stirred for an additional 0.4 hour at $-10°$, warmed to $25°$, and filtered under inert gas. Any of several methods of filtration can be used.[96] The simplest technique is to use a reaction flask with a fritted disk and two-way stopcock attached to the bottom; the solution can be forced through the frit with positive pressure into a second flask which is also under inert gas. The residue is washed with 50 ml of ether, filtered, and the combined ether filtrates are concentrated under aspirator vacuum at $-30°$. Pentane (125 ml) is added, the resulting slurry is filtered under inert gas, most of the pentane is removed from the filtrate at aspirator vacuum, and the remaining brown solution is cooled to $-78°$. Impure brown crystals (mp $35°$) of bis-(π-2-methylallyl)nickel are isolated by decanting the supernatant liquid under inert gas. It is convenient to store the complex at this stage at $-20°$ and then sublime the requisite quantity into the vessel for further reaction. Sublimation at $25°/0.1$ torr affords yellow crystals collecting in a $-78°$ receiver. No precise yield has been reported for this method of preparation.

Bis-(π-2-carbethoxyallyl)nickel (from Zinc Reduction of the Corresponding π-Allylnickel Bromide).[74] A solution of π-(2-carbethoxyallyl)nickel bromide[8] (0.61 g, 1.20 mmols) in 7 ml of dimethylformamide is added by syringe to a flask containing 2.5 g of zinc-copper granules[98] in 15 ml of pentane at $25°$. The solution is stirred for 0.5 hour, during which time the color changes from deep red to yellow. The pentane layer is transferred to an argon-filled flask through a length of 18-gauge

[96] D. F. Shriver, *Manipulation of Air-Sensitive Compounds*, McGraw-Hill, New York, 1969, pp. 139–205.
[97] J. J. Eisch and R. B. King, *Organometallic Synthesis*, Academic Press, 1965, pp. 55–60.
[98] E. LeGoff, *J. Org. Chem.*, **29**, 2048 (1964).

stainless steel tubing under a positive pressure of argon. The dimethylformamide layer is washed with two 10-ml portions of pentane, the combined pentane solutions are shaken with 10 ml of water, and the pentane is removed at aspirator vacuum. The residue is π-(2-carbethoxyallyl)nickel; 0.12 g (57%).

4,5-Dimethyl-1,4,7-*cis*,*cis*-*trans*-cyclodecatriene.[*37d] A mixture of anhydrous nickel acetylacetonate (2.19 g, 8.53 mmols), triphenylphosphine (2.24 g, 8.53 mmols), ethoxydiethylaluminum (2.3 g, 17.3 mmols), 1,3-butadiene (93.8 g, 1.74 mmols), 2-butyne (15.7 g, 0.29 mmol), and 50 ml of toluene is kept in a glass autoclave at 20° for 8 hours. The mixture is cooled to $-10°$ and transferred to a flask at $-78°$. The volatile components are flash-distilled below 25° at 12 torr into a trap at $-190°$. The residue is exposed to air to destroy the catalyst and is analyzed by glpc.

The dimethylcyclodecatriene, present in 86% yield (42.6 g) based on unrecovered 2-butyne, is isolated by careful fractional distillation. The fraction with bp 24–27°/0.1 torr (n^{20}D 1.5131) is 94.7% pure, contaminated with 1,5,9-cyclododecatriene. An analytically pure sample is obtained by crystallization from methyl alcohol at low temperature; n^{20}D 1.514, mp -30 to $-25°$.

π-(2-Methylallyl)nickel Bromide (from 2-Methylallyl Bromide and Nickel Carbonyl).[7,99] A 1-l three-necked flask is fitted with a reflux condenser, a pressure-equalizing dropping funnel, a three-way stopcock, and a large magnetic stirring bar. Benzene (380 ml) is placed in the flask, and the system is alternately evacuated to aspirator vacuum and filled with inert gas (argon or nitrogen) three or more times. From an inverted lecture cylinder, 38.5 ml (50.8 g, 0.298 mol) of nickel carbonyl is introduced into the top of the addition funnel while a slow flow of argon is allowed to exit from the same opening of the addition funnel. Argon is the best choice as inert atmosphere for this operation because it is more dense than air and forms a protective blanket at the opening which prevents introduction of oxygen. The nickel carbonyl is then added to the benzene under a positive pressure of inert gas, and the mixture is immersed in an oil bath at 50°. By means of a syringe, 10.04 g (0.0745 mol) of 2-methylallyl bromide is added during 10 minutes. After a short induction period, evolution of carbon monoxide becomes rapid and a deep red color appears. The exit gases are led from the top of the condenser through a gas bubbler in order to monitor the rate of gas evolution. As the gas evolution becomes vigorous, the bath temperature is raised to 70° and maintained at this temperature for 30 minutes after gas evolution ceases (total time after

---
* More detailed procedures are not available at present.
[99] M. F. Semmelhack and P. Helquist, *Org. Syntheses*, **52**, in preparation.

addition of 2-methylallyl bromide: 1.5 hours). After the solution cools to 25°, the benzene and excess nickel carbonyl are removed by applying aspirator vacuum. The solid red residue is dissolved in 150 ml of oxygen-free anhydrous diethyl ether, the solution is filtered under argon, and the filtrate is concentrated at aspirator vacuum and 25° until crystals appear. Crystallization is completed by cooling at −78° for 12 hours. The complex is isolated by drawing off the supernatant liquid through a syringe needle attached to an aspirator. The yield is 12.1 g (85%) of dark crystals.

Bis-(1,5-cyclooctadiene)nickel(0).[100] Into a 1-l three-necked round-bottomed flask equipped with a three-way stopcock, distilling head, a rubber serum stopper, and a heavy magnetic stirring bar are placed 57.0 g of commercial anhydrous bis(acetylacetonato)nickel(II) (Alfa Inorganics) and 500 ml of reagent grade toluene. Under a slow flow of argon the toluene is distilled from the flask at atmospheric pressure to remove the last traces of water from the nickel salt. With a slow flow of argon exiting from the opening, the distilling head is replaced by a 250-ml pressure-equalizing graduated addition funnel. After removal of all volatiles at 0.01 torr, the residue weighs 51.8 g [0.201 mol of bis(acetylacetonato)nickel(II)]. The apparatus is alternately evacuated (0.01 torr) and filled with argon several times. Toluene (155 ml) is distilled into the addition funnel under argon after discarding a 50-ml forerun, and then the toluene is added to the flask. Similarly, 123 ml (109 g, 1.00 mol) of 1,5-cyclooctadiene is distilled into the addition funnel and then added to the flask. A rubber septum is placed on the top opening of the addition funnel and the system is again alternately evacuated and filled with argon several times to remove traces of air. By means of a syringe needle, 1,3-butadiene (Matheson) is bubbled into the reaction mixture until a gain in weight of 5 g is observed.

By a special technique a solution of triethylaluminum in toluene (113 ml, 95.4 g of a 25% solution 0.209 mole of triethylaluminum, Texas Alkyls) is drawn into the addition funnel. The triethylaluminum solution is purchased in a gas cylinder bearing a valve with a dip tube that allows one to draw liquid directly from the upright cylinder by application of pressure with an inert gas through a second valve. A syringe needle is attached to a Luer-Lok syringe connector that is soldered on the tube leading from the cylinder. In this way the triethylaluminum solution can be forced out through the syringe needle into the addition funnel. With a slow flow of argon into the open three-way stopcock and exiting through a gas bubbler tube, the solution is cooled in ice, and the triethylaluminum solution is

[100] M. F. Semmelhack and P. Helquist, unpublished modifications of the procedure of Wilke, ref. 70.

added over 4 hours. Stirring is continued for an additional 5 hours at 0° and then 11 hours at 25°. The initially deep green solution changes to a yellow-brown suspension shortly after addition is complete. Essentially pure bis-(1,5-cyclooctadiene)nickel(0) is present as a yellow precipitate and is isolated by filtration under inert atmosphere.[96,97]

A simple filtration procedure is to use a dry bag (I²R Company) or dry box containing a lead for positive argon pressure, a fritted disk filtration funnel constructed so that a pressure of argon can be applied to the top, a flask to receive the filtrate and maintain inert atmosphere after being removed from the dry box, a flask to receive the residue and maintain an inert atmosphere, a spatula, and the reaction mixture. The mixture is poured into the funnel and filtered using argon pressure. The yellow residue is transferred to a flask and dried at 0.01 torr; 38.4 g (69%) of crude product is obtained. The filtrate is collected and cooled to $-78°$ slowly to afford more yellow crystals which are isolated by filtration as before. The combined yield of yellow crystalline bis-(1,5-cyclooctadiene)nickel(0) is 45.6 g (81%). It can be recrystallized from toluene, but no simple criteria for purity have been reported. The crystals taken directly from the reaction appear satisfactory for the applications described below.

π-(2-Carbethoxyallyl)nickel Bromide [using Bis-(1,5-cyclooctadiene)nickel(0)].[8] Bis-(1,5-cyclooctadiene)nickel (8.879 g, 32.4 mmols) is transferred under argon or nitrogen to a tared flask (equipped with a three-way stopcock and rubber serum stopper) in a dry box or by using inert atmosphere techniques.[96,97] Benzene (50 ml, dried by distillation, purged of air by alternately evacuating and filling with inert gas) is added via syringe at 25° followed by addition of ethyl 2-bromomethylacrylate (4.40 ml, 6.26 g, 32.4 mmols) over 30 minutes. A deep red color appears rapidly. After an additional 30 minutes at 25°, the reaction mixture is filtered under inert gas through a glass frit (as in the last paragraph of the preceding preparation) into a flask fitted with a three-way stopcock and a rubber serum stopper, and the filtrate is concentrated at 25°/25 torr until crystals appear. Then 25 ml of n-pentane (or other dry hydrocarbon solvent, air-free) is added and the mixture is cooled to 0° to complete precipitation of the complex, which is isolated by filtration through a glass frit under inert gas as above. The residue is washed twice with 25-ml portions of n-pentane and dried at 25°/0.1 torr to afford 6.17 g (76%) of red, microcrystalline π-(2-carboethoxyallyl)nickel bromide.

4-(2-Methylallyl)cyclohexanol.[8] π-(2-Methylallyl)nickel bromide (0.55 g, 1.42 mmols) is weighed under nitrogen or argon into a tared flask equipped with a three-way stopcock and rubber serum stopper. Dimethylformamide (4.0 ml, distilled at 50°/30 torr from calcium hydride and rendered air-free by alternately evacuating and filling with inert gas several times) is added via syringe. Then a solution of *trans*-4-iodocyclohexanol (0.643 g, 2.84 mmols, prepared[101] by reaction of cyclohexane-1,4-oxide with hydriodic acid, mp 60–61°) in 2.0 ml of dimethylformamide (air-free as above) is added rapidly by syringe. After addition the solution is allowed to warm to 23° during 1 hour and stirred at this temperature for 22 hours under a positive pressure of inert gas. The reaction mixture, now green, is poured into ether and the ether solution is washed with four portions of water, dried over magnesium sulfate, and concentrated at aspirator vacuum to afford 388 mg (89%) of 4-(2-methylallyl)cyclohexanol as a mixture of epimers (38:62) of greater than 98% purity, as determined by glpc (10% Fluorosilicone 1265 on Diatoport S at 130°, 15 ft × 0.125 in. column, major isomer at 17.3 minutes and the minor isomer at 15.5 minutes).

4-Hydroxy-2-methyl-4-phenyl-1-butene.[7,8] π-(2-Methylallyl)nickel bromide (1.77 mmols) is weighed (in a dry box or using inert gas techniques under argon or nitrogen) into a tared flask fitted with a three-way stopcock and a rubber serum stopper. Dimethylformamide (30 ml, made air-free by alternately evacuating and filling with inert gas) is added via syringe at 23° to give a deep red solution. This is followed by rapid addition of benzaldehyde (freshly distilled, 3.54 mmols). After 13 hours at 23° under a positive pressure of inert gas, the mixture is brown; it turns green only after an additional 24 hours at 50°. The reaction mixture is then cooled to 25°, poured into 100 ml of diethyl ether, and the ether solution is washed several times with 5% aqueous ammonium chloride. After having been dried over anhydrous magnesium sulfate, the ether layer is concentrated by rotary evaporation to give 492 mg (87%) of 4-hydroxy-2-methyl-4-phenyl-1-butene of >95% purity as determined by glpc analysis (10% Fluorosilicone 1265 on Diatoport S, 200°, 15 ft × 0.125 in. column, retention time: 4.4 minutes).

***cis*- and *trans*-Geranylcyclohexane.**[8] To a stirred solution of *trans*-geraniol (Aldrich puriss, 0.155 g, 1.00 mmol) in 10 ml of anhydrous diethyl ether is added phosphorus tribromide (0.271 g, 0.35 mmol) during several minutes. After 1 hour at 0° the mixture is shaken vigorously with two 15-ml portions of saturated aqueous sodium carbonate, dried over anhydrous magnesium sulfate, and concentrated at aspirator pressure to afford essentially pure *trans*-geranyl bromide (2.00 g, 0.94 mmol, 94%).[8,76] A sample of *trans*-geranyl bromide (4.68 g, 16.3 mmols) prepared in this way is dissolved in 100 ml of dry benzene in a 250-ml round-bottomed flask equipped with a three-way stopcock on top of a reflux condenser, a rubber serum stopper, and a magnetic stirring bar. The system is placed in an efficient hood and alternately evacuated to aspirator vacuum and filled

---

[101] E. J. Corey and L. Haefele, unpublished procedure.

with inert gas several times. Nickel carbonyl is added by syringe (8.2 ml, 11.0 g, 64 mmols), and the mixture is heated at 60–65° until the initially vigorous gas evolution becomes slow and constant (about 30 minutes). The deep red solution is concentrated at 25°/25 torr to constant weight (the benzene and excess nickel carbonyl are drawn into the aspirator stream). A vacuum trap cooled at −78° (dry ice) can be placed between the system and the vacuum to collect the volatiles; the unreacted nickel carbonyl which collects can be destroyed by adding it slowly to a solution of bromine in benzene.

To the red, viscous, liquid residue (crude π-geranylnickel bromide, not further purified) under inert gas is added via syringe 25 ml of dry, air-free dimethylformamide. To the deep red dimethylformamide solution of the complex at 25°, cyclohexyl iodide (2.14 ml, 16.3 mmols) is added by syringe over a period of two minutes, and the mixture is stirred at 23° for 5 hours. After an additional 17 hours at 60°, during which time the color changes from red to green, the solution is cooled to 25°, diluted with 100 ml of water, and washed with three 50-ml portions of n-pentane. The combined n-pentane solution is washed with two 50-ml portions of water, dried over anhydrous magnesium sulfate, and concentrated at aspirator vacuum to give a yellow residue. Distillation in a small apparatus without a fractionating column affords a center cut of bp 78–79°/2.5 torr (1.63 g). Glpc analysis (10% Fluorosilicone 1265 on Diatoport S, 160°, 15 ft × 0.125 in.) indicates the presence of two products, *cis*-geranylcyclohexane (40%, retention time: 12.2 minutes) and *trans*-geranylcyclohexane (60%, retention time: 13.4 minutes). The overall yield based on *trans*-geraniol is 58%.

cis-1-(*cis*-Cyclooct-2-en-1-yl)cyclooct-2-ene.[74]

Dimethylformamide (10 ml) is placed in a flask fitted with a three-way stopcock, magnetic stirring bar, and a rubber serum stopper. The system is placed under inert gas by alternately evacuating and filling with nitrogen or argon several times. Nickel carbonyl (5.64 mmols, 0.72 ml) is added in one portion by syringe, followed by 0.266 g (1.41 mmols) of 3-bromocyclooctene[102] added in the same manner. The solution (25°) turns red and carbon monoxide is evolved rapidly. After one hour at 25°, the color has changed to green, indicating complete reaction; the excess nickel carbonyl is removed by applying aspirator vacuum to the system for 30 minutes at 25°. The green solution is poured into 50 ml of n-pentane (or other convenient hydrocarbon solvent) and washed with two 50-ml portions of water, dried over magnesium sulfate, and concentrated by rotary evaporation to afford a colorless semisolid residue, 0.154 g (100%), which is a mixture of *meso*- and *d,l-cis*-1-(*cis*-cyclooct-1-en-2-yl)cyclooct-2-ene. Glpc analysis showed two

---

[102] A. C. Cope and L. L. Estes, Jr., *J. Amer. Chem. Soc.*, **52**, 1129 (1950).

peaks of retention times 17.5 and 18.3 minutes, of equal area (10% Fluorosilicone 1265 on Diatoport S, 200°, 15 ft × 0.125 in.).

trans, trans-**1,5-Cyclooctadecadiene.**[10] To a solution of nickel carbonyl (0.80 ml, 6.00 mmols) in dimethylformamide (45 ml, made air-free by alternately evacuating and filling with argon or nitrogen) at 50° in a flask equipped with a rubber serum stopper and a reflux condenser topped with a three-way stopcock is added a solution of *trans,trans*-1,18-dibromo-2,16-octadecadiene (0.41 g, 1.0 mmol) in 10 ml of dimethylformamide. Addition is made over 12 hours using a motor-driven syringe (Sage Instruments). Carbon monoxide forms slowly and is allowed to escape through a gas bubbler tube. The resulting green solution is cooled, poured into 150 ml of water, and extracted with four 20-ml portions of *n*-pentane. The combined *n*-pentane extracts are washed sequentially with two 50-ml portions of water and 50 ml of saturated aqueous sodium chloride, dried over anhydrous magnesium sulfate, and concentrated at aspirator vacuum. The residue is distilled at 115°/0.04 torr in a short-path apparatus to afford *trans, trans*-1,5-cyclooctadecadiene in 84% yield.

Methyl 2,5-Hexadienoate.[82] A 5-l round-bottomed flask is equipped with a stirrer, graduated dropping funnel, a solid-addition apparatus consisting of a glass test-tube connected to the flask with a flexible tube, a thermometer, a condenser cooled by liquid ammonia, and a gas inlet tube. The temperature is controlled with a water bath. The composition of the gaseous mixture (carbon monoxide and acetylene) is controlled by flow meters, and the rate of gas absorption is measured with flow meters before and after the reaction flask. The acetylene is generated from calcium carbide and the carbon monoxide from formic acid; the mixture also contains oxygen (0.15%), nitrogen (2–3%), and carbon dioxide (0.5–1%).

The apparatus is swept with the gas mixture (average composition: carbon monoxide, 53%; acetylene, 47%), and methyl alcohol (1.5 l) is introduced. The stirred solvent is saturated with the gas mixture at 25°. From the solid-addition apparatus is added a mixture of powdered manganese-iron alloy (10.0 g with the composition: Mn, 79; Fe, 16; C, 1.46; Si, 0.8; Ni, 0.2%. Particle size: 0.025–0.040 mm diameter), nickel chloride hexahydrate (10 g), thiourea (1.5 g), and magnesium oxide (33 g). Then a solution of allyl chloride (126 g), nickel chloride hexahydrate (10 g), thiourea (3 g), in methyl alcohol (300 ml) is added from the dropping funnel during 5 hours. After a few minutes the absorption of gases begins; the mixture of gases is introduced at a rate that keeps the rate of flow from the apparatus approximately constant at 2 l/hour. More powdered alloy is added at hourly intervals (2 g three times and 4 g once). When gas absorption ceases, the mixture is filtered under vacuum and the volatile components are distilled at atmospheric pressure into a receiver cooled in dry ice. The distillate may contain nickel carbonyl and is disposed of accordingly. The residue is diluted with water (300 ml), acidified with sulfuric acid, and extracted with several portions of ether and chloroform. After having been concentrated at reduced pressure, the mixture is fractionated to give a center cut of bp 47–50°/13 torr which weighs 152.3 g (74% yield of pure methyl 2,5-hexadienoate).

TABULAR SURVEY

An attempt has been made to include in Tables III–VII all synthetic applications of π-allylnickel complexes, either as discrete reagents or as transient intermediates, that were reported through August 1971. The tables are separated and ordered according to the presentation in the discussion.

Table III is concerned with synthesis by ligand displacement on discrete bis-(π-allyl)nickel complexes and is organized according to the number of carbon atoms in the allyl ligand. Table IV contains examples of syntheses for which bis-(π-allyl)nickel complexes have been characterized or implicated as intermediates; nearly all of these reactions involve 1,3-butadiene and therefore the table is organized according to the number of carbon atoms in the coreactant. The first entries in Table IV are for 1,3-butadiene alone.

Table V is a compilation of synthetic applications of π-allylnickel halides as discrete reagents and is divided into three sections: reaction with non-allylic halides, reaction with allylic derivatives, and reaction with oxygen functional groups. Within each section the entries are organized according to the number of carbon atoms in the π-allyl ligand. Table VI is concerned with the synthesis of 1,5-hexadiene derivatives via π-allylnickel halides as transient intermediates; it is divided into two sections: intermolecular coupling and intramolecular coupling (cyclization) reactions. Within each section the entries are in order of the number of carbon atoms in the carbon chain containing the allyl system; carbon atoms in ester and other functional groups are not counted in assigning priority.

Table VII consists of two sections concerned with the carbonylation of allylic derivatives via transient π-allylnickel halide complexes. The first section contains synthetic applications of simple carbonyl insertion, while the second section has examples of simultaneous carbon monoxide and acetylene insertion. The entries for each section are listed in order of increasing number of carbon atoms in the carbon chain containing the allyl system; carbon atoms in ester and other functional groups are not counted in assigning priority.

TABLE III. LIGAND DISPLACEMENT REACTIONS WITH BIS-(π-ALLYL)NICKEL COMPLEXES AS DISCRETE REAGENTS

Complex	Ligand	Coupling Product(s) and Yield(s) (%)	Refs.
Bis-(π-allyl)nickel	$(C_6H_5)_3P$	1,5-Hexadiene (high)	21b
	CO	1,5-Hexadiene (high)	9
	1,3-Butadiene	1,5-Hexadiene (high)	21b
	1,5-Cyclooctadiene	1,5-Hexadiene (high)	21b
	Cyclooctatetraene	1,5-Hexadiene (high)	21b
Bis-(π-2-methylallyl)nickel	CO	2,5-Dimethyl-1,5-hexadiene (high)	9
Bis-(π-crotyl)nickel	CO ($-40°$)	trans,trans-Octa-2,6-diene (96), 3-methylhepta-1,5-diene (2)	9
	CO ($> -40°$)	trans,trans-Octa-2,6-diene (58), 3-methylhepta-1,5-diene (38), 3,4-dimethylhexa-1,5-diene (4)	9
Bis-(π-3,3-dimethylallyl)nickel	CO	2,7-Dimethyl-2,6-octadiene (60), carbonyl insertion products (40)	74
Bis-(π-2-carbethoxyallyl)nickel	CO	2,6-Dicarbethoxy-1,5-hexadiene (high)	74
	I_2	2,6-Dicarbethoxy-1,5-hexadiene (high)	74
	$(C_6H_5)_3P$	1,5,9-Cyclododecatriene (high)	21b
	CO	(high)	21b
Bis-(π-cyclooctenyl)nickel	CO	Bis-(2-cyclooctenyl) (50, 100), bis-(2-cyclooctenyl) ketone (50, 0)	74, 21b

Note: References 103–110 are on p. 198.

TABLE IV. REACTIONS OF THE ZERO-VALENT NICKEL CATALYST WITH A 1,3-DIENE, A COREACTANT, AND A PHOSPHORUS LIGAND

(The diene is 1,3-butadiene unless otherwise specified.)

Coreactant	Ligand	Conditions	Product(s) and Yield(s) (%)	Refs.
—	—	—	1,5,9-Cyclododecatriene (high) (*cis,trans,trans*; *cis,cis,trans*; and all *trans*)	21a
—	(*o*-$C_6H_5C_6H_4O$)$_3$P	—	1,5-Cyclooctadiene (96), 4-vinylcyclohexene (3)	21a
—	"	85% conversion	1,5-Cyclooctadiene (41), 4-vinylcyclohexene (3), 1,2-divinylcyclobutane (39),	21a
—	(C_6H_{11})$_3$P	—	1,5-Cyclooctadiene (58), 4-vinylcyclohexene (39), cyclododecatriene (14)	28
CH_2=CH_2	—	80°	1,*trans*-4,9-Decatriene (69), *cis,trans*-1,5-cyclododecadiene (31)	29
	—	0°	*cis,trans*-1,5-Cyclododecadiene (96), 1,*trans*-4,9-decatriene (4)	29
CH_2=CHCO$_2$CH$_3$	(*o*-$C_6H_5C_6H_4O$)$_3$P	60°	Methyl 3,7-cyclodecadiene-1-carboxylate (63)	103
	(C_6H_5O)$_3$P	20°	Dimethyl 9-vinyl-4,9-undecadiendioate (92)	103
CH_2=C(CH$_3$)CO$_2$CH$_3$	—	—	Methyl 2-methylene-5,10-undecadienoate (high)	103
C_6H_5CH=CH_2	—	—	1-Phenyldodecatriene isomers (21), 1-phenyl-*cis,trans*-3,7-cyclodecatriene (14)	104
	(*o*-$C_6H_5C_6H_4O$)$_3$P	—	1-Phenyldodecatriene isomers (75)	104

TABLE IV. REACTIONS OF THE ZERO-VALENT NICKEL CATALYST WITH A 1,3-DIENE, A COREACTANT, AND A PHOSPHORUS LIGAND (*Continued*)

(The diene is 1,3-butadiene unless otherwise specified.)

Coreactant	Ligand	Conditions	Product(s) and Yield(s) (%)	Refs.
$CH_3C\equiv CCH_3$	$(C_6H_5)_3P$	Excess diene	4,5-Dimethyl-*cis,cis,trans*-1,4,7-cyclodecatriene (85)	38d
	"	Equimolar	1,2,3,4-Tetramethyl-1,3,6,10-cyclodecatetraene (20)	34
1-Methoxy-2-pentyne	$(C_6H_5)_3P$	Excess diene	4-Methoxymethyl-5-ethyl-1,4,7-cyclodecatriene (high)	34
	"	Equimolar	Mixture of di(methoxymethyl)-diethylcyclododecatetraene isomers (20)	34
1,4-Dimethoxy-2-butyne	$(C_6H_5)_3P$	Excess diene	4,5-Dimethoxymethyl)-1,4,7-cyclodecatriene (high)	34
	"	Equimolar	1,2,3,4-Tetra(methoxymethyl)-1,3,6,10-cyclododecatetraene (20)	34
Piperylene	$(o\text{-}C_6H_5C_6H_4O)_3P$	Equimolar	3-Methyl-1,5-cyclooctadiene (86)	105
Isoprene	$(o\text{-}C_6H_5C_6H_4O)_3P$	Equimolar	1-Methyl-1,5-cyclooctadiene (84)	105
1,3-Dimethyl-1,3-butadiene	$(o\text{-}C_6H_5C_6H_4O)_3P$	Equimolar	1,3-Dimethyl-1,5-cyclooctadiene (85)	106
2,3-Dimethyl-1,3-butadiene	$(o\text{-}C_6H_5C_6H_4O)_3P$	Equimolar	1,2-Dimethyl-1,5-cyclooctadiene (92)	104
1,3-Hexadiene	$(o\text{-}C_6H_5C_6H_4O)_3P$	Equimolar	3-Ethyl-1,5-cyclooctadiene (82)	106
2-Ethyl-1,3-butadiene	$(o\text{-}C_6H_5C_6H_4O)_3P$	Equimolar	1-Ethyl-1,5-cyclooctadiene (92)	106
Methyl 2,4-hexadienoate	$(o\text{-}C_6H_5C_6H_4O)_3P$	Equimolar	Methyl 4-methyl-2,5-cyclooctadiene-1-carboxylate (90)	103
1,3-*cis*-6-Octatriene	$(o\text{-}C_6H_5C_6H_4O)_3P$	Equimolar	2-(*cis*-2-Buten-1-yl)-1,5-cyclooctadiene (94)	106

Methyl 2,4-hexadieneoate and ethylene	$(o\text{-}C_6H_5C_6H_4O)_3P$	Equimolar	Methyl 5-methyl-2,7,10-undecatrieneoate (80)	4, 103
Morpholine	$(C_2H_5O)_3P$	—	cis,trans- and trans,trans-1,3,6-Octatriene (90), 1-morpholino-2,6-octatriene (75)[a]	40
[b]	$(o\text{-}C_6H_5C_6H_4O)_3P$	—	Dimethyl-1,5-cyclooctadiene isomers (91)	104
[c]	$(o\text{-}C_6H_5C_6H_4O)_3P$	—	Dimethyl-1,5-cyclooctadiene isomers (55)	104
[d]	$(o\text{-}C_6H_5C_6H_4O)_3P$	—	Tetramethyl-4-vinylcyclohexene isomers (86)	104

Note: References 103–110 are on p. 198.

[a] This yield is based on unrecovered amine.
[b] The diene in this experiment was piperylene.
[c] The diene in this experiment was isoprene.
[d] The diene in this experiment was 2,3-dimethyl-1,3-butadiene.

TABLE V. Reactions of π-Allylnickel Halides as Discrete Reagents

A. With Nonallylic Halides

π-Allyl Ligand, Halide	Coreactant	Product(s) and Yield(s) (%)	Refs.
π-2-Methylallyl, Br	CH_3Br	$CH_3CH_2C(CH_3)=CH_2$ (90)	7
	CH_3I	$CH_3CH_2C(CH_3)=CH_2$ (90)	7
	$CH_2=CHBr$	$CH_2=CHCH_2C(CH_3)=CH_2$ (70)	7
	$BrCH_2CH_2Br$	$CH_2=CH_2$, C_8 (100)	8
	CH_3SCH_2Cl	$CH_3SCH_2CH_2C(CH_3)=CH_2$ (20)	8
	▷—Br	C_6 (5%); C_7 (21%); C_8 (24%)[a] (50, total)	8
	CH_3COCH_2Cl	$CH_3COCH_2CH_2C(CH_3)=CH_2$ (46)	8
	$(CH_3)_3CBr$	$(CH_3)_3CCH_2C(CH_3)=CH_2$ (25)	7, 8
	$(CH_3)_3CCl$	$(CH_3)_3CCH_2C(CH_3)=CH_2$ (20)	7, 8
	▷—CH_2Br	$CH_2=CHCH_2CH_2CH_2C(CH_3)=CH_2$ (39)	8
	p-BrC_6H_4Br	(aromatic bis-methylallyl product) (97)	7
	C_6H_5I	$C_6H_5CH_2C(CH_3)=CH_2$ (98)	7
	$C_6H_{11}I$	$C_6H_{11}CH_2C(CH_3)=CH_2$ (91)	7
	I—⬡—OH (trans)	$CH_2=C(CH_3)CH_2$—⬡—OH (89)[b]	7
	$I(CH_2)_6I$	$CH_2=C(CH_3)(CH_2)_8C(CH_3)=CH_2$ (95)	7
	$C_6H_5CH_2Br$	$C_6H_5CH_2CH_2C(CH_3)=CH_2$ (91)	7
	$C_6H_5OCH_2Cl$	$C_6H_5OCH_2CH_2C(CH_3)=CH_2$ (50)	8
	$C_6H_5SCH_2Cl$	$C_6H_5SCH_2CH_2C(CH_3)=CH_2$ (24)	8

π-2-Carbethoxyallyl, Br	p-BrC$_6$H$_4$COCH$_2$Br	p-BrC$_6$H$_4$COCH$_2$CH$_2$CH$_2$C(CH$_3$)=CH$_2$	(75)	8
	C$_6$H$_5$CH$_2$CH$_2$CH$_2$Br	C$_6$H$_5$(CH$_2$)$_4$C(CH$_3$)=CH$_2$	(92)	7
	Cl(CH$_2$)$_3$I	Cl(CH$_2$)$_4$C(CO$_2$C$_2$H$_5$)=CH$_2$	(96)	8
π-1,1-Dimethylallyl, Br	CH$_3$I	CH$_3$CH$_2$CH=C(CH$_3$)$_2$	(90)	7, 8
	[bicyclic structure with I]	[structure]	(88)	7, 8
		α-Santalene		
	[bicyclic structure with exocyclic CH$_2$ and CH$_2$I]	[structure]	(88)	7, 8
		Epi-β-santalene		
	C$_6$H$_{11}$I	Geranylcyclohexane (cis, 24; $trans$, 36)		7, 8
π-Geranyl, Br	CH$_3$I	[steroid structures with OAc]	(35, total)	107
[steroid structure with OAc, Cl]				

Note: References 103–110 are on p. 198.

[a] C$_6$ is 1,5-hexadiene, C$_7$ is 2-methyl-1,5-hexadiene, C$_8$ is 2,5-dimethyl-1,5-hexadiene.
[b] The product is a mixture of epimers, 38:62.

TABLE V. Reactions of π-Allylnickel Halides as Discrete Reagents (*Continued*)

π-Allyl Ligand, Halide	Coreactant	Product(s) and Yield(s) (%)	Refs.
π-Geranyl, Br (*contd.*)			

A. *With Nonallylic Halides*

| | CH₃I | (good) [steroid structure with CH₃O and ketone] | 107 |

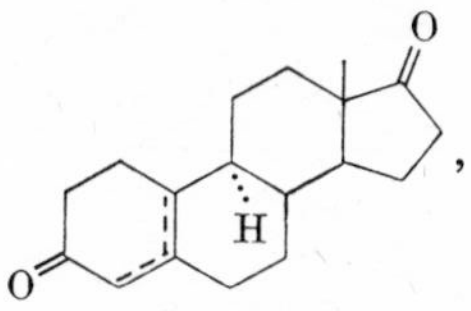

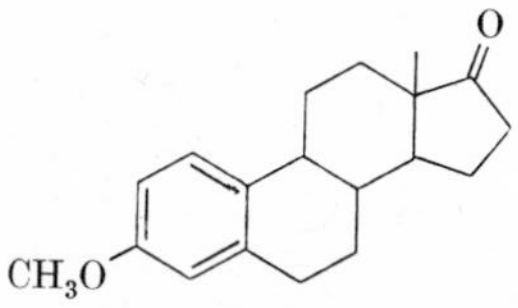

B. *With Allylic Halides*

π-Allyl, Br	CH₂=C(CH₃)CH₂Cl	C₆ (40%); C₇ (19%); C₈ (41%)[c] (95%)[d]	8
Br	CH₂=C(CH₃)CH₂Br	C₆ (35%); C₇ (24%); C₈ (41%)[c] (95%)[d]	8
I	CH₂=C(CH₃)CH₂Cl	C₆ (38%); C₇ (27%); C₈ (37%)[c] (79%)[d]	8
I	CH₂=C(CH₃)CH₂I	C₆ (33%); C₇ (37%); C₈ (30%)[c] (79%)[d]	8
	CH₂=CHCH₂Br	C₆ (22%); C₇ (53%); C₈ (25%)[c] (100%)[d]	62
	CH₂=CHCH₂OCOCH₃	C₆ (26%); C₇ (43%); C₈ (31%)[c] (34%)[d]	8
	CH₂=CHCH₂SCOCH₃	C₆ (3%); C₇ (39%); C₈ (58%)[c] (27%)[d]	8
	CH₂=CHCH₂OSO₂N(CH₃)₂	C₆ (13%); C₇ (51%); C₈ (36%)[c] (55%)[d]	8
	CH₂=CHCH₂SCN(C₂H₅)₂ (S=)	C₆ (0%); C₇ (70%); C₈ (30%)[c] (40%)[d]	8
	CH₂=CHCH₂SCN (S=, pyrrolidine)	C₆ (0%); C₇ (84%); C₈ (16%)[c] (52%)[d]	8
π-2-Methylallyl, Br	CH₂=CHCH₂OSO₂C₆H₄CH₃-*p*	C₆ (29%); C₇ (46%); C₈ (27%)[c] (95%)[d]	8
	CH₂=CHCH₂SCSC₆H₅	C₆ (8%); C₇ (45%); C₈ (47%)[c] (20%)[d]	8

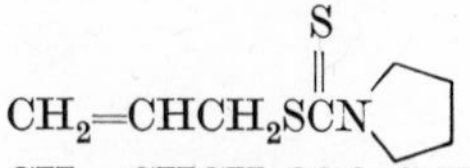

π-2-Carbethoxyallyl, Br	ClCH$_2$C(CH$_2$Cl)=CH$_2$	EtO$_2$C—[diene structure]—CO$_2$Et (15–25)	8, 79
π-Geranyl, Br	CH$_2$=CHC(CH$_2$Br)=CH$_2$	β-Farnesene (cis, 11; trans, 9)	62, 8
	[thiirane-S structure with CH$_2$SCN] CH$_2$=CHC=CH$_2$	β-Farnesene (cis, 18; trans, 28)	62, 8

C. With Oxygen Functional Groups

π-Allyl, Br	CH$_2$=CHCO$_2$CH$_3$	CH$_2$=CHCH$_2$CH$_2$CH$_2$CO$_2$CH$_3$ (—) CH$_2$=CHCH$_2$CH=CHCO$_2$CH$_3$ (—)	49
π-2-Methylallyl, Br	CH$_2$=CHCHO	CH$_2$=CHCH(OH)CH$_2$C(CH$_3$)=CH$_2$ (80)	7
	[cyclopentanone] C=O	[cyclopentyl]C(OH)CH$_2$CH(CH$_3$)=CH$_2$ (50)	7, 8
	C$_6$H$_5$CHO	C$_6$H$_5$CH(OH)CH$_2$C(CH$_3$)=CH$_2$ (87)	7, 8
	C$_6$H$_5$C—CH$_2$ (epoxide) O C$_6$H$_5$CO$_2$CH$_3$	C$_6$H$_5$CH(CH$_2$OH)CH$_2$C(CH$_3$)=CH$_2$ (60) No reaction	7, 8 8

Note: References 103–110 are on p. 198.

^{c} C$_6$ is 1,5-hexadiene, C$_7$ is 2-methyl-1,5-hexadiene, and C$_8$ is 2,5-dimethyl-1,5-hexadiene; the percentages indicate the proportion of each product, not the absolute yield.

^{a} This entry is the combined yield, C$_6$ + C$_7$ + C$_8$.

TABLE VI. COUPLING REACTIONS OF ALLYLIC COMPOUNDS WITH NICKEL CARBONYL

Allylic Reactant	Product(s) and Yield(s) (%)	Refs.
A. Intermolecular		
CH_2=$CHCH_2OCOCH_3$	CH_2=$CHCH_2CH_2CH$=CH_2 (50)	75
CH_2=$CHCHClCH_3$	$(CH_3CH$=$CHCH_2)_2$,	71
	CH_3CH=$CHCH_2CH(CH_3)CH$=CH_2 (78:22; total, 74)	
CH_3CH=$CHCH_2Cl$	CH_3CH=$CHCH_2CH(CH_3)CH$=CH_2,	71
	$(CH_3CH$=$CHCH_2)_2$ (79:21; total, 74)	
$NCCH$=$CHCH_2Cl$	$(NCCH$=$CHCH_2)_2$	108
	(*cis,cis*, 64; *trans,trans*, 30)	
CH_2=$C(CO_2C_2H_5)CH_2Br$	$[CH_2$=$C(CO_2C_2H_5)CH_2]_2$ (74)	79
CH_3O_2CCH=$CHCH_2Cl$	$(CH_3O_2CCH$=$CHCH_2)_2$	108
(*trans*)	(*trans,trans*, 38; *cis,trans*, 62)	
$(CH_3)_2C$=$CHCH_2Cl$	$[(CH_3)_2C$=$CHCH_2]_2$,	71
	$(CH_3)_2C$=$CHCH_2C(CH_3)_2CH$=CH_2 (67:33)	
$(CH_3)_2CClCH$=CH_2	$((CH_3)_2C$=$CHCH_2)_2$,	71
	$(CH_3)_2C$=$CHC(CH_3)_2CH$=CH_2 (62:38)	
$CH_3OCH_2CH_2CH$=$CHCH_2Cl$	$[CH_3O(CH_2)_2CH$=$CHCH_2]_2$ (main)	71
$CH_3OCH_2CH_2CHClCH$=CH_2	$[CH_3O(CH_2)_2CH$=$CHCH_2]_2$ (100)	71, 73
C_6H_5CH=$CHCH_2OCOCH_3$	$(C_6H_5CH$=$CHCH_2)_2$ (30)	75
B. Intramolecular		
[structure: 1,2-bis(chloromethyl) with =CH2]	[structure: cyclic triene] (62)	79

Starting material	Product(s) (yield %)	Reference
Cl–CH₂–C(=CH₂)–(CH₂)₂–C(=CH₂)–CH₂–Cl	1,4-bis(methylene)cyclohexane (62)[a]	8, 79
BrCH₂CH=CH(CH₂)₂CH=CHCH₂Br (mixture of isomers)	3-vinylcyclohexene (42)	10
BrCH₂CH=CH(CH₂)₄CH=CHCH₂Br (*cis,cis*)	1,2-divinylcyclohexane + 1,2-divinylcyclohexane (54, total)	10
BrCH₂CH=CH(CH₂)₄CH=CHCH₂Br (*trans,trans*)	1,2-divinylcyclohexane + 1,2-divinylcyclohexane (40, total)	10
(bromo diene methyl ester structure)	cyclohexane-CO₂CH₃ with isopropenyl and methyl (32) + divinylcyclohexane-CO₂CH₃ (23) + macrocyclic diene ester (11)	108a

Note: References 103–110 are on p. 198.
[a] The nickel reagent was triphenylphosphinenickel tricarbonyl.

TABLE VI. COUPLING REACTIONS OF ALLYLIC COMPOUNDS WITH NICKEL CARBONYL (*Continued*)

Allylic Reactant	Product(s) and Yield(s) (%)	Refs.
	B. Intramolecular (Continued)	
BrCH$_2$CH=CH(CH$_2$)$_6$CH=CHCH$_2$Br (*cis,cis*)	(59)	10
BrCH$_2$CH=CH(CH$_2$)$_6$CH=CHCH$_2$Br (*trans,trans*)	(57)	10
	(64)	79
BrCH$_2$CH=CH(CH$_2$)$_8$CH=CHCH$_2$Br (*cis,cis*)	(70)	10
BrCH$_2$CH=CH(CH$_2$)$_8$CH=CHCH$_2$Br (*trans,trans*)	(74)	10
	+ *cis,trans,trans* (68, total)	77, 78
	(5)	78

(structure)	(5)	(structure)	6, 78
Humulene			
(structure)	(10)	(structure)	6, 78
BrCH$_2$CH=CH(CH$_2$)$_{12}$CH=CHCH$_2$Br (cis,cis)		(structure) (84)	10
BrCH$_2$CH=CH(CH$_2$)$_{12}$CH=CHCH$_2$Br (trans,trans)		(structure) (70)	10

Note: References 103–110 are on p. 198.

TABLE VII. CARBONYLATION OF ALLYLIC DERIVATIVES WITH NICKEL CARBONYL

Allylic Reactant	Conditions	Product(s) and Yield(s) (%)	Ref.
A. With Carbon Monoxide in the Absence of Alkynes			
$CH_2=CHCH_2Cl$	CH_3OH	$CH_2=CHCH_2CO_2CH_3$	47
	$H_2O—CH_3OH$	$CH_2=CHCH_2CO_2H$	47
$CH_3CH=CHCH_2Cl$	CH_3OH	$CH_3CH=CHCH_2CO_2CH_3$	47
	$H_2O—CH_3OH$	$CH_3CH=CHCH_2CO_2H$	47
$NCCH_2CH=CHCH_2Cl$	CH_3OH	$NCCH_2CH=CHCH_2CO_2CH_3$	47
	$H_2O—CH_3OH$	$NCCH_2CH=CHCH_2CO_2H$	47
	Aprotic solvent	$NCCH_2CH=CHCH_2COCl$	47
B. With Carbon Monoxide in the Presence of Acetylene[a]			
$CH_2=CHCH_2Cl$		$CH_2=CHCH_2CH=CHCO_2CH_3$ (70)	47
		$CH_2=CHCH_2CH=CHCO_2CH_3$ (42)[b]	88
$CH_2=CHCH_2OCH_3$		$CH_2=CHCH_2CH=CHCO_2CH_3$ (84)[c]	87a
$CH_2=CHCH_2OCOCH_3$		$CH_2=CHCH_2CH=CHCO_2CH_3$ (58)[c]	87a
$CH_2=CHCH_2OC_4H_9\text{-}n$		$CH_2=CHCH_2CH=CHCO_2CH_3$ (84)[c]	87a
$CH_3CH=CHCH_2Cl$		$CH_3CH=CHCH_2CH=CHCO_2CH_3$ (81)	47
		$CH_3CH=CHCH_2CH=CHCO_2CH_3$ (70)[b]	88
$CH_3CH=CHCH_2OCH_3$		$CH_3CH=CHCH_2CH=CHCO_2CH_3$ (55)[c]	87a
$CH_3CH=CHCH_2OCOCH_3$		$CH_3CH=CHCH_2CH=CHCO_2CH_3$ (51)[c]	87a
$CH_2=CHCH(CH_3)Cl$		$CH_3CH=CHCH_2CH=CHCO_2CH_3$ (80)	47
$CH_2=C(CH_3)CH_2Cl$		$CH_2=C(CH_3)CH_2CH=CHCO_2CH_3$ (43)[b]	47
		$CH_2=C(CH_3)CH_2CH=CHCO_2CH_3$ (62)[d]	88
		$CH_2=C(CH_3)CH_2CH=CHCO_2CH_3$ (68)[c]	88
$CH_2=C(CH_3)CH_2OCH_3$		$CH_2=C(CH_3)CH_2CH=CHCO_2CH_3$ (40)	87a
$CH_3O_2CCH=CHCH_2Cl$		$CH_3O_2CCH=CHCH_2CH=CHCO_2CH_3$ (—)	47
$ClCH_2CH=CHCH_2Cl$ (*cis*)		$CH_2=CHCH=CH_2$ (—)	47
$ClCH_2CH=CHCH_2OCOCH_3$ (*cis*)		$CH_3O_2CCH_2CH=CHCH_2CH=CHCO_2CH_3$ (*cis,cis*) (—)	47
$ClCH_2CH=CHCH_2OH$		$(CH_3O_2CCH=CHCH_2CH=)_2$ (—)	87b

HOCH$_2$CH=CHCH$_2$OH (cis)	(CH$_3$O$_2$CCH=CHCH$_2$CH=)$_2$ (34)	87b
	HOCH$_2$CH=CHCH$_2$CH=CHCO$_2$CH$_3$	
CH$_3$CO$_2$CH$_2$CH=CHCH$_2$OCOCH$_3$	(CH$_3$O$_2$CCH=CHCH$_2$CH=)$_2$ (65)	87b
	HOCH$_2$CH=CHCH$_2$CH=CHCO$_2$CH$_3$ (70–77)	87b
CH$_2$—CHCH=CH$_2$	HOCH$_2$CH=CHCH$_2$CH=CHCO$_2$CH$_3$ (50)	87b
_O_		
HOCH$_2$CHClCH=CH$_2$	CH$_3$CH=CHCH(CH$_3$)CH=CHCO$_2$CH$_3$ (40)	47
CH$_3$CH=CHCH(CH$_3$)Cl	CH$_3$C(CH$_3$)=CHCH$_2$CH=CHCO$_2$CH$_3$ (35)	47
CH$_3$C(CH$_3$)=CHCH$_2$Cl	CH$_3$O$_2$CCH=CHCH$_2$CH=CHCO$_2$CH$_3$ (32)	47
CH$_3$O$_2$CCH$_2$CH=C(CH$_3$)CH$_2$Cl	CH$_3$O$_2$CCH=C(CH$_3$)CH$_2$CH=CHCO$_2$CH$_3$ (45)	47
NCCH$_2$CH=CHCH$_2$Cl	NCCH$_2$CH=CHCH$_2$CH=CHCO$_2$CH$_3$ (79)	47
	NCCH$_2$CH=CHCH$_2$CH=CHCOCl (—)	109
CH$_3$CO$_2$CH$_2$CH=CHCH$_2$Cl	CH$_3$CO$_2$CH$_2$CH=CHCH$_2$CH=CHCO$_2$CH$_3$ (35)	47
2-Cyclopentenyl bromide	Methyl 3-(cyclopent-2-en-1-yl)prop-2-enoate (50)	47
CH$_3$(CH$_2$)$_2$CH=CHCH$_2$Cl	CH$_3$(CH$_2$)$_2$CH=CHCH$_2$CH=CHCO$_2$CH$_3$ (62)	47
2-Cyclohexenyl bromide	Methyl 3-(cyclohex-2-en-1-yl)prop-2-enoate (50)	47
CH$_3$O$_2$CCH=CHCH$_2$CH=CHCH$_2$Cl	CH$_3$O$_2$CCH=CHCH$_2$CH=CHCH$_2$CH=CHCO$_2$CH$_3$ (80)	47
CH$_3$CO$_2$CH$_2$CH=CHCH$_2$CH=CHCH$_2$Cl (cis,cis)	(CH$_3$O$_2$CCH=CHCH$_2$CH=)$_2$ (—) (all cis)	47
CH$_3$(CH$_2$)$_4$CH=CHCH$_2$Cl	CH$_3$(CH$_2$)$_4$CH=CHCH$_2$CH=CHCO$_2$CH$_3$ (73)	47
(CH$_3$)$_3$CCH$_2$CH=CHCH$_2$Cl	(CH$_3$)$_3$CCH$_2$CH=CHCH$_2$CH=CHCO$_2$CH$_3$ (47)[b]	88
CH$_3$C(CH$_3$)$_2$CH$_2$CH=CHCH$_2$Cl	CH$_3$C(CH$_3$)$_2$CH$_2$CH=CHCH$_2$CH=CHCO$_2$CH$_3$ (55)	47
C$_6$H$_5$CH=CHCH$_2$Cl	C$_6$H$_5$CH=CHCH$_2$CH=CHCO$_2$CH$_3$ (54)	47
C$_6$H$_5$CH=CHCH$_2$NR$_3$Cl$^-$	C$_6$H$_5$CH=CHCH$_2$CH=CHCO$_2$CH$_3$ (—)	110
CH$_3$(CH$_2$)$_6$CH=CHCH$_2$Cl	CH$_3$(CH$_2$)$_6$CH=CHCH$_2$CH=CHCO$_2$CH$_3$ (64)	47
CH$_3$O$_2$C(CH=CHCH$_2$)$_3$Cl	CH$_3$O$_2$C(CH=CHCH$_2$)$_3$CH=CHCO$_2$CH$_3$ (19)	47
CH$_3$(CH$_2$)$_{14}$CH=CHCH$_2$Cl	CH$_3$(CH$_2$)$_{14}$CH=CHCH$_2$CH=CHCO$_2$CH$_3$ (48)	47

Note: References 103–110 are on p. 198.

[a] In Part B all products have the *cis* configuration about the α, β positions.
[b] The catalyst is Raney nickel.
[c] Hydrochloric acid is added.
[d] The catalyst is a mixture of iron alloy powder and nickel chloride.

REFERENCES TO TABLES III–VII

[103] C. Delliehausen, Dissertation, University of Bochum, 1968. See also reference 4.
[104] Unpublished results cited in ref. 4, p. 75.
[105] Unpublished results cited in ref. 4, p. 69.
[106] Unpublished results cited in ref. 4, p. 73.
[107] I. T. Harrison, E. Kimura, E. Bohme, and J. H. Fried, *Tetrahedron Lett.*, **1969**, 1589.
[108] G. P. Chiusoli and G. Cometti, *Chim. Ind.* (Milan), **45**, 404 (1963)
[108a] E. J. Corey and E. A. Broger, *Tetrahedron Lett.*, **1969**, 1779.
[109] G. P. Chiusoli, *Chim. Ind.* (Milan), **41**, 503 (1959).
[110] G. P. Chiusoli, *Atti Accad. Naz. Lincei, Rend., Cl. Sci. Fis. Mat. Nat.*, **26**, 790 (1959) [*C.A.*, **54**, 8709g (1960)].

CHAPTER 3

THE THIELE-WINTER ACETOXYLATION OF QUINONES

J. F. W. McOmie

School of Chemistry, University of Bristol, England

AND

J. M. Blatchly

Charterhouse, Godalming, Surrey, England

CONTENTS

	PAGE
Introduction	200
Mechanism	203
Scope and Limitations	206
General	206
Monosubstituted 1,4-Benzoquinones	211
Disubstituted 1,4-Benzoquinones	212
Trisubstituted 1,4-Benzoquinones	213
1,2-Quinones	213
1,4-Naphthoquinones	214
t-Butylquinones and Methylquinones	215
Use of Anhydrides Other than Acetic Anhydride	217
Related Reactions	218
Experimental Conditions	219
Experimental Procedures	221
Thiele-Winter Acetoxylations	221
1,2,4-Triacetoxybenzene	221
1.2.4-Triacetoxy-3-bromo-5-methoxybenzene	221

1,3,4-Triacetoxy-2-methylnaphthalene 221
1,2,3-Triacetoxy-5-*t*-butylbenzene 222
1,2,4-Triacetoxynaphthalene 222
Conversion of Triacetates into Related Compounds 222
2′,3′,5′-Trihydroxyterphenyl 222
2,4,5-Trimethoxybiphenyl 222
2-Hydroxy-1,4-naphthoquinone 222
5-Methoxy-2-phenyl-1,4-benzoquinone 223

TABULAR SURVEY 223

Table I. 1,4-Benzoquinone and Monosubstituted 1,4-Benzoquinones . 224
Table II. 2,3-Disubstituted 1,4-Benzoquinones 228
Table III. 2,5-Disubstituted 1,4-Benzoquinones 230
Table IV. 2,6-Disubstituted 1,4-Benzoquinones 235
Table V. Trisubstituted 1,4-Benzoquinones 238
Table VI. 1,4-Benzoquinones with One or More Electronegative Substituents 243
Table VIIA. 1,2-Benzoquinones 245
Table VIIB. Other 1,2-Quinones, Excluding 1,2-Naphthoquinones . . 248
Table VIIIA. 1,4-Naphthoquinones Unsubstituted in the Quinonoid Ring . 252
Table VIIIB. 2-Substituted 1,4-Naphthoquinones 256
Table VIIIC. 1,2- and 2,6-Naphthoquinones 259
Table IX. Anthraquinones 261
Table XA. Diphenoquinones 265
Table XB. Stilbenoquinones 266
Table XI. Miscellaneous Quinones 267
Table XII. Reaction of Quinones with Other Anhydrides and with Diacetyl Sulfide 269
Table XIII. Reaction of Quinones with Acid Chlorides and Acid Bromides 271

REFERENCES TO TABLES 276

INTRODUCTION

In a short paper, entitled "Ueber die Einwirkung von Essigsäureanhydrid auf Chinon und auf Dibenzoylstyrol" published in 1898, Johannes Thiele[1] described the reaction of *p*-benzoquinone with acetic anhydride in the presence of a small quantity of concentrated sulfuric acid to give 1,2,4-triacetoxybenzene in 80% yield.[2] He pointed out that the reaction

[1] For a biography of Johannes Thiele (1865–1918) with portrait and references, see F. Straus, *Ber.*, **60**, 75–132 (1927).
[2] J. Thiele, *Ber.*, **31**, 1247 (1898).

was completely analogous to the formation of chlorohydroquinone diacetate from acetyl chloride and p-benzoquinone[3] and to the formation of 2-benzenesulfonylhydroquinone from benzenesulfinic acid and p-benzoquinone.[4] He also mentioned that, if the sulfuric acid was replaced by zinc chloride, then chlorohydroquinone diacetate was formed. These studies constituted a small part of an extensive investigation by Thiele and his students on addition reactions of unsaturated compounds. Two years later, Thiele and Ernst Winter described further experiments in a paper entitled "Ueber die Einwirkung von Essigäureanhydrid und Schwefelsäure auf Chinone."[5] They showed that both 1,2- and 1,4-naphthoquinone gave 1,2,4-triacetoxynaphthalene (Eq. 1) and moreover that the same product was obtained when the catalytic amount of sulfuric acid was replaced by

(Eq. 1)

zinc chloride used in approximately the same molecular amount as the quinone. They found that toluquinone readily gave 5-methyl-1,2,4-triacetoxybenzene and 5-hydroxytoluquinone gave a tetraacetate whose orientation as 2,3,4,6-tetraacetoxytoluene was not established until 1967.[6] In contrast, 5,8-dihydroxy-1,4-naphthoquinone (naphthazarin) readily gave a diacetate which reacted only very slowly (several weeks) to give a pentaacetoxy naphthalene.

Many chemists used the reaction as an incidental part of their work, but Erdtman in 1934 was the first to use it systematically in order to study the reactivity of quinones.[7] Sulfuric acid was always used as the catalyst until 1948, when Fieser found that boron trifluoride etherate would bring about the reaction,[8] and a few years later Burton and Praill showed that perchloric acid is probably the most effective catalyst.[9]

The reaction has been carried out on a large number of quinones, but until now there has been no review of it. From a synthetic point of view the most important feature is the introduction of an oxygen atom into an aromatic nucleus, and for this reason we prefer to name the reaction the

[3] H. Schulz, *Ber.*, **15**, 652 (1882).
[4] O. Hinsberg, *Ber.*, **27**, 3259 (1894).
[5] J. Thiele and E. Winter, *Ann.*, **311**, 341 (1900).
[6] H. S. Wilgus and J. W. Gates, *Can. J. Chem.*, **45**, 1975 (1967).
[7] H. Erdtman, *Proc. Roy. Soc., Ser. A*, **143**, 177 (1934).
[8] L. F. Fieser, *J. Amer. Chem. Soc.*, **70**, 3165 (1948).
[9] H. Burton and P. F. G. Praill, *J. Chem. Soc.*, **1952**, 755.

Thiele-Winter *acetoxylation* of quinones rather than the more commonly used expression Thiele or Thiele-Winter *acetylation*. The series of reactions, illustrated in Eq. 2 with *p*-benzoquinone, constitutes a valuable procedure for the preparation of trihydroxyaromatics from hydroquinones or

$$\text{hydroquinone} \rightarrow \text{p-benzoquinone} \rightarrow \text{1,2,4-triacetoxybenzene} \rightarrow \text{1,2,4-trihydroxybenzene} \quad \text{(Eq. 2)}$$

catechols. *Para*-quinones give 1,2,4-triacetoxy compounds, whereas *ortho*-quinones can give either 1,2,3- or 1,2,4-triacetoxy products. However, the acid-catalyzed reaction of quinones with acetic anhydride does not invariably yield the acetate of a trihydric phenol, *e.g.*, side-chain, instead of nuclear, acetoxylation sometimes occurs. We therefore suggest that the Thiele-Winter acetoxylation reaction be defined as the acid-catalyzed reaction of quinones with acetic anhydride to give triacetoxyaromatic compounds in which two of the acetoxy groups are derived from the two quinonoid oxygen atoms and one is inserted during the reaction. Tables I–XI include all instances known to us where a quinone has been treated with acetic anhydride and an acid catalyst.

Very little work has been done on the reaction of quinones with anhydrides other than acetic anhydride. The few results obtained have been summarized in Table XII. Acetyl sulfide and acetyl chloride may be regarded as mixed anhydrides of acetic acid with thioacetic acid and with hydrochloric acid, respectively. Although the reactions of these mixed anhydrides with quinones do not fit into this definition, they are closely related to the Thiele-Winter acetoxylation reaction and they are included in the discussion and summarized in Tables XII and XIII.

The reaction of quinones with acetic anhydride in the presence of sodium acetate proceeds quite differently from the acetoxylation reaction. For example, 2,6-dimethoxyquinone does not undergo the Thiele-Winter reaction but, when an alkaline catalyst is used, the diacetoxy acid **1** is the main product and the diacetates **2** and **3** are minor products.[10] This type of

1: 2,6-dimethoxy-4-acetoxyphenyl-AcOCHCO$_2$H
2: 2,6-dimethoxy-4-acetoxyphenyl-OAc
3: 2,6-dimethoxy-4-acetoxyphenyl-CH$_2$OAc

[10] M. Lounasmaa, *Acta Chem. Scand.*, **22**, 70 (1968).

reaction has been extensively studied by Lounasmaa who has also suggested plausible reaction mechanisms.[10,11] The reaction is not discussed further in this chapter.

MECHANISM

The Thiele-Winter reaction may be represented in the simplest case by Eq. 3. A study of the kinetics of the reaction with p-benzoquinone and

$$\text{[p-benzoquinone]} + 2\text{Ac}_2\text{O} \longrightarrow \text{[1,2,4-triacetoxybenzene]} + \text{AcOH} \qquad \text{(Eq. 3)}$$

p-toluquinone, catalyzed by perchloric acid, has been made by Mackenzie and E. R. S. Winter.[12] They proposed the mechanism shown in Scheme 1.

SCHEME 1. R = H or CH$_3$
$$\text{Ac}_2\text{O} + \text{H}^+ \rightleftharpoons \text{Ac}^+ + \text{AcOH}$$

During the reaction acetic acid is produced, and it is probable that protonated acetic acid can catalyze the reaction, though probably not so effectively as the acetylium cation. This is borne out by the results of kinetic studies made with mixtures of acetic acid and acetic anhydride. When the catalyst is protonated acetic acid, the reaction may proceed normally, as in Eqs. 4 and 5, or the intermediate acetoxyhydroquinone may be oxidized by a molecule of the original quinone to give an acetoxyquinone (Eq. 6) which, when R = H or CH$_3$, can undergo a further

[11] M. Lounasmaa, *Acta Chem. Scand.*, **22**, 3191 (1968).
[12] H. A. E. Mackenzie and E. R. S. Winter, *Trans. Faraday Soc.*, **44**, 159, 171, 243 (1948).

Thiele-Winter reaction. In this way a mixture of di- and tetra-acetates may be produced in addition to the triacetate. For example, in the reaction of p-benzoquinone catalyzed by perchloric acid, 100% acetic anhydride gave

(Eq. 4)

(Eq. 5)

(Eq. 6)

more than 90% of 1,2,4-triacetoxybenzene, whereas 50:50 v/v acetic anhydride-acetic acid gave nearly 70% of triacetate accompanied by 17% each of hydroquinone diacetate and 1,2,4,5-tetraacetoxybenzene.[12]

The mechanisms proposed by Mackenzie and Winter are acceptable as far as they go. However, they do not enable one to predict whether a given quinone will undergo the reaction, nor do they enable one to predict the structure of the product when more than one product is possible.*

Schweizer has made some molecular orbital calculations on polycyclic quinones.[14,15] A consideration of the localization energy for the various carbon atoms and for the carbonyl oxygen atoms gives a quantitative measure of the reactivity of each position in the quinone toward nucleophilic reagents (e.g., amines) and in acid-catalyzed reactions (e.g., the Thiele-Winter reaction and the addition of hydrogen halides). For the quinones which have been studied experimentally, the degree of reactivity and the structure of the products are in agreement with those predicted. As

---

* Fieser has rationalized a few of the known results, but it is very difficult to apply his rationalizations to other quinones.[13] His statements that 2-methyl-1,4-naphthoquinone, 2,6-dichloro- and 2,3,5-trimethyl-1,4-benzoquinone do not undergo the Thiele–Winter reaction are misleading, for these quinones do react under appropriate conditions.

[13] L. F. Fieser and M. Fieser, *Advanced Organic Chemistry*, Reinhold, New York, 1961, p. 855.
[14] H. Hopff and H. R. Schweizer, *Helv. Chim. Acta*, **45**, 312, 1044 (1962).
[15] H. R. Schweizer, *Helv. Chim. Acta*, **45**, 1934 (1962).

far as the authors are aware, no calculations have been made on the reactivity of quinones substituted with alkyl, halogen, or methoxy groups.

The reactivity of a given quinone depends partly on the catalyst and partly upon steric and electronic effects within the molecule. The small amount of experimental evidence indicates that different catalysts exert a great effect on the rate of the reaction and usually a much smaller effect on the composition of the products obtained. The effect and choice of catalyst are dealt with in the section on "Experimental Conditions." Again, there is little evidence concerning the effect of steric hindrance but, since the inserted acetoxy group can enter next to large substituents such as iodine,[16] phenyl,[17] or cyclohexyl,[18] it appears that the electronic effect must be the main factor influencing the orientation of the product in quinones where more than one isomeric product is possible. The t-butyl group appears to be an exception to the preceding statement since an acetoxy group is rarely inserted next to it; instead, if the only vacant positions are adjacent to a t-butyl group, the t-butyl group is displaced. The behavior of quinones containing one or more t-butyl groups is described on p. 215.

When two different alkyl groups are present in a benzoquinone, their electronic effects are almost the same and either a single product is obtained or a mixture of the two possible isomers is formed, the proportions of which are probably determined mainly by steric effects. For example, 2-t-butyl-5-methyl-1,4-benzoquinone gives only the less hindered triacetate (Eq. 7),[19] while Thiele-Winter acetoxylation of the quinone **4** gives, after hydrolysis and oxidation, a mixture of the hydroxyquinones **5** and **6** in the ratio 4:1.[20] There is one apparent exception to this generalization, namely, the report that thymoquinone gave more (57%) of the hindered product **7** than of the less hindered one **8** (38%).[7] Earlier, Bargellini had carried out the same reaction and had obtained the isomers in 27% and 66% yield, respectively.[21] Bargellini used a more concentrated solution of thymoquinone in acetic anhydride but less sulfuric acid than Erdtman. Nevertheless it is surprising that the relatively small change in reaction conditions should have led to such a large change in the ratio of products. (Equations on p. 206).

A survey of the published results (see Tables I–XIII) together with many published[16,17,19,22] and unpublished results obtained by the authors of this

---

[16] J. M. Blatchly, J. F. W. McOmie, and J. B. Searle, *J. Chem. Soc.*, C, **1969**, 1350.
[17] J. M. Blatchly and J. F. W. McOmie, *J. Chem. Soc.*, **1963**, 5311.
[18] W. Flaig, T. Ploetz, and H. Biergans, *Ann.*, **597**, 196 (1955).
[19] J. M. Blatchly and J. F. W. McOmie, *J. Chem. Soc.*, C, **1972**, in press.
[20] D. A. Archer and R. H. Thomson, *J. Chem. Soc.*, C, **1967**, 1710.
[21] G. Bargellini, *Gazz. Chim. Ital.*, **53**, 235 (1923).
[22] J. M. Blatchly, R. J. S. Green, J. F. W. McOmie, and J. B. Searle, *J. Chem. Soc.*, C, **1969**, 1353.

review and their co-workers has revealed some useful generalizations concerning the reactivity of quinones and the probable orientation of the

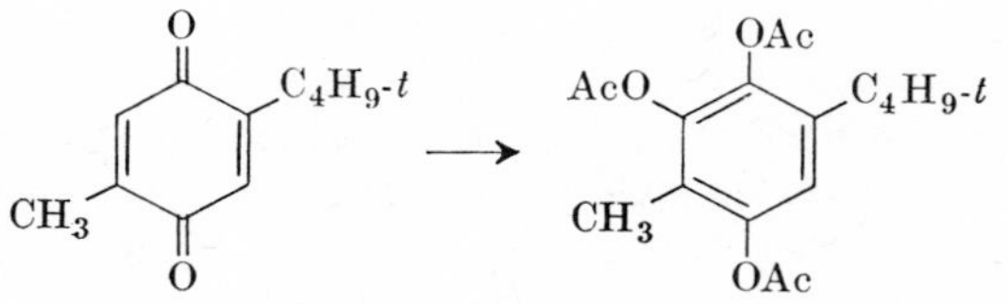

(Eq. 7)

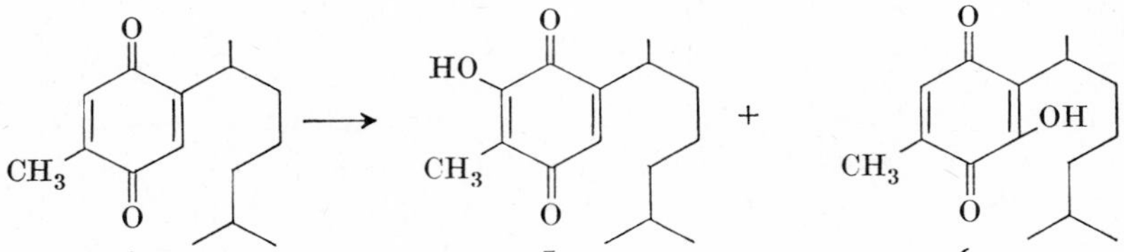

4 → **5** + **6**

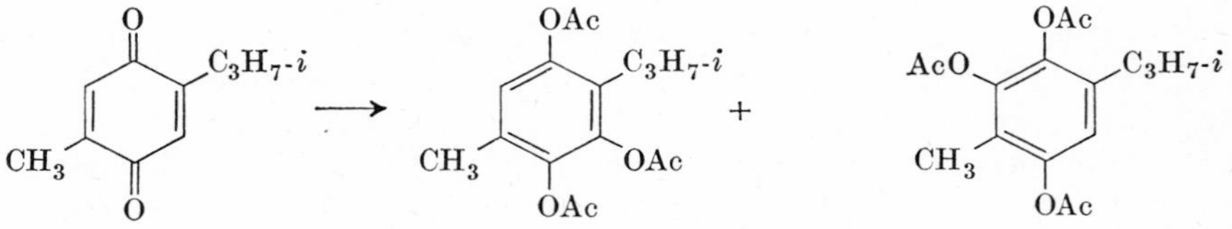

Thymoquinone → **7** (57%) + **8** (38%)

products; they are given in the next section. It is hoped that the studies now in progress in the authors' laboratories will provide information on which a general mechanism can be based.

SCOPE AND LIMITATIONS

General

Although many references to the Thiele-Winter acetoxylation of quinones are scattered throughout the literature, the reaction usually forms only an incidental part of the paper in which it occurs. However, as mentioned in the preceding section, the reaction is an important step in an efficient sequence for converting a catechol or hydroquinone into a trihydric phenol or into derivatives of a trihydric phenol. Some of the most useful interrelations are shown in Scheme 2. The ready oxidation of phenols to quinones by potassium nitrosodisulfonate (Fremy's salt)[23] followed by Thiele-Winter acetoxylation permits the conversion of a monohydric phenol into a trihydroxybenzene in three stages. Other

[23] H.-J. Teuber and S. Benz, *Chem. Ber.*, **100**, 2918 (1967), and earlier papers.

oxidative procedures for the preparation of benzoquinones have been reviewed in an earlier volume of this series.[23a]

The Thiele-Winter reaction has been used to correlate the structures of natural products; for example, fumigatin (**9**, R = H) has been converted

SCHEME 2

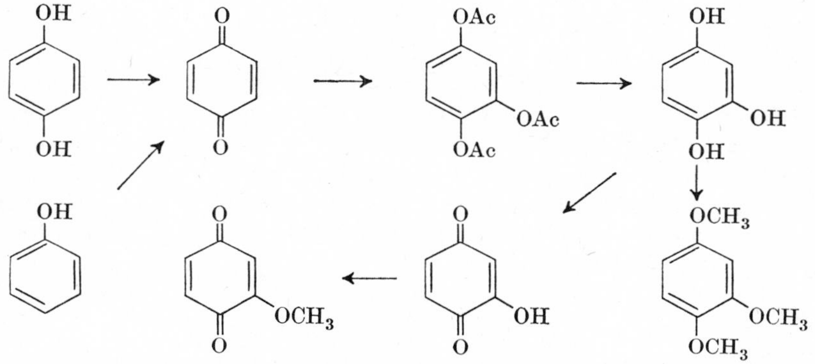

into spinulosin (**9**, R = OH),[24] and dermoglaucin (**10**, R = H), into dermocybin (**10**, R = OH) via the diquinone **11**.[25] The reaction was

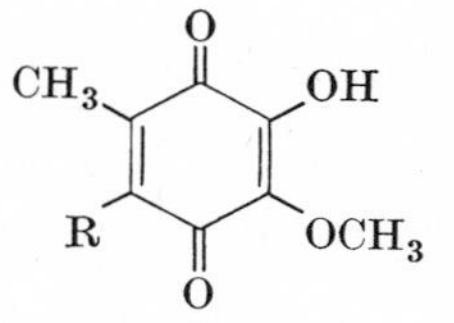

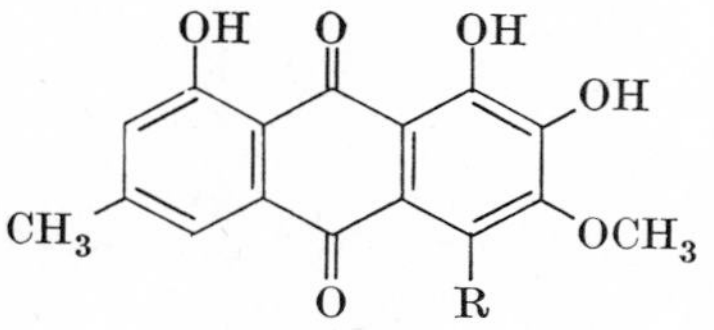

used by Posternak for a synthesis of phoenicin (**12**, R = OH) from the diquinone **12**, R = H,[26] and more recently the quinone **13** was converted into the hydroxy-quinone **14**, R = H, R' = CHO, in an overall yield of 67%.[27] This sequence formed an important part of the synthesis of

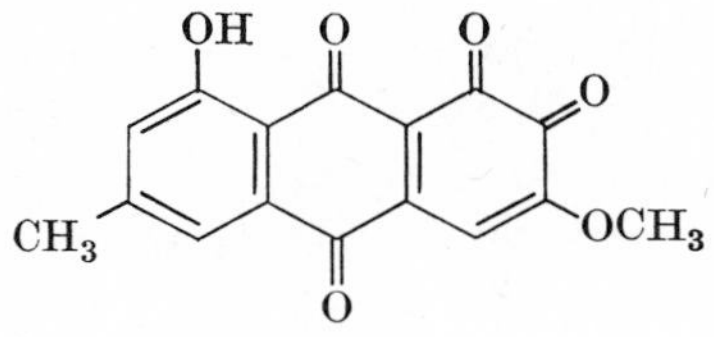

[23a] J. Cason, *Org. Reactions*, **4**, 305 (1948).
[24] W. K. Anslow and H. Raistrick, *Biochem. J.*, **32**, 687 (1938).
[25] W. Steglich, W. Lösel, and V. Austel, *Chem. Ber.*, **102**, 4104 (1969).
[26] T. Posternak, *Helv. Chim. Acta*, **21**, 1326 (1938).
[27] G. R. Allen, J. F. Poletto, and M. J. Weiss, *J. Org. Chem.*, **30**, 2897 (1965).

7-methoxymitosene (**14**, R = CH$_3$, R' = CH$_2$OCONH$_2$). It is noteworthy that the quinone **15**, R = CHO, which is closely related to the *o*-quinone **13**, did not undergo Thiele-Winter acetoxylation when treated with acetic anhydride and sulfuric acid at room temperature, but was simply converted into the corresponding acylal **15**, R = CH(OAc)$_2$.[28] This difference in behavior of closely related quinones illustrates that *o*-quinones are more reactive than *p*-quinones.

Thiele-Winter acetoxylation, followed by hydrolysis and oxidation, has been used in the determination of the structure of the natural product miltirone (**16**), which was thereby converted into the hydroxy-1,4-naphthoquinone **17**.[29] Very much earlier, Dimroth and his co-workers had investigated the structure of carminic acid, the coloring matter of cochineal. As part of their studies they converted deoxycarminic acid (**18**, R = H) back into carminic acid (**18**, R = OH). The hydroxyl groups in the glucosyl (C$_6$H$_{11}$O$_5$) side chain were first protected by acetylation, and the pentaacetate was then oxidized by lead tetraacetate to the corresponding 1,4-9,10-bisquinone which, without isolation, was immediately submitted to Thiele-Winter acetoxylation. Hydrolysis of the octaacetyl compound then gave carminic acid (p. 209).[30]

In a series of papers on the oxidation and the alkaline degradation of hydroxybenzoquinones, Corbett used the Thiele-Winter reaction, followed

---

[28] W. A. Remers and M. J. Weiss, *J. Amer. Chem. Soc.*, **88**, 804 (1966).
[29] T. Hayashi, H. Kakisawa, Hong-yen Hsü, and Yuh Pan Chen, *Chem. Commun.*, **1970**, 299.
[30] O. Dimroth and H. Kämmerer, *Ber.*, **53**, 471 (1920).

by hydrolysis and oxidation, to prepare many of the quinones investigated.[32]

The above examples give a good idea of the scope of the Thiele-Winter reaction. The limitations will now be considered. Thiele-Winter acetoxylation occurs with most of the quinones listed in the tables. However, many

unsuccessful attempts may not have been recorded. At present it is not possible to predict which quinones will fail to react. As a useful generalization it may be noted that, with one exception, an acetoxy group never enters adjacent to a methoxy group in a quinone. Thus, for example, 2,5- and 2,6-dimethoxy-1,4-benzoquinone fail to react,[7] and 2,3,5-trimethoxy-1,4-benzoquinone is decomposed.[22] With 5- or 6-substituted 2-methoxyquinones the acetoxy group enters adjacent to the 5- or 6-substituent, e.g., the bromoquinone **19** gives the triacetate **20**.[22] The one exception, the quinone **11**, mentioned above can probably be rationalized as resulting from activation of the o-quinonoid ring by the adjacent p-quinonoid ring. 3-Methoxy-1,2-benzoquinone gives a glassy solid,[33] and 4-methoxy-5-phenyl-1,2-benzoquinone undergoes reduction and acetylation to give 1,2-diacetoxy-4-methoxy-5-phenylbenzene.[34]

Hydroxyquinones are rapidly converted into acetoxyquinones some of which then undergo Thiele-Winter acetoxylation. The formation of some 2,3,4,6-tetraacetoxytoluene from toluquinone is assumed to be due to the intermediate formation of 5-acetoxytoluquinone when the reaction is carried out in a mixture of acetic anhydride and acetic acid.[6,12] However, under the usual conditions, 5-hydroxy-2-t-octyl-1,4-benzoquinone simply gives the corresponding acetate.[35] Not many quinones have been investigated in which the inserted acetoxy group would have to enter adjacent to an acetoxy group. 2,5-Dihydroxybenzoquinone gave the diacetoxyquinone

---

* Formula **18**, R = OH, shows the correct structure of carminic acid;[31] in Dimroth's paper the carboxyl group was assigned to position 5 instead of position 7.

[31] J. C. Overeem and G. J. M. Van der Kerk, *Rec. Trav. Chim. Pays-Bas*, **83**, 1023 (1964).
[32] J. F. Corbett, *J. Chem. Soc.*, *C*, **1966**, 2308; **1967**, 611; **1967**, 2408; **1970**, 2101.
[33] L. Horner and S. Göwecke, *Chem. Ber.*, **94**, 1267 (1961).
[34] J. M. Blatchly, J. F. W. McOmie, and co-workers, unpublished results.
[35] J. Pospíšil and V. Ettel, *Coll. Czech. Chem. Commun.*, **24**, 729 (1959).

which did not react further,[36] but 2-acetoxy-6-methoxybenzoquinone gave 1,2,3,4-tetraacetoxy-6-methoxybenzene.[34] 2-Hydroxy-5,6-dimethylbenzoquinone underwent reaction to give a product[37] now known to have the structure **21**.[34] Further hydroxylation of 2,5- and 2,8-dihydroxy-1,4-naphthoquinone failed,[38] but it is possible that the quinones would have undergone Thiele-Winter acetoxylation if the reaction had been carried out for a sufficiently long time. We have found that 2-acetoxy-1,4-naphthoquinone itself reacts if the mixture is kept for 40 days at room temperature.[34] The product (18% yield) has been shown by its nuclear magnetic resonance (nmr) spectrum to be the tricyclic compound **22**. When

21   **22**   **23**

it is heated just above its melting point for a short time, it isomerizes to 1,2,3,4-tetraacetoxynaphthalene. Similarly 2-acetamido-1,4-naphthoquinone has been found to react equally slowly to give the tricyclic compound **23**.[34] Probably neighboring group participation occurs in the Thiele-Winter acetoxylation of some 2-acetoxy- and 2-acetamido-1,4-quinones.

Another limitation of the Thiele-Winter reaction is that sometimes a mixture of products is formed or the quinone may polymerize (cf. the formation of derivatives of biphenyl and terphenyl by the action of hydrogen chloride on some quinones).[39] In most publications only one product is reported, but with the increasing use of more efficient methods of analysis it is becoming clear that mixtures of isomers and by-products are often formed. Thus phenyl-1,4-benzoquinone has been shown by gas-liquid chromatography to give 62% of 2,4,5-triacetoxybiphenyl and 31% of the isomeric 2,3,6-triacetoxy compound.[6] Apart from the formation of isomers, the acetylated hydroquinone is often formed as well as, or occasionally instead of, the expected Thiele-Winter acetoxylation product. This occurs most frequently when perchloric acid is the catalyst. A striking contrast occurs in the Thiele-Winter acetoxylation of 2,5-dibromobenzoquinone. With sulfuric acid as catalyst this quinone gives the expected triacetate in 49% yield, whereas with perchloric acid as catalyst only 12%

[36] J. A. Barltrop and M. L. Burstall, J. Chem. Soc., **1959**, 2183.
[37] G. Pettersson, Acta Chem. Scand., **18**, 2309 (1964).
[38] H. Singh, T. L. Folk, and P. J. Scheuer, Tetrahedron, **25**, 5301 (1969).
[39] F. M. Dean, A. M. Osman, and A. Robertson, J. Chem. Soc., **1955**, 11.

of the triacetate is formed, the major product (60%) being the acetylated hydroquinone, 1,4-diacetoxy-2,5-dibromobenzene.[16] Very reactive quinones, such as o-quinones and 2-iodo-3-methoxy-1,4-benzoquinone, which are easily oxidized or decomposed, tend to give mainly if not entirely the hydroquinone diacetate. Probably some of the quinone is polymerized (*cf.* ref. 39) and the polymer reduces other molecules of the quinone to the hydroquinone, which is then acetylated.

Attempted Thiele-Winter acetoxylation of 2-(2'-methoxyphenyl)-5-methoxy-1,4-benzoquinone in the presence of boron trifluoride gave a tar.[17] A similar result was obtained with 4'-methoxyphenyl-1,4-benzoquinone using sulfuric acid,[6] while under milder conditions with boron trifluoride as the catalyst a complex mixture was obtained from which no pure products could be isolated.[34] Blocking the positions *ortho* to the methoxy group with chlorine inhibits the Friedel-Crafts type of reaction on the methoxyl-substituted nucleus: (3',5'-dichloro-4'-methoxyphenyl)-1,4-benzoquinone reacted smoothly to give a mixture of 2,4,5- and 2,3,5-triacetoxybiphenyl in the ratio 65:35.[6]

The third limitation of the reaction, namely, the uncertainty in predicting the orientation of the product where more than one is possible, has already been mentioned. This topic is discussed separately for each of several types of quinones.

Monosubstituted 1,4-Benzoquinones (Tables I and VI)

The monosubstituted quinones usually undergo Thiele-Winter acetoxylation, but alkylthio quinones either remain unchanged or give tars. The orientation of the products is easily determined by nmr spectroscopy, since they contain two aromatic protons situated *ortho*, *meta*, or *para* to each other and these will show spin-spin coupling constants of 7–10, 2–3, 0–1 Hz, respectively. The orientations can be checked by converting the triacetates into the corresponding trimethoxy compounds and then remeasuring the nmr spectra.

In general, when the substituent is an electron-withdrawing group (*e.g.*, CN, COCH$_3$, CO$_2$CH$_3$), acetoxylation occurs adjacent to the substituent whereas, if the substituent is an electron-donating group, acetoxylation occurs mainly or exclusively at position 5. Thus 2-acetylquinone gives

2-acetyl-1,3,4-triacetoxybenzene,[6] 2-methoxyquinone gives 2-methoxy-1,4,5-triacetoxybenzene (98%),[7] and 2-phenylquinone gives a mixture of 2-phenyl-1,4,5-triacetoxybenzene (52–62%) and 2-phenyl-1,3,4-triacetoxybenzene (31–21%).[6]

Disubstituted 1,4-Benzoquinones (Tables II, III, and IV)

Almost all of the 2,3-, 2,5-, and 2,6-disubstituted 1,4-benzoquinones which have been studied have undergone the Thiele-Winter acetoxylation. The main exceptions are 2,5- and 2,6-dimethoxy-, 2,5- and 2,6-diacetoxy-, and, very curiously, 2,5-dichloro-quinone all of which remain unchanged. 2-Iodo-3-methoxyquinone yields the corresponding hydroquinone diacetate, 2-hydroxy-5-t-octylquinone gives the corresponding acetoxyquinone, and 2-methoxy-5-o-methoxyphenylquinone gives a tar. The lack of reactivity of the indoloquinone 15 has already been mentioned.

Acetoxylation during the Thiele-Winter reaction can occur adjacent to halogen, aryl, primary or secondary alkyl, and sometimes acetoxy groups, but never occurs adjacent to a methoxy group. It rarely occurs adjacent to a t-butyl group (see p. 215).

When the two substituents are different, two isomeric products are possible, and sometimes, e.g., with 2-iodo-5-methylquinone and 2-methyl-5-isopropylquinone, both isomers are formed. It is not possible to predict with certainty which isomer will predominate or be the exclusive product. On the basis of a limited number of examples, Wilgus and Gates made the following generalizations.[6] "Strong electron-donating groups direct primarily *para* and secondarily *meta;* the entering group will not enter the *ortho* position. Weak electron-donating groups direct both *para* and *meta*. When two electron-donating groups are present in the same quinone ring, the effect of the stronger electron-donating group predominates. Electron-withdrawing groups direct *ortho*. A bulky group will show some steric effect by inhibiting the reaction at the *ortho* position." These generalizations are useful, but Tables II, III, and IV contain some exceptions, and it is essential to confirm the orientation of the product or products by rigorous chemical or physical methods.

Unfortunately there is no general method for determining the orientation of the reaction product(s). However, when one of the substituents is halogen, the structure of the Thiele-Winter acetoxylation product can usually be determined simply by reductive removal of the halogen atom (preferably after the triacetate has been converted into the corresponding trimethoxy compound), followed by measurement of the nmr spectrum of the resulting tetrasubstituted benzene. Thus 2-bromo-5-phenylquinone on treatment with acetic anhydride gave a triacetate which, by the sequence of reactions in Scheme 3, gave a trimethoxybiphenyl whose nmr spectrum showed two aromatic protons with a coupling constant of 2.9 Hz, and hence the compound was 1,3,4-trimethoxy-5-phenylbenzene (2,3,5-trimethoxybiphenyl). Two of these methoxy groups (those at positions 1 and

SCHEME 3

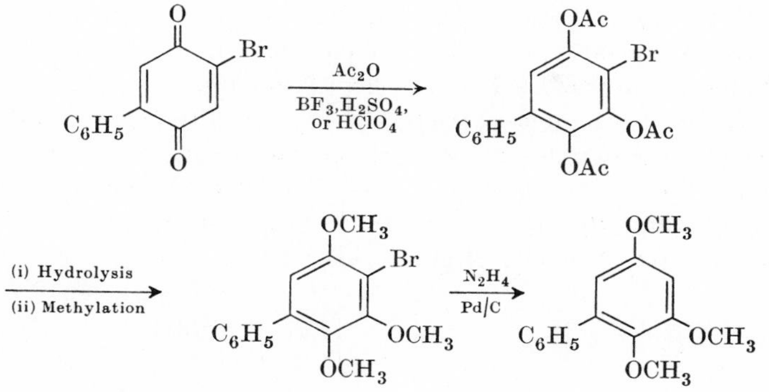

4) were derived from the original quinonoid oxygen atoms, the third (at position 3) had been introduced by acetoxylation, which thus occurred adjacent to the bromine atom in the original quinone.[34]

Trisubstituted Benzoquinones (Table V)

A study of Table V shows that most of the trisubstituted quinones that have been investigated do undergo the reaction. In most of the exceptions the vacant position in the quinone is adjacent to a methoxy group. 3,6-Dibromo-2-methoxyquinone is one of the exceptions, but here the rate of decomposition of the quinone exceeds the rate of acetoxylation and so no triacetoxy product can be obtained. Acetoxylation *ortho* to an acetoxy group is probably possible provided that the reaction is allowed to proceed for several weeks, preferably with boron trifluoride as the catalyst; *cf.* 2-acetoxy-1,4-naphthoquinone, which has already been discussed (p. 210).

There is unlikely to be any ambiguity in the structure of the reaction product. Nevertheless it is wise to confirm it by nmr spectroscopy or by chemical methods since, as with 2-acetoxy-1,4-naphthoquinone, the product may not always have the normal triacetoxybenzene structure.

1,2-Quinones (Tables VII and VIIIC)

1,2-Benzoquinone and its simpler derivatives are very readily polymerized by acids and do not undergo the Thiele-Winter reaction. Stable 1,2-quinones usually react rapidly and cleanly, the most satisfactory catalyst being a few drops of concentrated sulfuric acid. The 4 and 5 positions are more electron-deficient than the 3 and 6 positions; accordingly 1,2-quinones normally give 1,2,4-triacetoxybenzenes. However, if both the 4 and 5 positions are already substituted or if the 4-substituent is very bulky, e.g., t-butyl or t-octyl, 1,2,3-triacetates are formed instead. The formation of 1,2,4-triacetates involves the 1,4 addition of acetic anhydride (or of acetic acid followed by O-acetylation), whereas the formation of 1,2,3-triacetates involves a 1,6 addition.

1,2-Naphthoquinones and polycyclic 1,2-quinones are generally more stable than 1,2-benzoquinone itself, and they readily undergo acetoxylation. The reaction of purpurogallin quinone (24) to give the tetraacetate is particularly interesting since it involves a 1,8 addition reaction.[40] It is also noteworthy that the rather sterically hindered, internally chelated hydroxyl group does not undergo acetylation.

The reaction of methyl substituted o-quinones as the tautomeric methylenequinones is described on pp. 215–216.

The 1,2-quinones that have been submitted to the conditions of the Thiele-Winter reaction are recorded in Tables VIIA, VIIB, and VIIIC. Two examples of substituted 1,2-9,10-anthrabisquinones are given in Table IX.

1,4-Naphthoquinones (Tables VIIIA and VIIIB)

Almost all 1,4-naphthoquinones that have no substituent in the quinonoid ring undergo the Thiele-Winter reaction. If the other ring is substituted unsymmetrically, then both possible products are obtained. Thus 5-hydroxy-1,4-naphthoquinone (juglone) gives a difficultly separable mixture of 1,2,4,5- and 1,3,4,5-tetraacetoxynaphthalene.[38]

Early workers reported that 2-methyl-1,4-naphthoquinone was unreactive toward Thiele-Winter acetoxylation in the presence of sulfuric acid[41] or zinc chloride.[42] The latter result was confirmed by Burton and Praill, but these workers found that perchloric acid is an effective catalyst and that the quinone gives 1,3,4-triacetoxy-2-methylnaphthalene in 54% yield.[9] More recently it has been shown that the reaction can be brought

[40] L. Horner, S. Göwecke, and W. Dürckheimer, Chem. Ber., **94**, 1276 (1961).
[41] R. J. Anderson and M. S. Newman, J. Biol. Chem., **103**, 405 (1933).
[42] J. Madinaveita, Rev. Acad. Cienc. Madrid, **31**, 617 (1934) [C.A., **29**, 5438 (1935)].

about using boron trifluoride[6] or sulfuric acid as catalyst.[34] The reaction proceeds very slowly when sulfuric acid is the catalyst. Other 2-substituted 1,4-naphthoquinones have been reported to be unreactive, but the authors have found that prolonged treatment of 2-acetoxy- and 2-acetamido-1,4-naphthoquinone with acetic anhydride containing boron trifluoride gives the tricyclic products **22** and **23**, respectively.[34] It is probable that most of the quinones listed in Table VIIIB as unreactive would in fact react if perchloric acid were used as the catalyst or if the reaction were carried out for several weeks at room temperature using sulfuric acid or boron trifluoride as the catalyst.

t-Butylquinones and Methylquinones

As mentioned earlier, acetoxylation rarely occurs adjacent to a *t*-butyl group; instead, if no alternative position is vacant, one of the *t*-butyl groups is usually displaced by an acetoxy group. Replacement of a *t*-butyl group by hydrogen followed by acetoxylation does not seem to be a feasible pathway, since both 2,5-di-*t*-butyl- and 2,6-*t*-butyl-quinone do not yield the same triacetate(s) as they should if the common intermediate, 2-*t*-butylquinone, were formed. Instead, 2,5-di-*t*-butylquinone **(25)** yields the two triacetates shown in Eq. 8, while 2,6-di-*t*-butylquinone **(26)** yields a

third and different triacetate.[19] Many other examples of the displacement of *t*-butyl groups are known in *o*- and *p*-benzoquinones, and in 3,5,3′,5′-tetra-*t*-butyldiphenoquinone.

3-*t*-Butyl-5-methyl-1,2-benzoquinone is interesting because it reacts as such to give the triacetate and (presumably) in the tautomeric methylenequinone form **27** to give the acetoxymethyl compound (Eq. 9).[19] A few other examples of methyl- or substituted methyl-quinones reacting in the tautomeric methylenequinone form are known. For example, the 1,2-naphthoquinones **28**, R = H, C_6H_5, p-ClC_6H_4, react exclusively in the p-methylenequinone form to give the corresponding acetoxymethyl compounds of type **29**.[43]

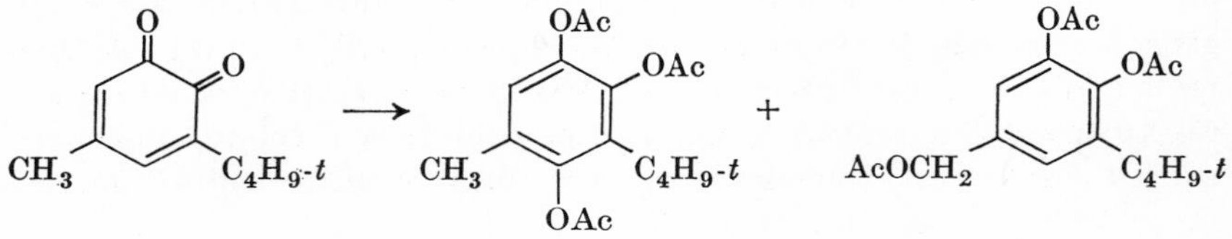

[structures for Eq. 9: o-quinone with CH_3 and C_4H_9-t → triacetoxy product (20%) + diacetoxy acetoxymethyl product (14.5%)]

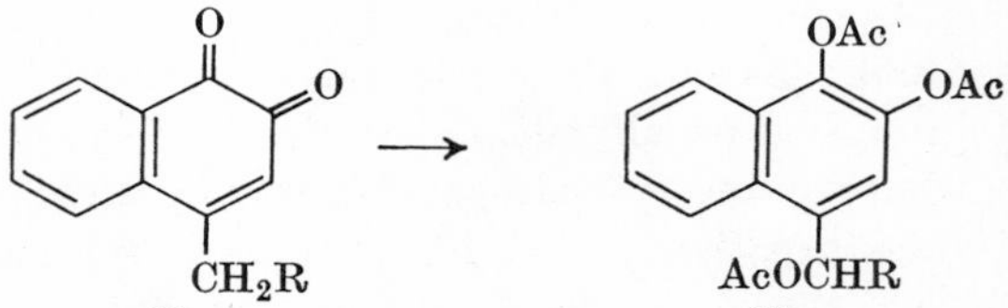

(Eq. 9)

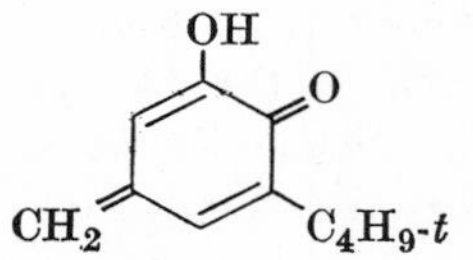

[structures: **27** (methylenequinone tautomer with OH); **28** (1,2-naphthoquinone with CH_2R) → **29** (naphthalene with OAc, OAc, AcOCHR)]

In general, the order of stability of methyl- or substituted methyl quinones is probably that shown in the accompanying formulas, although there may be some exceptions depending upon the exact nature of the

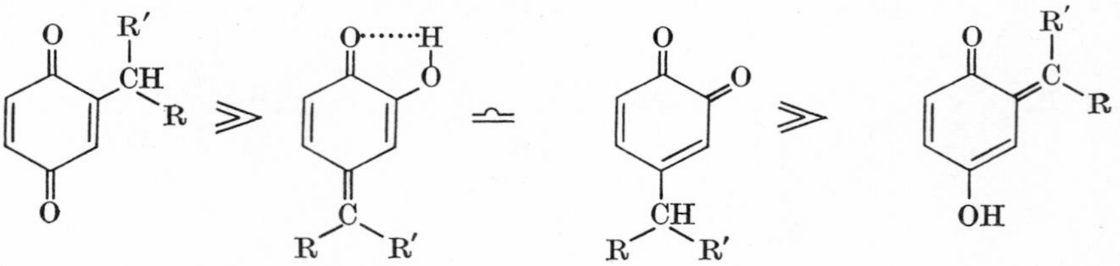

[stability series of quinone tautomers]

substituents R and R′. Thus the reaction of methylquinones in the tautomeric methylenequinone form is unlikely to occur except with suitably substituted o-quinones.

When methylquinones react in their tautomeric forms the product will, of course, be isomeric with the normal Thiele-Winter acetoxylation product, so it is not sufficient to rely on elemental analysis. This again emphasizes the necessity of confirming the structure of the product by physical or by chemical methods. Usually the nmr spectrum of the product is sufficient to confirm the structure; for example, the structure of the acetoxymethyl product in Eq. 9 was immediately apparent from its nmr spectrum.

---

[43] L. F. Fieser and C. K. Bradsher, *J. Amer. Chem. Soc.*, **61**, 417 (1939); L. F. Fieser and M. Fieser, *ibid.*, **61**, 596 (1939).

Incidentally, tautomerism is also possible in suitable hydroxyanthraquinones. The limited evidence (see Table IX) indicates that 9-hydroxy-1,4-anthraquinone reacts in its tautomeric form, 4-hydroxy-1,10-anthraquinone (**30**), to give the addition product **31**,[44] whereas 1-hydroxy-9,10- anthraquinones, *e.g.*, compound **32**, R = H, behave as normal 1,4-quinones (rather than as the tautomeric 9-hydroxy-1,10-anthraquinones) to give the corresponding esters (*e.g.*, compound **32**, R = Ac).[45]

Use of Anhydrides Other than Acetic Anhydride

Very few studies (see Table XII) have been made of the use of anhydrides other than acetic anhydride in the Thiele-Winter reaction. Thioacetic anhydride[46,47] reacts with quinones in the presence of boron trifluoride or sulfuric acid to give products analogous to those from acetic anhydride, *e.g.*, 3,5-dimethyl-1,2-benzoquinone gives the thioacetate **33**, R = SAc,[46] while Thiele-Winter acetoxylation gives the triacetate **33**, R = OAc.[46]

1,4-Benzoquinone reacts with benzoic anhydride[34] to give a mixture of hydroquinone dibenzoate (22%) and 1,2,4-tribenzoyloxybenzene (52%), but the quinone decomposes or polymerizes when it is treated with trifluoroacetic anhydride.[34] No doubt the quinones which undergo acetoxylation would react analogously with simple anhydrides such as propionic or butyric, but from a practical point of view these anhydrides are unlikely to have any advantage over acetic anhydride.

Acid halides may be regarded as mixed anhydrides of acetic acid and the hydrohalide acids. They react readily with quinones, but the products are

---

[44] K. Zahn, *Ber.*, **67**, 2063 (1934).
[45] R. Eder, C. Widmer, and R. Butler, *Helv. Chim. Acta*, **7**, 341 (1924).
[46] H. Budzikiewicz, W. Metlesics, and F. Wessely, *Monatsh. Chem.*, **91**, 117 (1960).
[47] W. Metlesics, *Monatsh. Chem.*, **88**, 804 (1957).

often mixtures. For example, acetyl bromide reacts with 1,4-benzoquinone to give a mixture of 2-bromo- and 2,5-dibromo-hydroquinone diacetate.³ The reaction of acetyl chloride with 2,5-dimethoxy-1,4-benzoquinone is very complex and gives mixtures of the partially chloro-deoxygenated compounds 34, and 35, R = H and Cl (p. 217).[48–51] Because, unfortunately, most investigators did not record yields, the utility of the reaction for the preparation of chloro- and bromo-hydroquinone diacetates is uncertain. The available information is collected in Table XIII.

RELATED REACTIONS

The addition of alcohols, thiols, or amines to quinones, followed by hydrolysis to give hydroxyquinones represents a reaction sequence closely related to the Thiele-Winter acetoxylation reaction. These addition reactions have not been reviewed, but there are many examples of them in the literature, and occasionally the reaction sequences have been used for the synthesis of some natural products or related compounds. In general, the orientation of the addition products is the same as in the Thiele-Winter reaction, but there is a tendency, especially with alcohols and amines, for 2 molecules of reagent to be added in a series of redox reactions. Thus the zinc chloride-catalyzed addition of methanol to 1,4-benzoquinone gives 2,5-dimethoxy-1,4-benzoquinone in high yield (allowing for the fact that 3 moles of benzoquinone are required to give 1 mole of product).[52] Although an acidic type of catalyst is usually necessary to bring about the addition of alcohols to quinones, Wanzlich and Jahnke have found that methanol adds very readily to o-benzoquinones generated *in situ*.[53]

The failure of the indoloquinone 15, R = CHO, to undergo the Thiele-Winter reaction has already been mentioned. However, Remers and Weiss were able to prepare both the 5- and 6-hydroxy derivatives of the quinone by the following method.[28] Addition to the quinone of 1 mol of p-toluenethiol in the presence of 1 mol of ferric chloride gave a mixture of the isomeric p-toluenethioquinones, Eq. 10. The separated isomers were then hydrolyzed by dilute sodium hydroxide to give the corresponding monohydroxy derivatives of the quinone.

The addition of methylamine to quinones proceeds very readily, and the resulting mono- or bis-(methylamino)quinones are easily hydrolyzed to

[48] A. Oliverio, *Gazz. Chim. Ital.*, **78**, 105 (1948).
[49] A. Oliverio and G. Castelfranchi, *Gazz. Chim. Ital.*, **80**, 267 (1950).
[50] L. Asp and B. Lindberg, *Acta Chem. Scand.*, **4**, 1192 (1950).
[51] G. Werber and A. Arcoleo, *Ric. Sci.*, **24**, 115 (1954) [*C.A.*, **49**, 3055 (1955)].
[52] F. Benington, R. D. Morin, and L. C. Clark, *J. Org. Chem.*, **20**, 102 (1955).
[53] H.-W. Wanzlich and U. Jahnke, *Chem. Ber.*, **101**, 3744, 3753 (1968).

hydroxyquinones. This sequence has been used, for example, to convert fumigatin (**9**, R = H) into spinulosin (**9**, R = OH),[54,55] and phoenicin (**36**, R = H) into oosporein (**36**, R = OH).[56]

EXPERIMENTAL CONDITIONS

Some account of the range of catalysts which may be used has already been given. Acetic acid itself is not a catalyst for the reaction even with the most reactive quinones. Zinc chloride is probably the mildest catalyst, but it is not very satisfactory because it has to be used in large quantities and chloro compounds are usually formed as well as triacetates. Boron trifluoride, sulfuric acid, and perchloric acid generally seem to constitute the order of increasing potency, but there are cases in the tables where the weakest catalyst gives the best yield, perhaps in a longer time. Recently it has been shown that acetic-phosphoric anhydride reacts with 1,4-benzoquinone to give 1,2,4-triacetoxybenzene (55% yield).[57] The reagent, made by mixing 100% phosphoric acid with acetic anhydride, has the advantage that phosphoric acid is nonoxidizing. The use of this reagent merits further investigation.

The quinone is usually added to acetic anhydride to give a concentration of 0.3–1.0 molar. It is best to cool the solution (or suspension) while adding the catalyst. The amount of catalyst is not critical, and generally the following concentrations are used: 0.5–1.0% v/v perchloric acid (commercial 70% solution); 1–5% v/v concentrated sulfuric acid; 5–10% v/v

[54] W. K. Anslow and H. Raistrick, *Biochem. J.*, **32**, 803 (1938).
[55] W. K. Anslow and H. Raistrick, *J. Chem. Soc.*, **1939**, 1446.
[56] F. Kögl and G. C. van Wessem, *Rec. Trav. Chim. Pays-Bas*, **63**, 5 (1944).
[57] A. J. Fatiadi, *Carbohyd. Res.*, **6**, 237 (1968) [*C.A.*, **69**, 27650y (1968)].

boron trifluoride (the commercial solutions of this reagent in ether or in acetic acid). A 5% v/v solution of sulfuric acid in acetic anhydride is sometimes referred to as "the Thiele Reagent."

The reaction should be carried out in glass vessels using acetic anhydride of reliable purity (a sample that darkens on treatment with sulfuric acid should be redistilled). For reactions that require a week or longer, it is preferable to keep the reaction mixture in the dark. The course of the reaction is conveniently followed by thin-layer chromatography. A spot of the reaction mixture may be applied directly to the plate (*i.e.*, there is no need to hydrolyze the product). In benzene on silica gel the R_f values of most quinones are greater than those of the Thiele-Winter acetoxylation products, and the latter may be located quickly by exposure to iodine vapor or by spraying the plate with 2.5% ceric sulfate or alcoholic alkali. Air oxidation of the polyhydroxy compounds formed by hydrolysis gives bright red, green, or purple spots. These colors also develop on keeping the unsprayed plate exposed to the air for a day or more.

As it is unlikely that any acetoxylation product would be decomposed by long exposure to the conditions of the reaction, there is no disadvantage in using long reaction times. When, after a long period, no quinone is recovered and no product is isolated, it is probably safe to conclude that decomposition reactions are faster than acetoxylation. High temperatures appear to speed the former more than the latter, giving darker products that are harder to work up. Room temperature or 25° is usually best, but with quinones of low solubility, where stirring for a long time would be impracticable, 35–45° is safe, particularly where halogen substituents are present to stabilize the quinones. Some workers have used temperatures as high as 70 or 100°, but often the products are tarry and the yields are low. Perchloric acid should not be used for lengthy reactions because it slowly reacts with acetic anhydride to give dark tarry, product-contaminating materials, the catalytic activity deteriorating at the same time.[12]

The reaction mixture is usually poured onto ice and stirred to hydrolyze the excess acetic anhydride. The use of ice is essential when sulfuric and perchloric acids have been used as catalyst. When boron trifluoride has been used, the rate of hydrolysis of acetic anhydride is usually much slower, and there is less risk of overheating. Very sensitive products, *e.g.*, some tetraacetates, are rapidly hydrolyzed by the acidic hydrolysis mixture to form water-soluble products. With these reactions, ether extraction should follow the hydrolysis promptly.

When all the quinone does not react, it can often be removed by taking advantage of its solubility in light petroleum. Methanol is a good cleaning solvent for the acetate products, and charcoaling is seldom necessary. Column and gas-liquid chromatography may, as a last resort or where an isomeric mixture is present, be used to isolate the pure products. Because alumina and, to a lesser extent, silica gel react slowly with quinones and hydrolyze polyacetates, chromatography should be carried out as rapidly as possible.

EXPERIMENTAL PROCEDURES

The examples of procedures given below have been chosen to illustrate the use of boron trifluoride, sulfuric acid, and perchloric acid as catalysts for the acetoxylation of o- and p-quinones. These five illustrative reactions are followed by examples of reaction sequences which often carried out on the Thiele-Winter reaction products, namely, hydrolysis, combined hydrolysis and methylation, hydrolysis and oxidation to give hydroxy quinones, and methylation of the hydroxy group in a hydroxy quinone.

Thiele-Winter Acetoxylation of Quinones

1,2,4-Triacetoxybenzene.[58] Full details for the Thiele-Winter acetoxylation of 1,4-benzoquinone to give 1,2,4-triacetoxybenzene in 86–87% yield are given in *Organic Syntheses*.

1,2,4-Triacetoxy-3-bromo-5-methoxybenzene.[22] Boron trifluoride in acetic acid (40% solution, 0.5 ml) was added to a solution of 1 g (4.6 mmols) of 2-bromo-6-methoxy-1,4-benzoquinone in 15 ml of acetic anhydride, and the mixture was kept at 35° for 16 hours. The mixture was then added to 200 ml of water with stirring and, after 0.5 hour, the precipitated product was collected by filtration. The crude product, after two recrystallizations from methanol, gave 1,2,4-triacetoxy-3-bromo-5-methoxybenzene (1.5 g, 90%) as prisms, mp 140°.

1,3,4-Triacetoxy-2-methylnaphthalene.[6] A suspension of 6.0 g (35 mmols) of 2-methyl-1,4-naphthoquinone in 30 ml of acetic anhydride containing 3 ml of boron trifluoride etherate was allowed to stand at room temperature for 5 days, with occasional shaking. The solid gradually dissolved during the first 2 days. The solution was poured onto ice and stirred for 1 hour, then the dark solid was extracted with diethyl ether. The ethereal extracts were combined, washed successively with water, aqueous sodium bicarbonate, and water, then dried and treated with charcoal. The solvent was removed and the residue was recrystallized from ethanol to give 1,3,4-triacetoxy-2-methylnaphthalene (5.2 g, 52%), mp 142–144°. The mp was raised to 144–145° by recrystallization from ethanol-ligroin.

[58] E. B. Vliet, *Org. Syn. Coll. Vol.*, **1**, 317 (1932).

When the reaction mixture was kept for only 18 hours instead of 5 days, the yield was 27%.

1,2,3-Triacetoxy-5-*t*-butylbenzene.[19] Concentrated sulfuric acid (0.3 ml) was added to a solution of 2.4 g (15 mmols) of 4-*t*-butyl-1,2-benzoquinone in 30 ml of acetic anhydride at 20°. After 6 minutes the almost colorless solution was poured onto crushed ice (200 g). The mixture was stirred for about 1 hour, and the solid was collected by filtration. Purification by recrystallization from methanol gave 1,2,3-triacetoxy-5-*t*-butylbenzene (3.5 g, 77%) as white crystals, mp 105–106°.

1,2,4-Triacetoxynaphthalene.[9] To a solution of 5 g (32 mmols) of 1,4-naphthoquinone in 30 ml of acetic anhydride at room temperature 0.6 ml of 72% perchloric acid was added. After 30 minutes the mixture was decomposed by ice water, and the resulting solid (yield 95%), mp 133.5–134.5°, was collected by filtration. Crystallization of the solid from ethanol gave pure 1,2,4-triacetoxynaphthalene, mp 135°.

In an experiment with 79 g (0.5 mol) of 1,4-naphthoquinone, Fieser and Fieser obtained the triacetoxynaphthalene in 80% yield.[59]

Conversion of Triacetates into Related Compounds

2′,3′,5′-Trihydroxyterphenyl (Hydrolysis of a Triacetate).[53] 2′,3′,5′-Triacetoxyterphenyl (20 g, 50 mmols) (prepared from 2,5-diphenyl-1,4-benzoquinone) was heated under reflux with ethanolic hydrochloric acid for 3 hours in an atmosphere of nitrogen. The mixture was concentrated, then diluted with water. The trihydroxy compound (13.2 g, 96%) was collected by filtration and recrystallized from ethanol to give material with mp 270–275° (dec).

2,4,5-Trimethoxybiphenyl (Combined Hydrolysis and Methylation).[17] Sodium hydroxide (25.2 g) in 30 ml of water was added dropwise, with stirring, to a solution of 3.0 g (9.2 mmols) of 2,4,5-triacetoxybiphenyl and 0.15 g of sodium dithionite in 30 ml of methanol and 30 ml of dimethyl sulfate at 60° during 20 minutes. The solution was stirred for a further 20 minutes at 60°, then diluted with water, cooled to 0°, and the crude product was collected by filtration. Recrystallization from petroleum (bp 60–80°) afforded 2,4,5-trimethoxybiphenyl (1.7 g, 72%) as granular crystals, mp 88°.

2-Hydroxy-1,4-naphthoquinone (Hydrolysis and Oxidation).[8] The reaction must be carried out in an inert atmosphere. The apparatus

---

[59] L. F. Fieser and M. Fieser, *Reagents for Organic Synthesis*, Vol. 1, John Wiley & Sons, 1967, p. 716.

consists of a three-necked flask with a gas-inlet tube in the central neck extending to the bottom, a dropping funnel, and a gas-exit tube connected to a water bubbler.

Nitrogen was bubbled through a suspension of 30.2 g (0.1 mol) of 1,2,4-triacetoxynaphthalene in 200 ml of ethanol, and the exit gas was allowed to bubble through 75 ml of 25% aqueous sodium hydroxide placed in the dropping funnel and covered with a layer of light petroleum. After 30 minutes the alkali was run in to hydrolyze the triacetate. Hydrolysis was complete in 20 minutes, after which the deep-yellow solution was acidified by the addition of 65 ml of 36% hydrochloric acid in 150 ml of water. The flask was opened and the contents warmed to 85°. The hydroxynaphthohydroquinone was oxidized by treatment with a solution of 65 g of ferric chloride hexahydrate and 22 ml of 36% hydrochloric acid in 100 ml of water. A precipitate formed in less than 1 minute. The mixture was then cooled in ice, and the yellow crystals of 2-hydroxy-1,4-naphthoquinone (16.8 g, 97%), m.p. 190–192° were collected by filtration.

This method of preparation is generally applicable but, if no product precipitates owing to high solubility, ether extraction may be employed. For certain sensitive quinones, e.g., 2-hydroxy-5,6-dimethyl-1,4-benzoquinone, ferric sulfate should be used in place of ferric chloride. For unstable quinones such as 2-hydroxy-1,4-benzoquinones that must be prepared under anhydrous conditions, the best procedure is to oxidize the pure hydroxyhydroquinone with silver oxide in dry ether.[32]

5-Methoxy-2-phenyl-1,4-benzoquinone (Methylation of a Hydroxyquinone).[8,17]

A mixture of 0.48 g (2.4 mmols) of 5-hydroxy-2-phenyl-1,4-benzoquinone, 5 ml of methanol, and 3 drops of boron trifluoride etherate was kept at room temperature for 10 days. The orange-yellow methoxyquinone (0.4 g, 78%) separated as needles, mp 199–200°.

TABULAR SURVEY

Compounds that have been submitted to the conditions of the reaction are collected in the tables. Within each table the compounds are arranged in order of molecular formula as in *Chemical Abstracts*. The literature has been surveyed up to the end of March 1971 but, since there is no systematic method of searching the literature for the reaction, we cannot guarantee the completeness of the tables.

A dash (—) in the product or in the yield column means that the yield was not reported. No entry under a product means that this product was not reported.

TABLE I. 1,4-Benzoquinone and Monosubstituted 1,4-Benzoquinones

Formula	2-Substituent	Catalyst	Product(s) or Substituent(s) in 1,4,5-Triacetoxybenzene and Yield(s) (%)	Refs.
$C_6H_4O_2$	Nil	$ZnCl_2$	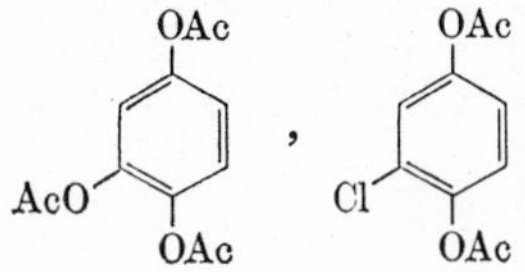 [structures: 1,2,4-triacetoxybenzene and 2-chloro-1,4,5-triacetoxybenzene] (—), (41), (6–9)	5
		$ZnCl_2$	(5)	9
		$ZnCl_2$	(5–6)	9
		H_2SO_4	(80, 87, 94)	2, 58, 60
		H_2SO_4	(—)	32
		$HClO_4^a$	(70)	12
		$HClO_4$	(96, 100)	9, 12
		$Ac^+ClO_4^-$	(>80)	12
		$AcOPO_3H_2$	(55)	57
$C_6H_3BrO_2$	Bromo	BF_3	2-Bromo (42)	16
		$HClO_4$	2-Bromo (28)	16
$C_6H_3ClO_2$	Chloro	H_2SO_4	2-Chloro (—)[b]	61, 62
$C_6H_3IO_2$	Iodo	BF_3	2-Iodo (35)	16
$C_7H_6O_2$	Methyl	$ZnCl_2$	[structures shown] (2), (trace), (>54)	9

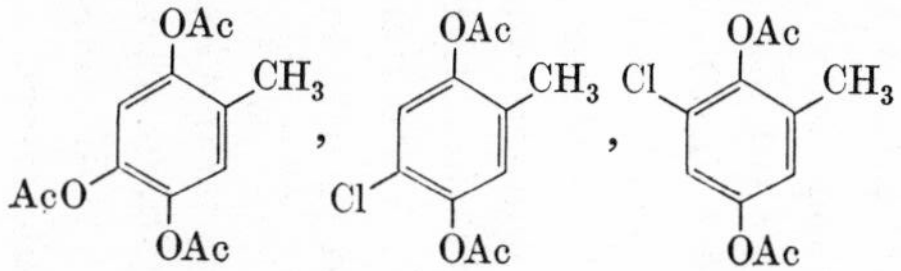

		BF$_3$	![structure: AcO-benzene-CH3-OAc-OAc] (89)	![structure: AcO-benzene-CH3-OAc-OAc] (11)	6
		H$_2$SO$_4$	(—)		5, 32
		,,	(78)	(15)	6
		HClO$_4$	(70)		12
		HClO$_4^c$			12
C$_7$H$_6$O$_2$S	Methylthio	BF$_3$, H$_2$SO$_4$, or HClO$_4$	2,5-Diacetoxytoluene (—), 2,4,5-triacetoxytoluene (30), 2,3,4,6-tetraacetoxytoluene (13) Starting material or tars		63
C$_7$H$_6$O$_3$	Methoxy	H$_2$SO$_4$	2-Methoxy (48, 98, 100)		65, 7, 64
C$_8$H$_8$O$_2$S	Ethylthio	BF$_3$, H$_2$SO$_4$, or HClO$_4$	Starting material or tars		63
C$_9$H$_{10}$O$_2$S	Propylthio	BF$_3$, H$_2$SO$_4$, or HClO$_4$	Starting material or tars		63
C$_{10}$H$_{12}$O$_2$	t-Butyl	BF$_3$	![structure: OAc-C4H9-t-OAc-OAc benzene] (6)	![structure: OAc-C4H9-t-OAc-AcO] (52) , ![structure: OAc-C4H9-t-OAc-AcO] (46)	19
		BF$_3$		![structure: AcO-OAc-C4H9-t-OAc] (3)	66

Note: References 60–148 are on pp. 276–277.

[a] This reaction was conducted in a 1:1 mixture of acetic acid and acetic anhydride. Hydroquinone diacetate (17%) and 1,2,4,5-tetraacetoxybenzene (17%) were obtained as well as the triacetate.
[b] The orientation of the product was not stated in ref. 61, but its structure was proved in ref. 62.
[c] This reaction was conducted in a 1:1 mixture of acetic acid and acetic anhydride.

TABLE I. 1,4-Benzoquinone and Monosubstituted 1,4-Benzoquinones (*Continued*)

Formula	2-Substituent	Catalyst	Product(s) or Substituent(s) in 1,4,5-Triacetoxylbenzene and Yield(s) (%)	Refs.
$C_{10}H_{12}O_2S$	*n*-Butylthio	BF_3, H_2SO_4, or $HClO_4$	Starting material or tars	63
$C_{12}H_7BrO_2$	*p*-Bromophenyl	BF_3	2-*p*-Bromophenyl (86)	34
$C_{12}H_7ClO_2S$	*p*-Chlorophenylthio	H_2SO_4 or $HClO_4$	2-*p*-Chlorophenylthio (good)	63
$C_{12}H_7NO_4$	*p*-Nitrophenyl (= Ar)	BF_3	[Ar, OAc, OAc, OAc isomer] (18), [Ar, OAc, OAc, OAc isomer] (82), [Ar, OAc, OAc, OAc isomer] (56), [Ar, OAc, OAc, OAc isomer] (19)	34
		H_2SO_4		6
$C_{12}H_8O_2$	Phenyl	BF_3	[C₆H₅ isomer] (31), [C₆H₅ isomer] (40, 41), (21), (62), (52)	17, 8
		BF_3		6
		H_2SO_4		6
	Phenylthio	H_2SO_4 or $HClO_4$	2-Phenylthio (good)	63
$C_{12}H_8O_3S$	*p*-Hydroxyphenyl	BF_3	2-*p*-Acetoxyphenyl (48)	34
$C_{12}H_{14}O_2$	Cyclohexyl	BF_3	2-Cyclohexyl (*ca.* 30)	8
		H_2SO_4	2-Cyclohexyl (—, 58)	18, 67

Formula	Ar group	Catalyst	Product	Ref.
$C_{13}H_8Cl_2O_3$	3′,5′-Dichloro-4′-methoxyphenyl (= Ar)	H_2SO_4	![structures with OAc, AcO, Ar groups]	6
$C_{13}H_{10}O_2S$	p-Methylphenylthio	H_2SO_4 or $HClO_4$	Ratio of isomers 65:35 2-p-Methylphenylthio (good)	63
$C_{13}H_{10}O_3$	p-Methoxyphenyl	BF_3 or H_2SO_4	Mixture of products or tars	6, 34
$C_{14}H_{10}O_4$	p-Acetoxyphenyl	BF_3	2-p-Acetoxyphenyl (45)	34
$C_{16}H_{22}O_2$	4′-Cyclohexylbutyl	H_2SO_4	2-(4′-Cyclohexylbutyl) (62)	67
$C_{24}H_{40}O_2$	n-Octadecyl	H_2SO_4	2-n-Octadecyl (—)	68

Note: References 60–148 are on pp. 276–277.

TABLE II. 2,3-DISUBSTITUTED 1,4-BENZOQUINONES

Formula	Substituents or Compound	Catalyst	Product(s) or Substituents in 1,4,5-Triacetoxybenzene and Yield(s) (%)	Refs.
$C_6H_2BrO_2$	Dibromo	BF_3	2,3-Dibromo (47)	16
$C_7H_5BrO_3$	2-Bromo-3-methoxy	BF_3	2-Bromo-3-methoxy (36)	22
$C_7H_5IO_3$	2-Iodo-3-methoxy	BF_3	2-Iodo-1,3,4-trimethoxybenzene[a] (8)	34
$C_8H_4O_2S$	(structure)	H_2SO_4	(structure) (33)	69
$C_8H_8O_2$	Dimethyl	BF_3 / H_2SO_4	2,3-Dimethyl (73, 90) / 2,3-Dimethyl (—, —, 49)	70, 6 / 32, 37, 71
$C_8H_8O_3$	2-Methoxy-3-methyl	H_2SO_4	(structure) (71)	72
$C_8H_8O_4$	Dimethoxy	H_2SO_4	2,3-Dimethoxy (63, 63, 77)	7, 55, 72a
$C_9H_8O_2$	(structure)	H_2SO_4	(structure) (—)	73
$C_{10}H_8O_2$	(structure)	BF_3	(structure) (60)	8

$C_{10}H_{10}O_2$	(tetrahydronaphthoquinone structure)	H_2SO_4	(diacetate structure, OAc/AcO/OAc) (—)	74
$C_{12}H_7BrO_2$	2-Bromo-3-phenyl	BF_3		34
$C_{12}H_{11}NO_3$	(indole-quinone with CHO, CH₃, C₂H₅)	H_2SO_4	2-Bromo-3-phenyl or 3-bromo-2-phenyl[b] (33) (indole-quinone with CH(OAc)₂, CH₃, C₂H₅) (—)	28
$C_{13}H_{10}O_3$	2-Methoxy-3-phenyl	BF_3 H_2SO_4 $HClO_4$	(triacetate with OCH₃, C₆H₅, OAc) (38) (54) (42)	34 34 34
$C_{20}H_{12}O_2$	(triptycene quinone structure)	BF_3	(triptycene diacetate structure) (95)	34

Note: References 60–148 are on pp. 276–277.

[a] This product was obtained after hydrolysis and methylation of the initial product.
[b] The orientation of the substituents is unknown.

TABLE III. 2,5-Disubstituted 1,4-Benzoquinones

Formula	Substituents or Compound	Catalyst	Product(s) or Substituents in 1,3,4-Triacetoxybenzene and Yield(s) (%)	Refs.
$C_6H_2Br_2O_2$	Dibromo	BF_3	2,5-dibromo-1,3,4-triacetoxybenzene (60); and tribromo isomer (49), (12)	16
$C_6H_2Cl_2O_2$	Dichloro	$HClO_4$, H_2SO_4	No reaction	16; 34, 61
$C_6H_4O_4$	Dihydroxy	$ZnCl_2$, BF_3, H_2SO_4, $HClO_4$	2-acetoxy-1,4-benzoquinone (75), (45), (60), (65)	75, 76, 35, 36
$C_7H_5BrO_3$	2-Bromo-5-methoxy	BF_3	2-Bromo-5-methoxy (33)	22
$C_7H_5IO_2$	2-Iodo-5-methyl	H_2SO_4	2-iodo-5-methyl-1,3,4-triacetoxybenzene (14, crude); and diiodo isomer (14)	77
$C_7H_6O_3$	2-Hydroxy-5-methyl	H_2SO_4, "	2-acetoxy-5-methyl-1,3,4-triacetoxybenzene (60); diacetoxy isomer (—)	5; 6[a], 7

Formula	Substrate	Reagent	Product	Yield (%)	Refs.
$C_7H_6O_4$	2-Hydroxy-5-methoxy	H_2SO_4	2-Acetoxy-5-methoxy-1,4-benzoquinone	(16)	55
$C_8H_8O_2$	Dimethyl	BF_3	2,5-Dimethyl	(91)	70
		H_2SO_4	2,5-Dimethyl	(—)	32, 64, 78
		"	2,5-Dimethyl	(94–96)	7, 18, 79
$C_8H_8O_3$	2-Methoxy-5-methyl	H_2SO_4	AcO—[OCH$_3$/CH$_3$]—OAc	(—, 86)	80, 54
		"		(95, 100)	72, 7[b]
$C_8H_8O_4$	Dimethoxy	H_2SO_4	No reaction		7, 64
		$HClO_4$	No reaction		22
$C_{10}H_8O_6$	Diacetoxy	$HClO_4$	No reaction		34, 36
$C_{10}H_{11}BrO_2$	2-Bromo-5-t-butyl	H_2SO_4	t-C$_4$H$_9$—[Br, OAc, OAc]—OAc , t-C$_4$H$_9$—[Br, OAc, OAc]—AcO	(4) (13), (10) (1)	19
		"	Decomposed		19
		$HClO_4$	No reaction		19
		BF_3			19
$C_{10}H_{12}O_2$	5-Isopropyl-2-methyl (thymoquinone)	H_2SO_4	AcO—[CH$_3$, i-C$_3$H$_7$]—OAc , [CH$_3$, OAc, OAc, i-C$_3$H$_7$, OAc]	(27) (67), (57) (38)	21
		"			7

Note: References 60–148 are on pp. 276–277.

[a] The orientation of the product was correctly established for the first time in ref. 6.
[b] An incorrect orientation was given in ref. 7.

TABLE III. 2,5-DISUBSTITUTED 1,4-BENZOQUINONES (Continued)

Formula	Substituents or Compound	Catalyst	Product(s) or Substituents in 1,3,4-Triacetoxybenzene and Yield(s) (%)	Refs.
$C_{10}H_{12}O_3$	2-t-Butyl-5-hydroxy	H_2SO_4	[quinone with C_4H_9-t and AcO], [triacetoxybenzene with OAc, OAc, OAc, AcO] (—) (5)	18, 19
$C_{11}H_{14}O_2$	2-t-Butyl-5-methyl	BF_3	[triacetoxybenzene with C_4H_9-t, OAc, OAc, AcO, CH_3] (low)	19
		H_2SO_4		(71) 19
$C_{12}H_7BrO_2$	2-Bromo-5-phenyl	BF_3	2-Bromo-5-phenyl (77)	34
		H_2SO_4	2-Bromo-5-phenyl (80)	34
		$HClO_4$	2-Bromo-5-phenyl (76)	34
$C_{12}H_8O_3$	2-Hydroxy-5-phenyl	BF_3	[triacetoxybenzene with OAc, OAc, AcO, OAc, C_6H_5] (56)	34
$C_{13}H_{10}O_3$	2-Methoxy-5-phenyl	BF_3	[triacetoxybenzene with OCH_3, OAc, AcO, OAc, C_6H_5] (38)	17

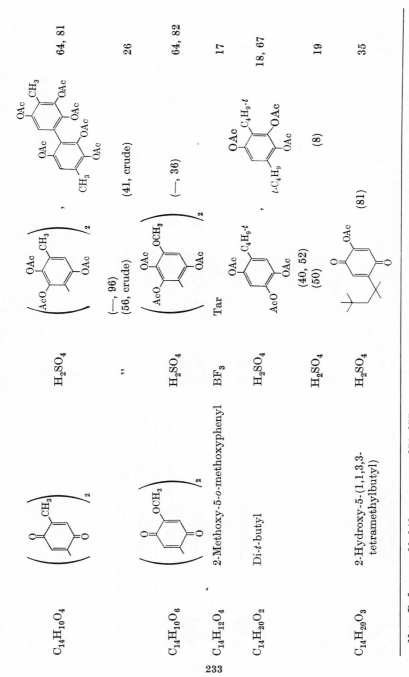

Note: References 60–148 are on pp. 276–277.

TABLE III. 2,5-DISUBSTITUTED 1,4-BENZOQUINONES (Continued)

Formula	Substituents or Compound	Catalyst	Product(s) or Substituents in 1,3,4-Triacetoxybenzene and Yield(s) (%)	Refs.
$C_{15}H_{22}O_2$		BF_3	(Ratio 4:1)[c]	20
$C_{17}H_{24}O_2$	2-(4'-Cyclohexylbutyl)-5-methyl	H_2SO_4	(49)	67
$C_{18}H_{12}O_2$	Diphenyl	$HClO_4$	2,5-Diphenyl (62, 73)	83, 53
$C_{18}H_{24}O_2$	Dicyclohexyl	H_2SO_4	2,5-Dicyclohexyl (—)	18
$C_{22}H_{16}O_6$	Bis-(2-methoxycarbonylphenyl)	$HClO_4$	2,5-Bis-(2-methoxycarbonylphenyl) (20)	84

Note: References 60–148 are on pp. 276–277.

[c] These products were obtained after hydrolysis and oxidation of the initial mixture.

TABLE IV. 2,6-Disubstituted 1,4-Benzoquinones

Formula	Substituents or Compound	Catalyst	Product(s) or Substituents in 1,3,4-Triacetoxybenzene and Yield(s) (%)	Refs.
$C_6H_2Br_2O_2$	Dibromo	BF_3	2,6-Dibromo (33)	16
		$HClO_4$	2,6-Dibromo (40)	16
$C_6H_2Cl_2O_2$	Dichloro	BF_3	2,6-Dichloro (62)	16
		H_2SO_4, $HClO_4$	2,6-Dichloro (—, —)	16
		H_2SO_4	No reaction	70
$C_6H_2I_2O_2$	Diiodo		(48) , 2-Bromo-6-methoxy (90)	
$C_7H_5BrO_3$	2-Bromo-6-methoxy	BF_3 $HClO_4$ BF_3	(10) (23) 2-Bromo-6-methoxy (90)	16 16 22
$C_7H_6O_4$	2-Hydroxy-6-methoxy	BF_3	(33)	34
		$HClO_4$	Tetraacetoxyanisole (38)[a]	34

Note: References 60–148 are on pp. 276–277.

[a] The orientation is unknown.

TABLE IV. 2,6-DISUBSTITUTED 1,4-BENZOQUINONES (Continued)

Formula	Substituents or Compound	Catalyst	Product(s) or Substituents in 1,3,4-Triacetoxybenzene and Yield(s) (%)	Refs.
$C_8H_8O_2$	Dimethyl	BF_3	2,6-Dimethyl (—, 87)	70, 34
		H_2SO_4	2,6-Dimethyl (—, —, 92)	32, 64, 7
$C_8H_8O_3$	2-Methoxy-6-methyl	BF_3	[2-CH₃, 3-OAc, 4-AcO, 5-OCH₃, 6-OAc benzene] (80)	34
		H_2SO_4	(89)	55
$C_8H_8O_4$	Dimethoxy	H_2SO_4, $HClO_4$	No reaction	7, 12, 22
$C_{11}H_{14}O_2$	2-t-Butyl-6-methyl		[2-CH₃, 3-OAc, 4-AcO, 5-C_4H_9-t, 6-OAc benzene], [2-OH, 3-C_4H_9-t, 4-OH, 6-$C_2H_5OCH_2$ benzene]	
$C_{12}H_7BrO_2$	2-Bromo-6-phenyl	H_2SO_4	(8) (3)^b	19
	2-Bromo-6-phenyl	BF_3	2-Bromo-6-phenyl (62)	34
	2-Bromo-6-phenyl	H_2SO_4	2-Bromo-6-phenyl (73)	34
	2-Bromo-6-phenyl	$HClO_4$	2-Bromo-6-phenyl (65)	34

Formula	Substituent	Catalyst	Product(s) (% yield)	Refs.
$C_{12}H_8O_3$	2-Hydroxy-6-phenyl	BF_3	[structure: C_6H_5-substituted tetraacetoxybenzene] (6), [structure: C_6H_5-substituted tetraacetoxybenzene] (30)	34
$C_{12}H_{16}O_3$	2-Methoxy-6-n-pentyl (primin)	H_2SO_4	[structure: n-C_5H_{11}, OCH_3-substituted triacetoxybenzene] (—)	84a
$C_{14}H_{20}O_2$	Di-t-butyl	BF_3	[structure: C_4H_9-t-substituted diacetoxybenzene] (low)	19
		H_2SO_4	(61)	19
		H_2SO_4	(40,[c] 50[c])	85, 86

Note: References 60–148 are on pp. 276–277.

[b] The reaction mixture was fractionally crystallized from ethanol and gave the two products shown.
[c] The overall yield is given, after hydrolysis and oxidation of the initial product to the corresponding hydroxyquinone.

TABLE V. TRISUBSTITUTED 1,4-BENZOQUINONES

Formula	Substituents or Compound	Catalyst	Product(s) or Substituents in 1,4,6-Triacetoxybenzene and Yield(s) (%)	Refs.
$C_6HBr_3O_2$	Tribromo	BF_3	AcO/OAc/Br/Br/OAc with Br (4); OAc/Br/Br/OAc/Br (9), (18)	16
$C_7H_4Br_2O_3$	2,3-Dibromo-5-methoxy	$HClO_4$	No reaction	16
	2,3-Dibromo-5-methoxy	BF_3, H_2SO_4, or $HClO_4$	No reaction	22
	2,5-Dibromo-3-methoxy	BF_3, H_2SO_4	No reaction	22
	2,5-Dibromo-3-methoxy	$HClO_4$	Quinone is decomposed	22
	2,6-Dibromo-3-methoxy	BF_3	OAc/Br/OCH$_3$/OAc/Br/AcO (29)	22
$C_7H_6O_4$	2,5-Dihydroxy-3-methyl	$HClO_4$	OAc/CH$_3$/OAc/OAc/AcO (—)[a]	87
$C_8H_7BrO_4$	2-Bromo-3,5-dimethoxy	BF_3 or H_2SO_4 or $HClO_4$	No reaction	22, 87a
	2-Bromo-3,6-dimethoxy	H_2SO_4 or $HClO_4$	No reaction	22, 87a

	2-Bromo-5,6-dimethoxy	BF$_3$ H$_2$SO$_4$ HClO$_4$	[structure: Br, OAc, OCH$_3$, CH$_3$O, OAc]	(6.5) (37) (11.5)	22
C$_8$H$_8$O$_3$	2-Hydroxy-3,5-dimethyl	H$_2$SO$_4$	[structure with OAc, CH$_3$, AcO, CH$_3$, OAc]	(90)	32
	2-Hydroxy-5,6-dimethyl	H$_2$SO$_4$	[structure with OAc, CH$_3$, CH$_3$, CH$_3$, OAc]	(—)	37 (see p. 210)
C$_8$H$_8$O$_4$	2-Hydroxy-3-methoxy-6-methyl (fumigatin)	H$_2$SO$_4$	[structure with OAc, OCH$_3$, CH$_3$, AcO, OAc]	(78)	24
	2-Hydroxy-5-methoxy-3-methyl	H$_2$SO$_4$	[quinone structure with OAc, CH$_3$, CH$_3$O, O, O]	(48)	55
	2-Hydroxy-6-methoxy-3-methyl	H$_2$SO$_4$	[quinone structure with OAc, CH$_3$, CH$_3$O, O, O]	(72)	72

Note: References 60–148 are on pp. 276–277.

[a] The compound shown was isolated after reductive acetylation of the initial product. This shows that Thiele-Winter acetoxylation did not occur.

TABLE V. Trisubstituted 1,4-Benzoquinones (Continued)

Formula	Substituents or Compound	Catalyst	Product(s) or Substituents in 1,4,6-Triacetoxybenzene and Yield(s) (%)	Refs.
$C_8H_8O_5$	2-Hydroxy-3,6-dimethoxy	H_2SO_4	OAc, OCH₃, CH₃O (88)	88
$C_9H_{10}O_2$	Trimethyl	BF_3 / H_2SO_4	AcO/CH₃/CH₃/OAc/CH₃ (62) / (—, 37)	6 / 32, 34
$C_9H_{10}O_3$	2-Methoxy-5,6-dimethyl	H_2SO_4 / H_2SO_4	No reaction / No reaction	70 / 37
$C_9H_{10}O_4$	2,3-Dimethoxy-5-methyl	H_2SO_4	HO/OCH₃/OCH₃/CH₃ (55)[b]	55
$C_9H_{10}O_5$	2,6-Dimethoxy-3-methyl / Trimethoxy	H_2SO_4 / $HClO_4$	No reaction / Tar	80 / 22
$C_{10}H_{12}O_2$	2-Ethyl-3,5-dimethyl	H_2SO_4	AcO/C₂H₅/CH₃/OAc/CH₃/OAc (62)	89
	2-Ethyl-3,6-dimethyl	H_2SO_4	OAc/C₂H₅/CH₃/OAc/CH₃/AcO (29)	89

	2-Ethyl-5,6-dimethyl	H_2SO_4	(42)	89
$C_{10}H_{12}O_3$	2-Hydroxy-3-isopropyl-6-methyl	H_2SO_4	(low)	21
	2-Hydroxy-6-isopropyl-3-methyl	H_2SO_4	(low)	21
	(phoenicin)	H_2SO_4	(80)	26
$C_{14}H_{10}O_6$	(isophoenicin)	H_2SO_4	(—)	26

Note: References 60–148 are on pp. 276–277.

[b] The quinone shown was obtained after hydrolysis and oxidation of the initially formed, oily triacetoxybenzene.

TABLE V. Trisubstituted 1,4-Benzoquinones (*Continued*)

Formula	Substituents	Catalyst	Product(s) or Substituents in 1,4,6-Triacetoxybenzene and Yield(s) (%)	Refs.
$C_{14}H_{12}O_4$	2,6-Dimethoxy-3-phenyl	BF_3	No reaction	17
$C_{18}H_{14}O_8$	(phoenicin diacetate)	H_2SO_4	(—)	26
$C_{18}H_{26}O_2$	2-(4-Cyclohexyl-*n*-butyl)-3,5-dimethyl	H_2SO_4	(96)	67
	2-(4-Cyclohexyl-*n*-butyl)-5,6-dimethyl	H_2SO_4	(27)	67

Note: References 60–148 are on pp. 276–277.

TABLE VI. 1,4-BENZOQUINONES WITH ONE OR MORE ELECTRONEGATIVE SUBSTITUENTS

Formula	Substituents	Catalyst	Product(s) or Substituent(s) in 1,3,4-Triacetoxybenzene and Yield(s) (%)	Refs.
$C_7H_3NO_2$	2-Cyano	H_2SO_4	2-Cyano (40)	34
$C_7H_5NO_5$	2-Methoxy-3-nitro	H_2SO_4	(10) structure with OAc, OCH$_3$, NO$_2$, OAc	34
$C_8N_2N_2O_2$	2,3-Dicyano	H_2SO_4	(—) structure with OAc, CN, CN, OAc, AcO	61
$C_8H_6N_2O_6$	2,5-Dimethyl-3,6-dinitro	H_2SO_4	(34) structure with CH$_2$OAc, NO$_2$, OAc, O$_2$N, CH$_3$, OAc	90
$C_8H_6O_3$	2-Acetyl	BF$_3$ H_2SO_4	2-Acetyl (78) ,, (53, 79, 92)	6 34, 91, 6

Note: References 60–148 are on pp. 276–277.

TABLE VI. 1,4-Benzoquinones with One or More Electronegative Substituents (*Continued*)

Formula	Substituents	Catalyst	Product(s) or Substituent(s) in 1,3,4-Triacetoxybenzene and Yield(s) (%)	Refs.
$C_8H_6O_4$	2-Methoxycarbonyl	BF_3	2-Methoxycarbonyl (57)	34
		H_2SO_4	,, (—, 34, 68)	92, 6, 34
$C_8H_6O_5$	2-Hydroxy-3-methoxy-carbonyl	H_2SO_4	(63) structure: benzene ring with OAc, OAc, CO_2CH_3, OAc, AcO substituents	34
$C_{12}H_8O_4S$	2-Phenylsulfonyl	H_2SO_4	2-Phenylsulfonyl (26)	34
$C_{13}H_8O_3$	2-Benzoyl	H_2SO_4	2-Benzoyl[a] (33)	93
$C_{20}H_{12}O_4$	2,5-Dibenzoyl	H_2SO_4	2,5-Dibenzoyl (62)	93

Note: References 60–148 are on pp. 276–277.

[a] In the original paper this product was thought to be the 6-benzoyl isomer. Here we assume it to be the 2 isomer by analogy with the products from 2-acetyl- and 2-methoxycarbonyl-1,4-benzoquinone.

TABLE VIIA. 1,2-Benzoquinones

Formula	Substituents	Catalyst	Product(s) and Yield(s) (%)	Refs.
$C_6H_3BrO_2$	4-Bromo	H_2SO_4 or $HClO_4$	Glassy solid	34
$C_6H_4O_2$	Nil	BF_3 or H_2SO_4	Polymeric material	34
$C_7H_6O_2$	3-Methyl	H_2SO_4	Polymeric material	34
	4-Methyl	H_2SO_4	[triacetoxy benzene structure] (—)	46
$C_7H_6O_3$	3-Methoxy	H_2SO_4	Glassy solid	33
$C_8H_8O_2$	3,5-Dimethyl	H_2SO_4	[tetrasubstituted structure] (40)	46
	4,5-Dimethyl	H_2SO_4	[tetrasubstituted structure] (60)	94
$C_{10}H_{11}BrO_2$	3-Bromo-5-t-butyl	H_2SO_4	[bromo triacetoxy structure] (5)	19

TABLE VIIA. 1,2-Benzoquinones (*Continued*)

Formula	Substituents	Catalyst	Product(s) and Yield(s) (%)	Refs.
$C_{10}H_{12}O_2$	4-t-Butyl	H_2SO_4	AcO–C₆H₂(OAc)₂–C_4H_9-t (77)	19
$C_{11}H_{14}O_2$	3-t-Butyl-5-methyl	H_2SO_4	OAc,OAc,C_4H_9-t,CH₃ (27), OAc,OAc,C_4H_9-t,CH₂OAc (20), and trisubstituted isomer (14.5)	95
	5-t-Butyl-3-methyl	H_2SO_4	OAc,CH₃,OAc,t-C_4H_9 (12)	19
	3-t-Butyl-5-methoxy	BF_3, H_2SO_4, or $HClO_4$	Red tar	19
$C_{11}H_{14}O_3$	4-t-Butyl-5-methoxy	H_2SO_4	OAc,OAc,CH₃O (6)	19
		BF_3	Black tar	19
$C_{13}H_{10}O_3$	4-Methoxy-5-phenyl	H_2SO_4	OCH₃, C₆H₅ quinone (22)[a]	34

Formula	Substituent	Reagent	Product	Yield	Ref.
$C_{14}H_{20}O_2$	3,5-Di-t-butyl	H_2SO_4	(structure: OAc, OAc, C_4H_9-t, AcO)	(86–93)	18
	4-(1,1,3,3-Tetramethylbutyl)	H_2SO_4	(structure: OAc, OAc, OAc)	(—)	35
$C_{14}H_{20}O_3$	3,5-Di-t-butyl-6-hydroxy	H_2SO_4	(quinone structure with t-C_4H_9, C_4H_9-t, AcO)	(45)	18
$C_{21}H_{20}O_4$	(diquinone structure)	H_2SO_4	(acetoxy arene structure)	(34)	96
	(diquinone structure)	H_2SO_4	(acetoxy arene structure) and an unsymmetrical isomer	(—) (—)	19, 97

Note: References 60–148 are on pp. 276–277.

[a] This compound was obtained when the initial product was successively hydrolyzed, methylated, and oxidized.

TABLE VIIB. OTHER 1,2-QUINONES, EXCLUDING 1,2-NAPHTHOQUINONES

Formula	Compound	Catalyst	Product(s) and Yield(s) (%)	Refs.
$C_{11}H_6O_5$		H_2SO_4	(10–24)	40
$C_{12}H_4Br_2O_2$		BF_3	No reaction	98
$C_{12}H_6O_2$		$HClO_4$	No reaction	34
$C_{13}H_{11}NO_3$		BF_3	(77)	27, 99

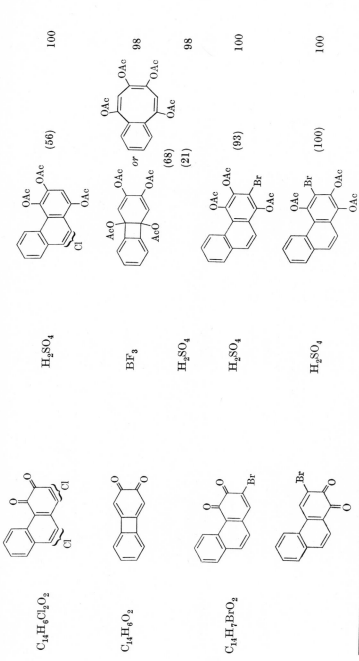

Note: References 60–148 are on pp. 276–277.

[a] This product was obtained by hydrolysis and oxidation of the initial triacetoxy product, which could not be obtained pure.

TABLE VIIB. OTHER 1,2-QUINONES, EXCLUDING 1,2-NAPHTHOQUINONES (*Continued*)

Formula	Compound	Catalyst	Product(s) and Yield(s) (%)	Refs.
$C_{14}H_8O_2$	(phenanthrene-1,2-quinone)	H_2SO_4	1,3,4-triacetoxyphenanthrene (36)	101
	(phenanthrene-3,4-quinone)	H_2SO_4	2-hydroxyphenanthrene-1,4-quinone (30)[a]	102
	(phenanthrene-9,10-quinone)	BF_3 H_2SO_4 $HClO_4$	9,10-diacetoxyphenanthrene (9) (—) (15)	34 14 34

$C_{14}H_8O_3$	[phenanthrene-dione with HO]	H_2SO_4	[phenanthrene triacetate with AcO] (72)	103
$C_{14}H_{10}O_4$	[dione with CH$_3$O, CH$_3$O]	ZnCl$_2$, BF$_3$, H$_2$SO$_4$, or HClO$_4$	No reaction	104
$C_{16}H_{14}O_4$	[dione with CH$_3$, CH$_3$, CH$_3$O, CH$_3$O]	H_2SO_4	No reaction	34
$C_{16}H_{14}O_6$	[dione with OCH$_3$, OCH$_3$, CH$_3$O, CH$_3$O]	BF$_3$, H$_2$SO$_4$, or HClO$_4$	No reaction	105

Note: References 60–148 are on pp. 276–277.

TABLE VIIIA. 1,4-NAPHTHOQUINONES UNSUBSTITUTED IN THE QUINONOID RING

Formula	Quinone	Catalyst	Product(s) and Yield(s) (%)	Refs.
$C_{10}H_{10}O_4$		H_2SO_4	(—, 56)	106, 107
$C_{10}H_5ClO_2$		BF_3		108
$C_{10}H_6O_2$		H_2SO_4 $ZnCl_2$ BF_3 H_2SO_4 $HClO_4$	(total yield, 63) (36) (49) (32, 87) (75, 81) (75, 97) (80, 95)	109 9, 5 59, 8 8, 109a 59, 9
$C_{10}H_6O_3$	(juglone)	H_2SO_4 "	(74) (approximately equal amounts) (total yield, 68)	110 111 107

$C_{10}H_6O_4$ (naphthazarin)

H_2SO_4	(37–42)[a]	(52)[a] 112
		(21–26)[a]
H_2SO_4		38
H_2SO_4		109
H_2SO_4	(52)	(8) 38
H_2SO_4	(69)[a]	(10)[a] 38
	(—)	5
H_2SO_4		113

Note: References 60–148 are on pp. 276–277.

[a] The initial products were hydrolyzed, then oxidized to the quinones shown.

TABLE VIIIA. 1,4-Naphthoquinones Unsubstituted in the Quinonoid Ring (*Continued*)

Formula	Quinone	Catalyst	Product(s) and Yield(s) (%)		Refs.
$C_{10}H_6O_5$	(naphthopurpurin)	H_2SO_4	(56)		107
$C_{11}H_6O_5$		H_2SO_4	(20)^a	(55)^a	114
$C_{11}H_8O_2$		H_2SO_4		(7)	109
$C_{11}H_8O_4$		H_2SO_4	(62)	(—)^{ab}	113
$C_{12}H_8O_6$		H_2SO_4	(—)^c		114

$C_{12}H_9NO_3$	[AcNH-naphthoquinone structure]	BF_3 or H_2SO_4 $HClO_4$	No reaction Unidentified product	109 109
$C_{12}H_{10}O_4$	[5,8-dihydroxy-6-ethyl-1,4-naphthoquinone structure]	H_2SO_4	[structure with OH, OH, C_2H_5] (—)^{ab}	113
$C_{13}H_{10}O_5$	[methoxy ethyl naphthoquinone structure]	H_2SO_4	[tetraacetoxy structure with CH_3O, C_2H_5] (—)^b	115
$C_{14}O_{10}O_6$	[diacetoxy naphthoquinone structure]	H_2SO_4	[hexaacetoxy naphthalene structure] (41)	116
$C_{16}H_{18}O_4$	[dihydroxy isohexyl naphthoquinone structure]	H_2SO_4	[dihydroxy isohexyl naphthoquinone structure] (—)^{ab}	113

Note: References 60–148 are on pp. 276–277.

^{a} The initial products were hydrolyzed, then oxidized to the quinones shown.
^{b} The orientation of the product is unknown.
^{c} The quinone shown was obtained after the initial product had been hydrolyzed.

TABLE VIIIB. 2-SUBSTITUTED 1,4-NAPHTHOQUINONES

Formula	Substituents in 1,4-Naphthoquinone	Catalyst	Product(s) or Substituent(s) in 1,4-Naphthoquinone and Yield(s) (%)	Refs.
$C_{10}H_3Br_3O_3$	2,3,6-Tribromo-5-hydroxy	H_2SO_4	2,3,6-Tribromo-5-acetoxy (—)	117
$C_{10}H_5BrO_3$	2-Bromo-8-hydroxy	H_2SO_4	2-Bromo-8-acetoxy (—)	117
$C_{10}H_5ClO_3$	2-Chloro-8-hydroxy	H_2SO_4	2-Chloro-8-acetoxy (—)	117
$C_{10}H_5ClO_4$	2-Chloro-5,8-dihydroxy	H_2SO_4	2-Chloro-5,8-diacetoxy (—)	106
$C_{10}H_6O_3$	2-Hydroxy (lawsone)	$ZnCl_2$, H_2SO_4	2-Acetoxy (—, —)	5
$C_{10}H_6O_4$	2,5-Dihydroxy	—	Starting material recovered after hydrolysis	38
	2,6-Dihydroxy	H_2SO_4	2,6-Diacetoxy	118
	2,8-Dihydroxy	—	Starting material recovered after hydrolysis	38
$C_{10}H_6O_5$	2,5,7-Trihydroxy (flaviolin)	$HClO_4$	2,5,7-Triacetoxy (good)	119
	2,5,8-Trihydroxy (naphthopurpurin)	H_2SO_4	5,6,8-Triacetoxy (—, —, 94)	109a, 113, 107
$C_{10}H_6O_6$	2,3,5,8-Tetrahydroxy (spinazarin)	H_2SO_4	2,3,5,8-Tetraacetoxy (—, 100)	120, 120a
$C_{11}H_6Br_2O_4$	5,7-Dibromo-2,6-dihydroxy-8-methyl	H_2SO_4	5,7-Dibromo-2,6-diacetoxy-8-methyl (34)	121
$C_{11}H_8O_2$	2-Methyl	BF_3	OAc, CH₃, OAc, OAc structure (52, 59)	6, 34
		H_2SO_4	(44)	34
		$HClO_4$	(54)	9
		$ZnCl_2$	No reaction[a]	9, 42
		H_2SO_4	No reaction	41

Formula	Substituents	Catalyst	Product (% yield)	Refs.
$C_{11}H_8O_3$	3-Hydroxy-2-methyl	H_2SO_4	3-Acetoxy-2-methyl (—)	109a
	5-Hydroxy-2-methyl (plumbagin)	H_2SO_4	(—), (—)	111
$C_{11}H_8O_6$	8-Hydroxy-2-methyl	H_2SO_4,	(—)	122
				123
	2,3,5,8-Tetrahydroxy-6-methyl[b]	H_2SO_4	8-Acetoxy-2-methyl (—)	122
		H_2SO_4	2,3,5,8-Tetraacetoxy-6-methyl (—, 100)	120, 120a
$C_{12}H_8O_3$	2-Acetyl	BF_3	(82)	123a
$C_{12}H_8O_4$	2-Acetoxy	BF_3	(18)	34

Note: References 60–148 are on pp. 276–277.

[a] When the reaction was done in nitromethane, a trace of 1,4-diacetoxy-3-chloro-2-methylnaphthalene was formed.[9]

[b] This quinone may have reacted in its tautomeric form, 5,6,7,8-tetrahydroxy-2-methyl-1,4-naphthoquinone, to give the corresponding tetraacetate.

TABLE VIIIB. 2-SUBSTITUTED 1,4-NAPHTHOQUINONES (*Continued*)

Formula	Substituents in 1,4-Naphthoquinone	Catalyst	Product(s) or Substituent(s) in 1,4-Naphthoquinone and Yield(s) (%)	Refs.
$C_{12}H_9NO_3$	2-Acetamido	BF_3	(18) [structure: naphthoquinone fused with oxazoline bearing CH_3; with OAc, OAc substituents]	34
$C_{12}H_{10}O_2$	2,5-Dimethyl	H_2SO_4	No reaction	124
	2,6-Dimethyl	H_2SO_4	No reaction	125
	2,8-Dimethyl	H_2SO_4	No reaction	124
$C_{12}H_{10}O_6$	2,5,8-Trihydroxy-3-methoxy-6 (or 7)-methyl	H_2SO_4	2,5,8-Triacetoxy-3-methoxy-6 (or 7)-methyl (—)	120a
$C_{13}H_{10}O_2$	2-Allyl	H_2SO_4	Unidentified, colorless product (—)	126
$C_{13}H_{12}O_6$	5,8-Dihydroxy-2,3-dimethoxy-6-methyl[c]	H_2SO_4	5,8-Diacetoxy-2,3-dimethoxy-6-methyl (—)	120a

Note: References 60–148 are on pp. 276–277.

[c] This quinone may have reacted in a tautomeric form to give a diacetate isomeric with the one shown.

TABLE VIIIC. 1,2- AND 2,6-NAPHTHOQUINONES

Formula	Quinone or Substituent in 1,2-Naphthoquinone	Catalyst	Product(s) or Substituent(s) in 1,2,4-Triacetoxynaphthalene and Yield(s) (%)	Refs.
1,2-Naphthoquinones				
$C_{10}H_5BrO_2$	6-Bromo	H_2SO_4	6-Bromo (62)	128
$C_{10}H_5ClO_2$	3-Chloro	H_2SO_4	3-Chloro (49)	43
	4-Chloro	H_2SO_4	No reaction	43
$C_{10}H_6O_2$	Nil	$ZnCl_2$ or H_2SO_4	Nil (—, 87)	5
$C_{10}H_6O_3$	6-Hydroxy	H_2SO_4	6-Acetoxy (—, —)	107, 118
	7-Hydroxy	H_2SO_4	7-Acetoxy (—, —)	118
$C_{11}H_8O_2$	4-Methyl	H_2SO_4	![structure with OAc, OAc, CH₂OAc] (79)	43
$C_{12}H_9NO_3$	6-Methyl	H_2SO_4	Unidentified colorless product (—)	128
	7-Acetamido	H_2SO_4	7-Acetamido (0–18)	129
$C_{12}H_{10}O_2$	3,7-Dimethyl	H_2SO_4	3,7-Dimethyl (—, 85)	125, 130

Note: References 60–148 are on pp. 276–277.

TABLE VIIIC. 1,2- AND 2,6-NAPHTHOQUINONES (*Continued*)

Formula	Quinone or Substituent(s) in 1,2-Naphthoquinone	Catalyst	Product(s) or Substituent(s) in 1,2,4-Triacetoxynaphthalene and Yield(s) (%)	Refs.
$C_{17}H_{11}ClO_2$	4-*p*-Chlorobenzyl	H_2SO_4	X = Cl (—)	43
$C_{11}H_{12}O_2$	4-Benzyl	H_2SO_4	X = H (51)	43
$C_{17}H_{16}O_6$	[structure with CH(CO₂C₂H₅)₂]	H_2SO_4	AcOC(CO₂C₂H₅)₂ (good)	43
$C_{19}H_{22}O_2$	[structure with C₃H₇-*i*]	H_2SO_4	[tetraacetoxy phenanthrene with C₃H₇-*i*] (—)	29
	2,6-Naphthoquinone			
$C_{10}H_4Cl_2O_2$	[dichloro-2,6-naphthoquinone]	H_2SO_4	Unidentified colorless product (—)	127

Note: References 60–148 are on pp. 276–277.

TABLE IX. ANTHRAQUINONES

Formula	Quinone	Catalyst	Product(s) and Yield(s) (%)	Refs.
$C_{14}H_2Cl_4O_4$		H_2SO_4	(—)	131
$C_{14}H_6Cl_2O_3$		H_2SO_4	(—)	44
$C_{14}H_6O_5$		H_2SO_4	(—)[a]	132
$C_{14}H_6O_6$		H_2SO_4	(—)[b]	132
$C_{14}H_6O_7$		H_2SO_4	(—)	132

Note: References 60–148 are on pp. 276–277.

[a] The product was a mixture of two isomers.
[b] This compound was obtained after hydrolysis of the initial product.
[c] This quinone may have reacted in a tautomeric form to give a product isomeric with the one shown.

TABLE IX. Anthraquinones (Continued)

Formula	Quinone	Catalyst	Product(s) and Yield(s) (%)	Refs.
$C_{14}H_6O_8$		H_2SO_4	(—)	132
$C_{14}H_7ClO_3$		H_2SO_4	(51)	133
$C_{14}H_8O_2$		H_2SO_4 or $HClO_4$	No reaction	14, 34
$C_{14}H_8O_3$		H_2SO_4	(—)	44
$C_{14}H_8O_4$	(quinizarin)	H_2SO_4	(—)	133

$C_{15}H_6O_6$		H$_2$SO$_4$	(—)[a]	134
$C_{15}H_8O_4$		H$_2$SO$_4$	(—)	106
$C_{15}H_{10}O_3$		H$_2$SO$_4$	(—)	45
		H$_2$SO$_4$	(—)	45
$C_{16}H_8O_6$		H$_2$SO$_4$	(—)	135

Note: References 60–148 are on pp. 276–277.

TABLE IX. Anthraquinones (Continued)

Formula	Quinone	Catalyst	Product(s) and Yield(s) (%)	Refs.
$C_{16}H_9ClO_4$	(structure)	H_2SO_4	(—)	44
$C_{16}H_{10}O_4$	(structure)	H_2SO_4	(—)	44
$C_{16}H_{10}O_6$	(structure)	H_2SO_4	(—) (dermocybin tetraacetate)	25
$C_{32}H_{28}O_7$	(structure)	H_2SO_4	(—) (octaacetylcarminic acid)[a]	30

Note: References 60–148 are on pp. 276–277.

[a] See footnote on p. 209.

TABLE XA. Diphenoquinones

Formula	Substituents in 4,4'-Diphenoquinone	Catalyst	Product(s) and Yield(s) (%)	Refs.
$C_{12}H_8O_2$	Nil	BF_3	(trace)	34
		H_2SO_4	(trace, —, —)	34, 64, 136
		$HClO_4$	(4)	34
$C_{16}H_{16}O_2$	3,3',5,5'-Tetramethyl	H_2SO_4	(—, —, 65)	7, 64, 34
$C_{16}H_{16}O_6$	3,3',5,5'-Tetramethoxy (coerulignone)	BF_3	(trace)	34
		H_2SO_4	No isolable product	64
$C_{22}H_{24}O_2$	3,3'-Di-t-butyl-5,5'-dimethyl	BF_3	(76)	19
$C_{28}H_{40}O_2$	3,3',5,5'-Tetra-t-butyl	BF_3	(90)	19

Note: References 60–148 are on pp. 276–277.

TABLE XB. STILBENOQUINONES

Formula	Substituents in [quinone-CH=CH-quinone structure]	Catalyst	Product or Substituents in [AcO-C6H4-CH(OAc)-CH(OAc)-C6H4-OAc structure] and Yield(s) (%)	Refs.
$C_{14}H_6Br_4O_2$	3,3′,5,5′-Tetrabromo	H_2SO_4	3,3′,5,5′-Tetrabromo (—)	137
$C_{14}H_6Cl_4O_2$	3,3′,5,5′-Tetrachloro	H_2SO_4	3,3′,5,5′-Tetrachloro (—)	138
$C_{18}H_{18}O_2$	3,3′,5,5′-Tetramethyl	H_2SO_4	3,3′,5,5′-Tetramethyl (—)	139
$C_{30}H_{42}O_2$	3,3′,5,5′-Tetra-t-butyl	H_2SO_4	3,3′,5,5′-Tetra-t-butyl (126),[a] [structure with AcO, t-C4H9, C≡C, C4H9-t, OAc] (48)[b]	139

Note: References 60–148 are on pp. 276–277.

[a] This is the yield of crude product when the reaction mixture was kept for 36 hours at room temperature.
[b] The diphenylacetylene was obtained when the reaction mixture was worked up after 6 hours at room temperature.

TABLE XI. Miscellaneous Quinones

Formula	Quinone	Catalyst	Product(s) and Yield(s) (%)	Refs.
$C_8H_6N_2O_6$	2-methyl-3-nitro-5-methyl-6-nitro-1,4-benzoquinone	H_2SO_4	tetrasubstituted benzene with OAc, CH$_2$OAc, NO$_2$, OAc, CH$_3$, O$_2$N (34)	90
$C_{10}H_6Br_2O_2$	2,3-dibromo-5,6-dimethyl-1,4-benzoquinone	H_2SO_4	No reaction	34
$C_{10}H_{12}O_2$	tetramethyl-1,4-benzoquinone	BF_3, H_2SO_4, or $HClO_4$	No reaction	34
$C_{20}H_{10}O_2$	dibenzo quinone	$HClO_4$	dihydroxy product $(—)^a$	136, 140

Note: References 60–148 are on pp. 276–277.

[a] The compound shown was obtained after hydrolysis and oxidation of the initial product.

TABLE XI. MISCELLANEOUS QUINONES (*Continued*)

Formula	Quinone	Catalyst	Product(s) and Yield(s) (%)	Refs.
$C_{30}H_{28}O_8$	(erythroaphin-*fb*)	$HClO_4$	pentaacetate (*ca.* 80)[b]	141, 142
	(erythroaphin-*sl*)	$HClO_4$	Mixture of pentaacetates (R = H, R' = OAc) and (R = OAc, R' = H) (—)	141, 142

Note: References 60–148 are on pp. 276–277.

[b] Hydrolysis of the pentaacetate, followed by oxidation, gave hydroxyerythroaphin-*fb* in an overall yield of 80%.

TABLE XII. REACTION OF QUINONES WITH OTHER ANHYDRIDES AND WITH DIACETYL SULFIDE

Formula	Quinone	Reagent	Catalyst	Product(s) and Yield(s) (%)	Refs.
$C_6H_4O_2$	1,4-Benzoquinone	$(C_6H_5CO)_2O$	H_2SO_4	1,4-bis(OCOC$_6$H$_5$)benzene (22), 1,2,4-tris(OCOC$_6$H$_5$)benzene (52)	34
		$(CF_3CO)_2O$	BF_3 or H_2SO_4	Decomposition occurs	34
		$(CH_3CO)_2S$	BF_3	2,4-(OAc)$_2$-SAc-benzene (70)	47
$C_7H_6O_2$	1,4-Toluquinone	$(CH_3CO)_2S$	BF_3	2,5-(OAc)$_2$-4-CH$_3$-AcS-benzene (17), 2,6-(OAc)$_2$-CH$_3$-SAc-benzene (45)	47
$C_8H_8O_2$	3,5-Dimethyl-1,2-benzoquinone	$(CH_3CO)_2S$	H_2SO_4	tetrasubstituted product (33)	46
$C_8H_8O_4$	2,6-Dimethoxy-1,4-benzoquinone	$(CF_3CO)_2O$	H_2SO_4	No reaction	34

Note: References 60–148 are on pp. 276–277.

TABLE XII. REACTIONS OF QUINONES WITH OTHER ANHYDRIDES AND WITH DIACETYL SULFIDE (*Continued*)

Formula	Quinone	Reagent	Catalyst	Product(s) and Yield(s) (%)	Refs.
$C_{10}H_6O_2$	1,2-Naphthoquinone	$(CH_3CO)_2S$	BF_3	[naphthalene with OAc, OAc, SAc substituents (42)], [naphthalene bis-(OAc, OAc) $)_2$ (25)]	47
	1,4-Naphthoquinone	$(CH_3CO)_2S$	BF_3	[naphthalene with OAc, SAc, SAc, OAc (3)], [naphthalene with OAc, SAc, OAc, OAc (80)]	47

Note: References 60–148 are on pp. 276–277.

TABLE XIII. REACTION OF QUINONES WITH ACID CHLORIDES AND ACID BROMIDES

Formula	Substituents in 1,4-Benzoquinone	Acid Halide	Catalyst	Product(s) and Yield(s) (%)	Refs.
$C_6HCl_3O_2$	Trichloro	CH_3COCl	—	[structure: tetrachloro diacetoxybenzene] (—)	143
$C_6H_2Cl_2O_2$	2,3-Dichloro	CH_3COCl	$AlCl_3$	[structure] (—)	48
	2,5-Dichloro	CH_3COCl	$AlCl_3$	[structure] (—)	48
	2,6-Dichloro	CH_3COCl	$AlCl_3$	[structure] (—), [structure] (small)	48
$C_6H_3ClO_2$	Chloro	CH_3COCl	—	[structure] (—)	3

Note: References 60–148 are on pp. 276–277.

TABLE XIII. Reaction of Quinones with Acid Chlorides and Acid Bromides (Continued)

Formula	Substituents in 1,4-Benzoquinone	Acid Halide	Catalyst	Product(s) and Yield(s) (%)	Refs.
$C_6H_4O_2$	Nil	CH_3COBr	—	[structure: OAc/Cl/Br/OAc benzene] (—)	3
		CH_3COCl	—	[structure: OAc/Cl/OAc benzene] (—), [structure: OAc/Cl/OAc benzene with Cl] (—)	3
		CH_3COCl	CH_3CO_2H	(—)	144
		CH_3COCl	H_2O[a]	(20)	145
		CH_3COCl		(13)	146
		CH_3COCl	$Zn(OAc)_2$[b]	(17), (7)	146
		CH_3COBr	—	[structure: OAc/Br/OAc benzene] (—), [structure: OAc/Br/OAc/Br benzene] (—)	3
$C_7H_6O_2$	Methyl	CH_3COCl	H_2O	[structure: CH_3/OAc/Cl/OAc benzene] (23), [structure: CH_3/OAc/Cl/OAc benzene with Cl] (—)	146
		"	$Zn(OAc)_2$	(—)	146

$C_8H_8O_3$	2-Methoxy-5-methyl	CH_3COCl	—	![structure with OAc, OCH$_3^c$, Cl, CH$_3$] (—)	7
$C_8H_8O_4$	2,5-Dimethoxy	CH_3COCl	$AlCl_3^d$	![structure with OAc, OCH$_3^c$, Cl, CH$_3$O] (—) , ![structure with OCH$_3$, Cl, Cl, CH$_3$O]	48, 49
			H_2SO_4	(66)	50
			$ZnCl_2$	$(46)^e$	51
			BF_3	$(7.5)^e$	
			$FeCl_3$	$(30)^e$	
			$SnCl_4$	$(31)^e$	(trace) (4.5) (18) (3.5) (5.5)
		CH_3COBr	—	![structure OAc, OCH$_3$, OAc, CH$_3$O] (83)	50
	2,6-Dimethoxy	CH_3COCl	$AlCl_3$ or H_2SO_4	No reaction	48, 50

Note: References 60–148 are on pp. 276–277.

[a] With this catalyst a small amount of 1,4-diacetoxybenzene was obtained also.
[b] With this catalyst, 1,4-diacetoxybenzene (3.6%) was obtained also.
[c] This compound was originally formulated as the alternative isomer, based on an analogy with an incorrect structure.
[d] With this catalyst a small amount of 1,2,4,5-tetrachloro-2,5-dimethoxybenzene was obtained also.
[e] The yield of this product is calculated from the yield of the phenol obtained from it by hydrolysis.

TABLE XIII. REACTION OF QUINONES WITH ACID CHLORIDES AND ACID BROMIDES (Continued)

Formula	Substituents in 1,4-Benzoquinone	Acid Halide	Catalyst	Product(s) and Yield(s) (%)		Refs.
		CH₃COBr	HgBr₂	structure with OAc, OCH₃, Br, OAc, CH₃O groups; Main product	structure with OAc, OCH₃, R, OAc, CH₃O, R groups; Small amounts of compound R = H and R = Br	50
C₁₀H₁₂O₂	5-Isopropyl-2-methyl	CH₃COCl	—	structure with CH₃, OAc, OAc, Cl, i-C₃H₇ (—)		3
		CH₃COBr	—	structure with CH₃, OAc, OAc, Br, i-C₃H₇ (—)		3
		C₆H₅COCl	—	structure with CH₃, OCOC₆H₅, OCOC₆H₅, Cl, i-C₃H₇ (—), structure with OCOC₆H₅, CH₃, Cl, OCOC₆H₅, i-C₃H₇ (—)		3

			Main product		
$C_{10}H_{12}O_4$	2,5-Diethoxy	CH_3COCl	$AlCl_3$ or H_2SO_4	[structure: AcO, OC₂H₅, Cl, Cl, C₂H₅O] , [structure: Cl, OC₂H₅, Cl, C₂H₅O] (trace)	147

Stilbenequinone

$C_{14}H_{10}O_2$	Nil	CH_3COBr	—	[structure: AcO–C₆H₄–CH(Br)–CH(Br)–C₆H₄–OAc] (—)	148
$C_{18}H_{18}O_2$	3,3′,5,5′-Tetramethyl	CH_3COCl	CH_3CO_2H	[structure with CH₃ groups, CHCl–CHCl, OAc] (14)	139
$C_{30}H_{42}O_2$	3,3′,5,5′-Tetra-t-butyl	CH_3COCl	CH_3CO_2H	[structure with t-C₄H₉ groups, CHCl–CHCl, OAc] (16.5)	139

Note: References 60–148 are on pp. 276–277.

REFERENCES TO TABLES

[60] G. Bargellini and G. Avrutin, *Gazz. Chim. Ital.*, [2] **40**, 347 (1910).
[61] J. Thiele and F. Günther, *Ann.*, **349**, 45 (1906).
[62] A. Oliverio and G. Castelfranchi, *Gazz. Chim. Ital.*, **80**, 276 (1950).
[63] Shigeo Ukai and Kazuo Hirose, *Chem. Pharm. Bull.* (Tokyo), **16**, 202 (1968).
[64] H. Erdtman, *Svensk Kem. Tidskr.*, **44**, 135 (1932) [*C.A.*, **26**, 4803 (1932)].
[65] S. L. Fries, A. H. Soloway, B. K. Morse, and W. C. Ingersoll, *J. Amer. Chem. Soc.*, **74**, 1305 (1952).
[66] H. Musso and D. Maassen, *Ann.*, **689**, 93 (1965).
[67] W. M. McLamore, *J. Amer. Chem. Soc.*, **73**, 2225 (1951).
[68] A. H. Cook, I. M. Heilbron, and F. B. Lewis, *J. Chem. Soc.*, **1942**, 659.
[69] L. F. Fieser and R. G. Kennelly, *J. Amer. Chem. Soc.*, **57**, 1611 (1935).
[70] L. F. Fieser and M. I. Ardao, *J. Amer. Chem. Soc.*, **78**, 774 (1956).
[71] J. A. Ballantine, C. H. Hassall, and G. Jones, *J. Chem. Soc.*, **1965**, 4672.
[72] W. K. Anslow, J. N. Ashley, and H. Raistrick, *J. Chem. Soc.*, **1938**, 439.
[72a] J. C. Catlin, G. D. Daves, and K. Folkers, *J. Med. Chem.*, **14**, 45 (1971).
[73] W. Baker, J. F. W. McOmie, and T. L. V. Ulbricht, *J. Chem. Soc.*, **1952**, 1825.
[74] A. Skita and W. Rohrmann, *Ber.*, **63**, 1473 (1930).
[75] G. Kehrmann and M. Sterchi, *Helv. Chim. Acta*, **9**, 859 (1926).
[76] A. H. Crosby and R. E. Lutz, *J. Amer. Chem. Soc.*, **78**, 1233 (1956).
[77] T. Posternak, H. W. Ruelius, and J. Tcherniak, *Helv. Chim. Acta*, **26**, 2031 (1943).
[78] Yasuhiko Asahina and Ei-Iti Ishibashi, *Ber.*, **62**, 1207 (1929).
[79] A. Sonn, *Ber.*, **64**, 1847 (1931).
[80] G. Aulin and H. Erdtmann, *Svensk Kem. Tidskr.* **50**, 42 (1938) [*C.A.*, **32**, 4552 (1938)].
[81] H. Erdtman, *Proc. Roy. Soc.*, *Ser. A*, **143**, 191 (1934).
[82] H. Erdtman, *Ann.*, **513**, 240 (1934).
[83] B. F. Cain, *J. Chem. Soc.*, **1961**, 936.
[84] H. Erdtman and M. Nilsson, *Acta Chem. Scand.*, **10**, 735 (1956).
[84a] G. B. Marini-Bettòlo, F. delle Monache, O. G. da Lima, and S. de B. Coelho, *Gazz. Chim. Ital.*, **101**, 41 (1971).
[85] H. Musso and R. Zunker, *Ann.*, **717**, 64 (1968).
[86] U. Cuntze, D. Maassen, and H. Musso, *Chem. Ber.*, **102**, 2851 (1969).
[87] J. C. Sheehan, W. B. Lawson, and R. J. Gaul, *J. Amer. Chem. Soc.*, **80**, 5536 (1958).
[87a] B. Lindberg, *Acta Chem. Scand.*, **6**, 1048 (1952).
[88] W. Baker, *J. Chem. Soc.*, **1941**, 668.
[89] I. Ljungcrantz and K. Mosbach, *Acta Chem. Scand.*, **18**, 638 (1964).
[90] G. Schill, *Ann.*, **693**, 182 (1966).
[91] N. N. Vorozhtsov and V. P. Mamaev, *Sb. Statei Obshch. Khim. Akad. Nauk SSSR*, **1**, 533 (1953) [*C.A.*, **49**, 925e (1955)].
[92] A. Kreuchunas, U.S. Pat., 2,849,480 [*C.A.*, **53**, 4219c (1959)].
[93] M. T. Bogert and H. P. Howells, *J. Amer. Chem. Soc.*, **52**, 837 (1930).
[94] L. Horner and K. Sturm, *Ann.*, **597**, 1 (1955).
[95] F. Takacs, *Monatsh. Chem.*, **95**, 961 (1964).
[96] L. Taimr and J. Pospíšil, *Chem. Ind.* (London), **1969**, 456.
[97] W. Baker and J. C. McGowan, *J. Chem. Soc.*, **1943**, 486.
[98] J. M. Blatchly, J. F. W. McOmie, and S. D. Thatte, *J. Chem. Soc.*, **1962**, 5090.
[99] G. R. Allen, J. F. Poletto, and M. J. Weiss, *J. Amer. Chem. Soc.*, **86**, 3877 (1964).
[100] L. F. Fieser and J. T. Dunn, *J. Amer. Chem. Soc.*, **59**, 1024 (1937).
[101] L. F. Fieser, *J. Amer. Chem. Soc.*, **51**, 1896 (1929).
[102] L. F. Fieser, *J. Amer. Chem. Soc.*, **51**, 940 (1929).
[103] M. S. Newman and R. L. Childers, *J. Org. Chem.*, **32**, 62 (1967).
[104] J. F. W. McOmie, M. L. Watts, and D. E. West, *J. Chem. Soc.*, *C*, **1969**, 646.
[105] W. Baker, N. J. McLean, and J. F. W. McOmie, *J. Chem. Soc.*, **1964**, 1067.
[106] K. Zahn and P. Ochwat, *Ann.*, **462**, 72 (1928).

[107] L. F. Fieser and J. T. Dunn, *J. Amer. Chem. Soc.*, **59**, 1016 (1937).
[108] L. F. Fieser and R. H. Brown, *J. Amer. Chem. Soc.*, **71**, 3615 (1949).
[109] J. M. Lyons and R. H. Thomson, *J. Chem. Soc.*, **1953**, 2910.
[109a] Chika Kuroda, *J. Sci. Res. Inst.* (Tokyo), **45**, 166 (1951) [*C.A.*, **46**, 6115 (1952)].
[110] R. H. Thomson, *J. Org. Chem.*, **16**, 1082 (1951).
[111] R. G. Cooke, H. Dowd, and W. Segal, *Aust. J. Chem.*, **6**, 39 (1953).
[112] R. H. Thomson, *J. Org. Chem.*, **13**, 870 (1948).
[113] Tika Kuroda and Mizu Wada, *Sci. Pap. Inst. Phys. Chem. Res.* (Tokyo), **34**, 1740 (1938) [*C.A.*, **33**, 2511 (1939)].
[114] I. Singh, R. E. Moore, C. W. J. Chang, R. T. Ogata, and P. J. Scheuer, *Tetrahedron*, **24**, 2969 (1968).
[115] K. Wallenfels, *Ber.*, **75**, 785 (1942).
[116] O. Dimroth and H. Roos, *Ann.*, **456**, 177 (1927).
[117] R. H. Thomson, *J. Org. Chem.*, **13**, 377 (1948).
[118] O. Dimroth and B. Kerkovius, *Ann.*, **399**, 36 (1913).
[119] B. D. Astill and J. C. Roberts, *J. Chem. Soc.*, **1953**, 3302.
[120] Tika Kuroda and Hama Osima, *Proc. Imp. Acad.* (Tokyo), **16**, 214 (1940) [*C.A.*, **34**, 6939 (1940)].
[120a] Chika Kuroda, *J. Sci. Res. Inst.* (Tokyo), **46**, 188 (1952) [*C.A.*, **48**, 6411 (1954)].
[121] G. Rohde and G. Dorfmüller, *Ber.*, **43**, 1363 (1910).
[122] L. F. Fieser and J. T. Dunn, *J. Amer. Chem. Soc.*, **58**, 572 (1936).
[123] Michizo Asano and Jun-ichi Hase, *J. Pharm. Soc. Japan*, **63**, 90 and 410 (1943) [*C.A.*, **46**, 93 (1952)].
[123a] D. J. Cram, *J. Amer. Chem. Soc.*, **71**, 3953 (1949).
[124] M. Tishler, L. F. Fieser, and N. L. Wendler, *J. Amer. Chem. Soc.*, **62**, 2866 (1940).
[125] L. F. Fieser and A. M. Seligman, *J. Amer. Chem. Soc.*, **56**, 2690 (1934).
[126] L. F. Fieser, W. P. Campbell, and E. M. Fry, *J. Amer. Chem. Soc.*, **61**, 2206 (1939).
[127] R. Willstätter and J. Parnas, *Ber.*, **40**, 3971 (1907).
[128] L. F. Fieser, J. L. Hartwell, and A. M. Seligman, *J. Amer. Chem. Soc.*, **58**, 1223 (1936).
[129] W. A. Remers, P. N. James, and M. J. Weiss, *J. Org. Chem.*, **28**, 1169 (1963).
[130] L. F. Fieser and M. Fieser, *Reagents For Organic Synthesis*, Vol. **2**, John Wiley and Sons, 1969, p. 9.
[131] H. Waldmann, *J. Prakt. Chem.*, [2] **147**, 331 (1936/1937).
[132] O. Dimroth and V. Hilcken, *Ber.*, **54**, 3050 (1921).
[133] A. Green, *J. Chem. Soc.*, **1926**, 1428.
[134] F. Kögl and W. B. Deijs, *Ann.*, **515**, 23 (1935).
[135] O. Dimroth, O. Friedemann, and H. Kämmerer, *Ber.*, **53**, 481 (1920).
[136] B. R. Brown and A. R. Todd, *J. Chem. Soc.*, **1954**, 1280.
[137] T. Zincke and K. Fries, *Ann.*, **325**, 19 (1902).
[138] T. Zincke and K. Fries, *Ann.*, **325**, 44 (1902).
[139] W. Bradley and J. D. Sanders, *J. Chem. Soc.*, **1962**, 480.
[140] A. Calderbank, A. W. Johnson, and A. R. Todd, *J. Chem. Soc.*, **1954**, 1285.
[141] D. W. Cameron, R. I. T. Cromartie, Y. K. Hamied, P. M. Scott, and A. R. Todd, *J. Chem. Soc.*, **1964**, 62.
[142] D. W. Cameron and Lord Todd in W. I. Taylor and A. R. Battersby, Eds. *Oxidative Coupling of Phenols*, Edward Arnold Ltd., 1967, Chapter 5, p. 222.
[143] C. Graebe, *Ann.*, **146**, 1 (1868).
[144] B. Scheid, *Ann.*, **218**, 195 (1883).
[145] J. Cason, R. E. Harman, S. Goodwin, and C. F. Allen, *J. Org. Chem.*, **15**, 860 (1950).
[146] H. Burton and P. F. G. Praill, *J. Chem. Soc.*, **1952**, 2546.
[147] A. Oliverio and G. Werber, *Ann. Chim.* (Rome), **42**, 145 (1952) [*C.A.*, **47**, 3264 (1953)].
[148] T. Zincke and S. Münch, *Ann.*, **335**, 157 (1904).

CHAPTER 4

OXIDATIVE DECARBOXYLATION OF ACIDS BY LEAD TETRAACETATE

Roger A. Sheldon*

Koninklijke/Shell Laboratorium, Amsterdam

AND

Jay K. Kochi*

Indiana University, Bloomington

CONTENTS

	PAGE
Introduction	281
Mechanism	284
Primary Decomposition, Formation of Alkyl Radicals	285
Secondary Processes, Oxidative Elimination and Substitution	287
Table I. Rates of Oxidation of Phenethyl Radicals by Cu^{II}	290
Catalysis by Cupric Acetate	291
Ligand Transfer Oxidation of Alkyl Radicals: Halodecarboxylation	292
Stoichiometry of Decarboxylations with Lead Tetraacetate	292
Solvent Effects	293
Stereochemistry of Oxidative Decarboxylation	293
Decarboxylation of Formic Acid, α-Hydroxy Acids, α-Amino Acids, and α-Keto Acids	295
Bisdecarboxylation	298
Scope and Limitations	299
Primary and Secondary Carboxylic Acids	299
Alkane Formation	300
Alkene versus Ester	301
Table II. Esters and Alkenes from Oxidative Decarboxylation of Acids by Lead Tetraacetate	302
Alkene Formation through Use of Lead Tetraacetate/Cupric Acetate	303
Aromatic Substitution	306
Table III. Aromatic Substitution during Oxidative Decarboxylation by Lead Tetraacetate	307
Alkyl Halides	308
Tertiary Carboxylic Acids	308

* We appreciate the assistance of Mrs. Annemarie R. Hamori and Miss Susan A. Vladuchick, Du Pont Experimental Station, in searching the chemical literature. J. K. K. also thanks the National Science Foundation for financial support.

α-Aralkanecarboxylic Acids 312
Unsaturated Acids 313
  α,β-Unsaturated Acids 313
  β,γ-Unsaturated Acids 314
  γ,δ-Unsaturated Acids 316
  Miscellaneous Unsaturated Acids 321
Aromatic Carboxylic Acids 322
β-Keto Acids, β-Cyano Acids, and Malonic Half-Esters 323
γ-Keto Acids 325
Halodecarboxylation 326
  Metal Halide 327
  Iodine (Photochemical) 331
  Iodine (Thermal) 333
Bisdecarboxylation of 1,2-Dicarboxylic Acids 334
  Alkene Synthesis by Lead Dioxide or Lead Tetraacetate . . . 335
  Rearrangements and Side Reactions Accompanying Bisdecarboxylation . 342
Decarboxylation of Malonic Acids and Glutaric and Adipic Acids . . 348
Decarboxylation of α-Amino and α-Hydroxy Acids and α-Keto Acids . 352
Summary and Comparison with Other Methods of Decarboxylation . . 353
  $RCO_2H \rightarrow RH$ 353
  $RCO_2H \rightarrow R(-H)$ 353
  $RCO_2H \rightarrow ROAc$ 353
  $RCO_2H \rightarrow RX$ (X = Cl, Br, I) 354
  $HO_2CC-CCO_2H \rightarrow C=C$ 354
  Decarboxylation by Other Metal Salts 354

EXPERIMENTAL CONDITIONS 354
  Preparation and Properties of Lead(IV) Salts 354
  Solvents 355
  Reaction Temperature and Time 355
  Catalysts and Inhibitors 355

EXPERIMENTAL PROCEDURES 356
  1-Hexene from n-Heptanoic Acid (Use of Pb^{IV}/Cu^{II} Reagent) . . 356
  Cyclobutene from Cyclobutanecarboxylic Acid (Use of Pb^{IV}/Cu^{II} Reagent) 356
  6-Heptenoic Acid from Suberic Acid (Use of Pb^{IV}/Cu^{II} Reagent) . . 357
  $\Delta^{22\text{-}24}$-Norcholene from Cholanic Acid (Use of Pb^{IV}/Cu^{II} Reagent) . 357
  5-Acetoxy-2-pentene and 5-Acetoxy-1-pentene from 2-Methyl-5-acetoxypentanoic Acid 357
  Triphenylcarbinol from Triphenylacetic Acid (Acetoxydecarboxylation) . 358
  exo-Norbornyl Acetate from exo-Norbornane-2-carboxylic Acid (Acetoxydecarboxylation) 358
  5-Nonanone from Di-n-butylmalonic Acid 358
  Ethyl 2-Acetoxyhexanoate from Ethyl Hydrogen n-Butylmalonate (Acetoxydecarboxylation) 359
  General Procedure for Iododecarboxylation with $Pb(OAc)_4\text{-}I_2$ . . 359
  11-Acetoxy-1-iodoheptadecane from 12-Acetoxystearic Acid (Iododecarboxylation) 359
  General Procedure for Chlorodecarboxylation with $Pb(OAc)_4$-LiCl . 359
  cis,cis- and trans,trans-3,5-Dimethyl-1-chlorocyclohexane from cis,cis-3,5-Dimethylcyclohexanecarboxylic Acid (Chlorodecarboxylation) . . 360

OXIDATIVE DECARBOXYLATION OF ACIDS BY Pb(OAc)$_4$ 281

1-Phenyl-2-chloro-2-cyanopropane from 2-Benzyl-2-cyanopropanoic Acid (Chlorodecarboxylation)	360
General Procedure for Bisdecarboxylation	360
trans-Stilbene from *d,l*-2,3-Diphenylsuccinic Acid (Bisdecarboxylation)	361
1-Ethoxycarbonylbicyclo[2,2,2]oct-2-ene (Bisdecarboxylation)	362
Bicyclo[2,2,2]oct-2-en-5-one	362
TABULAR SURVEY	362
Table IV. Decarboxylation of Primary Carboxylic Acids by Lead Tetraacetate	363
Table V. Decarboxylation of Secondary Carboxylic Acids by Lead Tetraacetate	366
Table VI. Decarboxylation of Tertiary Carboxylic Acids by Lead Tetraacetate	370
Table VII. Decarboxylation of α-Aralkanecarboxylic Acids by Lead Tetraacetate	374
Table VIII. Decarboxylation of Unsaturated Carboxylic Acids by Lead Tetraacetate	375
Table IX. Decarboxylation of *o*-Substituted Benzoic Acids by Lead Tetraacetate	379
Table X. Decarboxylation of Acids by Lead Tetraacetate/Cupric Acetate	381
Table XI. Decarboxylation of Malonic Acids, Half-Esters of Malonic Acids, β-Cyano Acids, and β-Keto Acids by Lead Tetraacetate	386
Table XII. Decarboxylation of γ-Keto Acids by Lead Dioxide	389
Table XIII. Halodecarboxylation of Acids by Lead Tetraacetate and Lithium Halide or Iodine	390
Table XIV. Comparisons of Halodecarboxylation Procedures	398
Table XV. Decarboxylation of 1,2-, 1,3-, and 1,4-Dicarboxylic Acids	400
Table XVI. Decarboxylation of α-Amino and α-Hydroxy Acids by Lead Tetraacetate	418
Table XVII. Decarboxylation of α-Keto Acids by Lead Tetraacetate	420
REFERENCES TO TABLES IV–XVII	420

INTRODUCTION

Lead tetraacetate, one of the most versatile oxidizing agents known in organic chemistry, is capable of reacting with a variety of common functional groups; its uses in organic chemistry have been reviewed.[1-8] The

---

[1] R. Criegee in *Oxidation in Organic Chemistry*, K. Wiberg, Ed., Chapter V, Academic Press, New York, 1965.

[2] L. F. Fieser and M. F. Fieser, *Reagents for Organic Synthesis*, Vol. I (1967), Vol. II (1969), John Wiley & Sons, New York.

[3] J. D. Bacha, Ph.D. Thesis, Case Institute of Technology, *Diss. Abstr.*, B28, 2329 (1967); Univ. Microfilms Order No. 67-16055.

[4] R. Criegee, *Angew. Chem.*, **70**, 173 (1958).

[5] J. B. Aylward, *Quart. Rev.*, **25**, 407 (1971).

[6] M. L. Mihailovic and Z. Cekovic, *Synthesis*, 209 (1970).

[7] K. Heusler and J. Kalvoda in *Steroid Synthesis*, J. Fried, Ed., Holden-Day, San Francisco, 1971.

[8] R. M. Moriarty, "Lead Tetraacetate Oxidation of Alkenes," in *Selective Organic Transformations*, B. S. Thyagarajan, Ed., Interscience, New York, 1972.

purpose of this chapter is to review the oxidative process involving the decarboxylation of acids, whereby carboxylic acids are converted to a variety of compounds depending on the experimental conditions, e.g., solvent, structure of substrate, presence of additives. The primary objective is to evaluate critically the usefulness of this reaction in organic synthesis. Reactions of lead tetraacetate with other functional groups are not treated here unless they are directly pertinent to the discussion of the decarboxylation reaction.

Many carboxylic acids and their derivatives are readily available organic compounds. They are prepared by very common condensation reactions such as the Diels-Alder, acetoacetic ester, cyanoacetic ester, and malonic ester syntheses. Hence, methods for converting the carboxyl group to a variety of other functional groups in high yield are most desirable synthetic objectives. In addition, lead tetraacetate is a readily available and reasonably inexpensive reagent. Where possible, other methods of carrying out decarboxylation of acids are compared to the method involving lead tetraacetate. In addition, alternative methods of synthesis of particular compounds, not involving decarboxylation, are presented.

In the crystal, lead tetraacetate has four acetate ligands showing eight-coordination around the lead nucleus with a distorted dodecahedral symmetry.[9] The strong absorption bands between 1520 and 1540 cm^{-1} and weaker ones between 1400 and 1410 cm^{-1} of various lead(IV) carboxylates have been assigned to the asymmetric and symmetric stretching modes, respectively, of the carboxylate groups.[10] According to the interpretation of the infrared spectra, both oxygen atoms of each carboxylate ligand are equivalent. Lead(IV) carboxylates are generally colorless but sometimes yellow. The ultraviolet absorption spectra of lead(IV) carboxylates have maxima at approximately 230–235 nm ($\epsilon \approx 1.5$–2.0×10^4 M^{-1} cm^{-1}), the tails of which extend beyond 300 nm ($\epsilon > 10^3$ M^{-1} cm^{-1}). The latter allow photolyses of lead(IV) carboxylates to be carried out conveniently with radiation at wavelengths of 350 nm or shorter.[11]

In common with other metal acetates, solutions of lead tetraacetate in acetic acid exhibit practically no electrical conductivity.[12,13] Ebulliometric measurements indicate little dissociation or association of the lead species in solution.[14] Furthermore, isotopic tracer studies show that ligand

[9] B. Kamenar, *Acta Cryst.*, **16**, A34 (1963).
[10] K. Heusler and H. Loeliger, *Helv. Chim. Acta*, **52**, 1495 (1969).
[11] V. Franzen and R. Edens, *Angew. Chem.*, **73**, 579 (1961).
[12] K. Heyman and H. Klaus in *Chemistry of Lower Fatty Acids*, G. Jander, Ed., Interscience, New York, 1963.
[13] R. Partch and J. Monthony, *Tetrahedron Lett.*, **1967**, 4427.
[14] G. Rudakoff, *Z. Naturforsch.*, **17B**, 623 (1962).

exchange of acetato groups coordinated to the Pb^{IV} nucleus with the solvent is too rapid to measure by conventional techniques.[15] The pre-equilibrium metathesis of lead tetraacetate with carboxylic acids (Eqs. 1, 2, and 3) therefore generally does not impose kinetic limitations on the decarboxylation reactions. For most acids the equilibrium constant for the first metathesis (Eq. 1) is 1, and it is probably close to 1 for the other equilibria as well.

$$RCO_2H + Pb(OAc)_4 \rightleftharpoons RCO_2Pb(OAc)_3 + HOAc \qquad (Eq.\ 1)$$

$$RCO_2H + RCO_2Pb(OAc)_3 \rightleftharpoons (RCO_2)_2Pb(OAc)_2 + HOAc \qquad (Eq.\ 2)$$

$$RCO_2H + (RCO_2)_2Pb(OAc)_2 \rightleftharpoons (RCO_2)_3Pb(OAc) + HOAc \qquad (Eq.\ 3)$$

$$RCO_2H + (RCO_2)_3Pb(OAc) \rightleftharpoons (RCO_2)_4Pb + HOAc \qquad (Eq.\ 4)$$

Lead tetraacetate is hygroscopic and turns brown on exposure to air. Acetic acid, however, inhibits oxide formation and, where practicable, it is desirable to use directly the moist material crystallized from acetic acid solutions containing small amounts of acetic anhydride.

Decarboxylation of α-hydroxy and α-amino acids by lead tetraacetate has been known for many years.[16-18] It was originally suggested that when lead tetraacetate itself is heated in solution it decomposes by first-order kinetics to yield carbon dioxide, lead diacetate, and acetyl peroxide or free acetoxy and/or methyl radicals.[19] Subsequently, Kharasch and co-workers reported exhaustive product studies of the decomposition of lead tetraacetate in acetic acid.[20] Sodium acetate accelerates the reduction of lead tetraacetate,[21, 22] but no carbon dioxide is liberated under these conditions.[23] Mosher and Kehr also examined the thermal decomposition of various organic acids by lead tetraacetate.[24] Since then, extensive studies of the mechanism and synthetic utility of this reaction and its many modifications, such as halodecarboxylation and bisdecarboxylation, have been reported. As a result the decarboxylation of acids with lead

---

[15] E. A. Evans, J. L. Huston, and T. H. Norris, *J. Amer. Chem. Soc.*, **74**, 4985 (1952).
[16] H. Oeda, *Bull. Chem. Soc. Jap.*, **9**, 8 (1934).
[17] J. Houben, *Die Methoden der Organischen Chemie*, 3rd ed., Vol. 2, G. Thieme, Liepzig, 1925, p. 147.
[18] O. Süs, *Ann.*, **564**, 137 (1949).
[19] W. A. Waters, *J. Chem. Soc.*, **1946**, 409; *Trans. Faraday Soc.*, **42**, 184 (1946); *The Chemistry of Free Radicals*, Oxford University Press, 1948.
[20] M. S. Kharasch, H. N. Friedlander, and W. H. Urry, *J. Org. Chem.* **16**, 533 (1951).
[21] D. Benson, L. H. Sutcliffe, and J. Walkley, *J. Amer. Chem. Soc.*, **81**, 4488 (1959).
[22] R. O. C. Norman and M. Poustie, *J. Chem. Soc.*, B, **1968**, 781.
[23] The formation of acetoxyacetic acid is consistent with carboxymethyl radicals as intermediates in the presence of acetate salts. E. I. Heiba, R. M. Dessau, and W. J. Koehl, Jr., *J. Amer. Chem. Soc.*, **90**, 1082, 2706 (1968).
[24] W. A. Mosher and C. L. Kehr, *J. Amer. Chem. Soc.*, **75**, 3172 (1953).

tetraacetate has been used extensively for organic synthesis and proof of structure.

MECHANISM

First a definition of the term *oxidative decarboxylation* is in order. By analogy with other decarboxylation reactions (*e.g.*, the copper-quinoline method), the decarboxylation of acids by lead tetraacetate might be expected to yield alkane and carbon dioxide. In this reaction there is no

$$RCO_2H \rightarrow RH + CO_2$$

formal change in the oxidation level and, in principle, an oxidant is not required to bring about this change. However, in only a few cases does the decarboxylation of acids by lead tetraacetate conform to this simple stoichiometry. More commonly the products consist of alkene $R(-H)$ and ester ROAc or RCO_2R in addition to alkane RH. Alkenes and esters are formally derived from the cation R^+ in which the alkyl radical R has undergone a one-electron oxidation. The overall process shown in Eq. 5 represents a two-electron oxidation of RCO_2H and is referred to as an *oxidative decarboxylation*.

$$RCO_2H \xrightarrow{-2e} CO_2 + H^+ + [R^+] \begin{matrix} \nearrow \text{Alkene} + H^+ \\ \searrow_{HOAc} \\ \text{Ester} + H^+ \end{matrix} \quad \text{(Eq. 5)}$$

The mechanism of oxidative decarboxylation of carboxylic acids by lead tetraacetate has received considerable attention in recent years.[25–34] There is now compelling evidence in support of a free-radical chain

[25] For a general discussion of the mechanism of the oxidation-reduction reactions in organic chemistry, see: J. K. Kochi, *Science*, **155**, 422 (1967); J. K. Kochi, *Rec. Chem. Prog.*, **27**, 246 (1966); and refs. 26–34.
[26] E. J. Corey and J. Casanova, Jr., *J. Amer. Chem. Soc.*, **85**, 165 (1963); also see E. J. Corey, J. Casanova, Jr., P. A. Vatakencherry, and R. Winter, *ibid.*, 169.
[27] (a) W. H. Starnes, *J. Amer. Chem. Soc.*, **86**, 5603 (1964) and references therein; (b) *J. Org. Chem.*, **33**, 2767 (1968).
[28] (a) J. K. Kochi, *J. Amer. Chem. Soc.*, **87**, 1811 (1965); (b) *ibid.*, 3609.
[29] J. K. Kochi, J. D. Bacha, and T. W. Bethea, *J. Amer. Chem. Soc.*, **89**, 6538 (1967).
[30] K. Torssell, *Ark. Kemi*, **31**, 401 (1970).
[31] D. I. Davies and C. Waring, *Chem. Commun.*, **1965**, 263.
[32] D. I. Davies and C. Waring, *J. Chem. Soc., C*, **1968**, 1865.
[33] E. G. Gream and D. Wege, *Tetrahedron Lett.*, **1967**, 503.
[34] J. K. Kochi, R. A. Sheldon, and S. S. Lande, *Tetrahedron*, **25**, 1197 (1969).

mechanism for this reaction.[28,29,31,32] One such scheme is shown in Eqs. 6–11,* and elaborated further in the following pages.

Metathesis:

$$Pb^{IV}(OAc)_4 + RCO_2H \longrightarrow Pb^{IV}(OAc)_3O_2CR + HOAc \qquad (Eq.\ 6)$$

Initiation:

$$Pb^{IV}O_2CR \xrightarrow[\text{uv light}]{\text{Heat or}} Pb^{III} + R\cdot + CO_2 \qquad (Eq.\ 7)$$

Propagation:

$$R\cdot + Pb^{IV}O_2CR \longrightarrow [R^+] + Pb^{III}O_2CR \qquad (Eq.\ 8)$$

$$Pb^{III}O_2CR \longrightarrow Pb^{II} + R\cdot + CO_2 \qquad (Eq.\ 9)$$

Termination:

$$R\cdot + Pb^{III}O_2CR \longrightarrow [R^+] + Pb^{II}O_2CR \qquad (Eq.\ 10)$$

$$R\cdot + SH \longrightarrow RH + S\cdot, \text{ etc.} \qquad (Eq.\ 11)$$

Primary Decomposition, Formation of Alkyl Radicals

Subsequent to fast metathesis (Eq. 6), the reaction proceeds via initial homolytic decomposition of the lead(IV) carboxylate into carbon dioxide, Pb^{III}, and an alkyl radical (Eq. 7). Strictly speaking, this should be regarded as an electron transfer from a carboxylate anion to a lead(IV) species, since the infrared absorption of lead(IV) carboxylates suggests that they are ionic.[10, 35-37] There is much experimental evidence in support of a homolytic mechanism. Thus the reaction may be initiated photochemically[11, 28, 38] or by the presence of free-radical initiators such as peroxides.[28, 38] The alkyl radicals formed in the homolytic decomposition have been trapped by efficient radical scavengers such as butadiene[28] and styrene.[39] Similarly, homolytic alkylation of benzene by cyclopropyl radicals was shown in an elegant experiment employing optically active

---

* In this and the following discussion the ligands around Pb are not included unless pertinent to the discussion.

[35] K. Heusler, H. Labhart, and H. Loeliger, *Tetrahedron Lett.*, **1965**, 2847.

[36] K. Heusler, *Chimia*, **21**, 557 (1967).

[37] H. Loeliger, *Helv. Chim. Acta*, **52**, 1516 (1969).

[38] N. A. Maier, L. I. Guseva, and Yu. A. Ol'dekop, *J. Gen. Chem. USSR*, **37**, 2528 (1967); N. A. Maier and Yu A. Ol'dekop, *Doklady Chemistry*, **172**, 39 (1967).

[39] R. O. C. Norman and C. B. Thomas, *J. Chem. Soc., B*, **1967**, 771.

cis- and trans-2-phenylcyclopropanecarboxylic acid.[40] Both alkyl and α-carboxyalkyl radicals have been observed by electron-spin resonance in the photolysis of lead(IV) carboxylates in a solid benzene matrix at $-196°$.[10, 35-37] Initial appearance of the spectrum of the alkyl radical was followed subsequently by the appearance of the α-carboxyalkyl radical spectrum. The results were explained as shown in the accompanying scheme. However, results obtained in a matrix are not directly applicable to reactions in solution.

$$(R'CO_2)_2Pb^{IV} \begin{matrix} -O-\overset{O}{\overset{\|}{C}}-CHR_2 \\ -O_2CCHR_2 \end{matrix} \longrightarrow (R'CO_2)_2Pb^{III} \quad CO_2 \quad \cdot CHR_2 \\ \quad\quad\quad\quad\quad\quad\quad\quad\quad\quad\quad\quad\quad\quad\quad -O_2C-\overset{H}{\overset{|}{C}}R_2 \longrightarrow$$

$$(R'CO_2)_2Pb^{III} + CO_2 + R_2CH_2 + R_2\dot{C}CO_2^-, \quad etc.$$

The homolytic decomposition represented by Eq. 7 could, in principle, proceed via two discrete steps: initial formation of the acyloxy radical $RCO_2\cdot$, followed by its subsequent rapid decomposition to an alkyl radical and carbon dioxide. Recent kinetic studies, employing a competitive technique, have shown that the decarboxylation does proceed via a multibond cleavage as shown (Eqs. 7, 9).[34] A consequence of this multibond fragmentation is that the rate of overall decarboxylation of an acid is markedly dependent on the structure of R. Thus, photochemically, pivalic acid (tertiary) decarboxylates at 30° approximately 100 times faster, and isobutyric acid (secondary) 20 times faster, than n-butyric acid (primary).[34] The same conclusion was reached by examination of the competition between the thermal decarboxylation of various lead(IV) carboxylates and the oxidation of alcohols.[41] These differences in rates could be useful in synthesis, as in the selective decarboxylation of a tertiary carboxyl group in the presence of a primary carboxyl group in the same molecule. (For examples see pp. 311, 351.) However, a concerted mechanism may not always be operative. Thus lead tetraacetate oxidation of many aromatic carboxylic acids gives products containing the intact carboxyl group,[42] and the reaction appears to have proceeded by way of aroyloxy radicals.

Homolytic decarboxylation of acids by lead tetraacetate is analogous to the well-known Kolbe electrolysis (anodic oxidation) of carboxylic acids

[40] T. Aratani, Y. Nakanisi, and H. Nozaki, *Tetrahedron Lett.*, **1969**, 1809.
[41] K. Heusler, *Helv. Chim. Acta*, **52**, 1520 (1969).
[42] D. I. Davies and C. Waring, *J. Chem. Soc., C*, **1967**, 1639.

which also involves the formation of alkyl radicals via a one-electron transfer. In this case kinetic studies suggest that acyloxy radicals are discrete intermediates.[43] The subsequent fate of the alkyl radicals depends on (among other factors) the composition of the electrode.

$$RCO_2^- \xrightarrow{-e} RCO_2\cdot \longrightarrow R\cdot + CO_2$$
$$R\cdot \longrightarrow R-R, \quad R(-H), \text{ ester, } etc.$$

The rate of decarboxylation by lead tetraacetate is enhanced in the presence of metal salts, such as lithium, sodium, or potassium acetates, and in the presence of organic bases, such as pyridine.[28] The effect of the metal acetates is attributed to the formation of thermally less stable anionic complexes such as $Pb(OAc)_5^-$ and $Pb(OAc)_6^{2-}$. A binuclear lead(IV) species has also been suggested.[22] The catalysis by pyridine is attributed to facilitation of the initial decomposition (Eq. 7) by way of coordination of the organic base to the lead(IV) nucleus. Hence an equivalent amount of pyridine is often added in decarboxylations carried out with lead tetraacetate. The pyridine adduct of lead tetraacetate has been isolated and characterized.[13]

Secondary Processes, Oxidative Elimination and Substitution

As shown in the reaction scheme, Eqs. 7, 8, and 9 represent a free-radical chain mechanism in which the first is the initiation and the last two are the propagation steps. Inhibition by radical scavengers such as phenols or oxygen is given as evidence in support of a free-radical chain mechanism.[28] Inhibition by oxygen leads to products derived from the alkoxy radical as an intermediate.

$$C_2H_5\underset{CO_2H}{C}HCH_3 \xrightarrow[O_2]{Pb(OAc)_4} C_2H_5\underset{OH}{C}HCH_3 + C_2H_5COCH_3$$

$$(CH_3)_3CCO_2H \xrightarrow[O_2]{Pb(OAc)_4} (CH_3)_3COH + CH_3COCH_3 + (CH_3)_3CO-OC(CH_3)_3$$

$$(C_6H_5)_3CCH_2CH_2CO_2H \xrightarrow[O_2]{Pb(OAc)_4}$$ [chroman-type product with C_6H_5 C_6H_5 substituents]

Chain termination may occur by dimerization of alkyl radicals or hydrogen transfer with solvent (Eq. 11). The latter is valid only if the solvent-derived radical (S·) is not readily oxidized and cannot participate

---

[43] P. H. Reichenbacher, M. Y.-C. Liu, and P. S. Skell, *J. Amer. Chem. Soc.*, **90**, 1816 (1968). However, see L. Eberson, *Acta Chem. Scand.*, **17**, 2004 (1963), and *J. Amer. Chem. Soc.*, **91**, 2402 (1969).

in a chain-propagating sequence similar to that in Eqs. 8 and 9. A third possibility for chain termination is oxidation of the alkyl radicals with Pb^{III}. However, the metastability of Pb^{III} makes its existence outside the solvent cage unlikely.[10,29] Hence oxidation of alkyl radicals by Pb^{III} would, of necessity, be a reaction only between the Pb^{III}-alkyl radical partners formed in the initial decarboxylation.

A combination of Eqs. 7 and 10 differs in timing from a direct two-equivalent oxidation, variations of which have been invoked in a number of mechanisms involving oxidations with lead tetraacetate. Such processes, however, cannot account for the chain nature of the decomposition. Recent work has indicated that under certain conditions the oxidation of alkyl radicals by Pb^{III} in the cage may compete with Pb^{IV} oxidation, and it may even be the major pathway in some photochemical reactions at lower temperatures.[44] Of course, the extent of oxidation by Pb^{III} will be influenced by factors which usually affect cage reactions, e.g., temperature, viscosity of the solvent, structure of $R\cdot$, rate of fragmentation of the Pb^{III} carboxylate.

The oxidations represented by Eqs. 7–10 are all examples of *electron-transfer oxidations*. A number of other oxidizing metal species can also effect oxidative decarboxylation of carboxylic acids.[45–50] The mechanisms of electron-transfer oxidations of alkyl radicals,[25] particularly by Cu^{II},[51,52] have been studied extensively. Two distinct processes have been delineated in the latter case, and they can be described as oxidative elimination (Eq. 12) and oxidative substitution (Eq. 13). In decarboxylation reactions oxidative elimination of R leads to alkenes, and oxidative substitution usually produces esters.

$$R\cdot + M^{n+} \xrightarrow{\text{Oxidative elimination}} R(-H) + M^{(n-1)+} + H^+ \quad \text{(Eq. 12)}$$

$$R\cdot + M^{n+} + R'CO_2H \xrightarrow{\text{Oxidative substitution}} RO_2CR' + M^{(n-1)+} + H^+ \quad \text{(Eq. 13)}$$

Electron-transfer oxidation of alkyl radicals by Pb^{IV} (or Pb^{III}) has analogies with anodic oxidation of carboxylic acids. Thus, under high

---

[44] R. A. Sheldon, unpublished results.
[45] S. S. Lande and J. K. Kochi, J. Amer. Chem. Soc., **90**, 5196 (1968).
[46] R. A. Sheldon and J. K. Kochi, J. Amer. Chem. Soc., **90**, 6688 (1968).
[47] J. M. Anderson and J. K. Kochi, J. Amer. Chem. Soc., **92**, 2450 (1970).
[48] J. M. Anderson and J. K. Kochi, J. Amer. Chem. Soc., **92**, 1651 (1970); J. Org. Chem., **35**, 986 (1970).
[49] J. K. Kochi and T. W. Bethea, J. Org. Chem., **33**, 75 (1968).
[50] L. F. Fieser and M. J. Haddadin, J. Amer. Chem. Soc., **86**, 2392 (1964).
[51] J. K. Kochi and R. V. Subramanian, J. Amer. Chem. Soc., **87**, 4855 (1965).
[52] J. K. Kochi, A. Bemis, and C. L. Jenkins, J. Amer. Chem. Soc., **90**, 4616 (1968).

anodic voltages and/or the use of carbon electrodes, further oxidation of alkyl radicals (R·) to carbonium ions (R⁺) has been shown to occur.[53-55]

Rates of oxidative elimination and oxidative substitution are dependent on several factors. For a particular metal oxidant the most important factor is the structure of R. The rates of electron-transfer oxidation of alkyl radicals are directly related to the stability of the carbonium ion, that is, inversely related to the ionization potential of the radical (R·), and they increase in the order R = methyl < primary < secondary < tertiary < allylic. In the reaction scheme given, the cation is written as [R⁺] since it is uncertain whether a free carbonium ion is an actual intermediate. Oxidation may proceed by way of an alkyl-lead species and may not actually involve a free carbonium ion.[34, 51, 52]

Oxidation of methyl radicals and primary alkyl radicals by lead tetraacetate is an inefficient process, and the principal product of decarboxylation of acetic and other primary carboxylic acids is the alkane (RH) formed by hydrogen transfer from the solvent. Decarboxylation of these acids by lead tetraacetate is also slow owing to the inefficiency of the chain process. When R is a secondary alkyl group, decarboxylation of acids by lead tetraacetate affords substantial amounts of alkene and ester in addition to alkane. When R is a tertiary alkyl group, alkenes and esters are formed to the virtual exclusion of alkane, and the rate of oxidative decarboxylation is fast. The importance of alkane as a side product thus is inversely related to the ease of oxidation of the alkyl radical, which decreases in the order: tertiary > secondary > primary > methyl.

The distribution of oxidation products between alkene (*oxidative elimination*) and ester (*oxidative substitution*) is also influenced by the structure of R.[52] An example is the competition between oxidative elimination and oxidative substitution in the Cu^{II} oxidation of the series of phenethyl radicals, shown in Table I.[52]

The rate of oxidative solvolysis as shown in Table I is facilitated by electron-releasing substituents in a manner reminiscent of the solvolysis of phenethyl tosylates (Table I, column 6).[56] These same substituents, on the other hand, exert an opposite polar effect on oxidative elimination which is more akin to the base-induced elimination of phenethylammonium salts (Table I, column 7).[57] Isotopic labeling studies show further that oxidative elimination involves direct loss of the β-hydrogen. Complete

---

[53] J. T. Keating and P. S. Skell, *J. Org. Chem.*, **34**, 1479 (1968).

[54] J. T. Keating and P. S. Skell, *J. Amer. Chem. Soc.*, **91**, 695 (1969).

[55] E. J. Corey, N. L. Bauld, R. T. LaLonde, J. Casanova, Jr., and E. T. Kaiser, *J. Amer. Chem. Soc.*, **82**, 2645 (1960).

[56] M. G. Jones and J. L. Coke, *J. Amer. Chem. Soc.*, **91**, 4284 (1969).

[57] A. F. Cockerill, S. Rottschaefer, and W. H. Saunders, Jr., *J. Amer. Chem. Soc.*, **89**, 901, 4985 (1967); W. H. Saunders, Jr., D. G. Bushman, and A. F. Cockerill, *ibid.*, **90**, 1775 (1968).

TABLE I. Rates of Oxidation of Phenethyl Radicals by Cu^{II}

Radical	$k_e/k_s{}^a$	$k_e{}^b$	$k_s{}^b$	$k_e + k_s$	$k_{solv}{}^c$	$k_{elim}{}^d$
		(M^{-1} sec^{-1} $\times 10^{-6}$)			(Relative)	
$m\text{-}CH_3OC_6H_4CH_2CH_2\cdot$	50	—	—	—	—	—
$C_6H_5CH_2CH_2\cdot$	33	1.6	0.05	1.7	1.0	1.0
$p\text{-}CH_3C_6H_4CH_2CH_2\cdot$	1.3	1.0	0.81	1.8	3.0	0.30
$p\text{-}CH_3OC_6H_4CH_2CH_2\cdot$	0.014	0.021	1.6	1.6	39.0	0.10

^a k_e/k_s measures the relative yields of styrenes compared to phenethyl esters.

^b k_e and k_s are the second-order rate constants for oxidative elimination and substitution respectively.

^c k_{solv} is the relative first-order titrimetric rate of acetolysis of phenethyl tosylates (Ref. 56).

^d k_{elim} is the relative second-order rate constant for elimination of phenethylammonium salts induced by sodium alkoxide in alcohol (Ref. 57).

$$C_6H_5CD_2CH_2\cdot + Cu^{II} \xrightarrow[\text{elimination}]{\text{Oxidative}} C_6H_5CD=CH_2 + Cu^{I} + D^+$$

scrambling of the α- and β-carbon atoms in the phenethyl acetates is consistent with the formation of carbonium ion intermediates in oxidative solvolysis.[52] Alkyl radicals are generally oxidized by cupric acetate to

$$\text{Ph-}CD_2CH_2\cdot + Cu^{II} + HOAc \xrightarrow{\text{Oxidative solvolysis}} \left[\text{Ph}^+ \begin{array}{c} CD_2 \\ | \\ CH_2 \end{array} \right]$$

$$+ Cu^{I} + H^+ \longrightarrow \text{Ph-}CD_2CH_2OAc + \text{Ph-}CH_2CD_2OAc$$

alkenes by oxidative elimination in preference to oxidative solvolysis. The latter is important only with alkyl radicals that are readily oxidized to relatively stable carbonium ions.

A similar consideration should be given to the factors involved in the oxidation of alkyl radicals by Pb^{IV}. Thus acids containing tertiary alkyl groups on oxidation with lead tetraacetate give mainly alkene, whereas acids which produce allylic radicals give, almost exclusively, ester. Alkyl radicals that form stable carbonium ions generally give esters by oxidative substitution. At this point, however, it is not completely clear that alkenes produced in the lead tetraacetate decarboxylation of acids are derived wholly by oxidative elimination since there are differences between Pb^{IV} and Cu^{II} as oxidants.[58] There is some evidence that both the alkenes and esters formed in the oxidation of alkyl radicals by lead tetraacetate may be derived from the carbonium ion in contrast to oxidations with cupric acetate.

Catalysis by Cupric Acetate

Rates of oxidative decarboxylation by lead tetraacetate are enhanced by the presence of catalytic amounts of cupric acetate. Moreover, the distribution of products is often changed completely. Thus primary carboxylic acids are decarboxylated in the presence of cupric acetate to afford alkenes (R—H) in high yield. In contrast, only low yields of alkenes are obtained in the absence of cupric acetate. This effect is attributed to the rapid scavenging of alkyl radicals by Cu^{II}. The rate of electron-transfer oxidation of alkyl radicals by Cu^{II} approaches a diffusion-controlled rate,[51,52] and it has been shown, for example, that isopropyl radicals are oxidized by Cu^{II} at least 100 times faster than by Pb^{IV}.[44] Primary and secondary alkyl radicals are generally oxidized by cupric acetate selectively to alkenes (oxidative elimination).[25] An overall reaction in the presence of cupric acetate is given in the accompanying scheme. As shown, it represents an alternative pathway for facile decomposition of those lead(IV) carboxylates whose chain decomposition (by Pb^{IV} oxida-

$$Pb^{IV}O_2CR \rightarrow Pb^{III} + R\cdot + CO_2$$
$$R\cdot + Cu^{II} \rightarrow R(-H) + H^+ + Cu^I$$
$$Cu^I + Pb^{IV}O_2CR \rightarrow Cu^{II} + Pb^{III}O_2CR$$
$$Pb^{III}O_2CR \rightarrow Pb^{II} + R\cdot + CO_2, \quad etc.$$

tion of alkyl radicals) is not facile. By the same reasoning the presence of cupric acetate has little effect on the decarboxylation of tertiary carboxylic acids by lead tetraacetate. In this process the ready oxidation of tertiary alkyl radicals by Pb^{IV} competes effectively with oxidation by Cu^{II}. Cu^{II} catalysis of lead tetraacetate decarboxylation[58,59] is particularly applicable to the oxidative decarboxylation of primary and secondary acids to alkenes.[60] The combination of lead tetraacetate and Cu^{II} will be referred to henceforth as the Pb^{IV}/Cu^{II} reagent.

[58] J. D. Bacha and J. K. Kochi, *J. Org. Chem.*, **33**, 83 (1968).
[59] J. K. Kochi and J. D. Bacha, *J. Org. Chem.*, **33**, 2746 (1968).
[60] J. D. Bacha and J. K. Kochi, *Tetrahedron*, **24**, 2215 (1968).

Ligand Transfer Oxidation of Alkyl Radicals: Halodecarboxylation

Addition of halide ion, usually in the form of metal halides such as lithium, potassium, or calcium chloride, also completely alters the course of the decarboxylation reaction involving lead tetraacetate.[61, 62] Under these conditions the alkyl halide RX is usually the sole product. Formation of alkyl halide in high yield, even in the case of methyl, is attributed to rapid ligand-transfer oxidation of alkyl radicals by halo-Pb^{IV} species, a reaction that competes effectively with the relatively slow electron-transfer oxidation.[63] By employing this variation of the decarboxylation

$$R\cdot + Pb^{IV}X \rightarrow RX + Pb^{III}, \quad etc.$$

$$X = Cl, Br, \quad etc.$$

of acids with lead tetraacetate, a variety of acids can be converted to alkyl halides.

The chlorodecarboxylation of *endo*- and *exo*-2-norbornanecarboxylic acid with lead tetraacetate and lithium chloride yields the same 2-norbornyl radical,[64] which on reaction with chloro-Pb^{IV} affords six times more *exo*-2-norbornyl chloride than *endo* isomer. The stereoselectivities of the reactions between the 2-norbornyl radical and a variety of chlorine atom donors have been compared. A similar stereochemical study has also been carried out with the related apobornyl radical.

Stoichiometry of Decarboxylations with Lead Tetraacetate[34]

From the discussion above it is apparent that the yield of carbon dioxide formed in the decarboxylation of acids by lead tetraacetate is not always stoichiometric. Many simultaneous reactions are possible, and they differ largely in the formal oxidation states of the products. The latter determine the stoichiometric requirements of lead tetraacetate for each mole of carbon dioxide formed. Formation of ester or alkene is an oxidative process which requires 1 mole of lead tetraacetate and generates 1 mole of carbon dioxide according to Eqs. 14 and 15. The decarboxylation of tertiary carboxylic acids conforms to this stoichiometry since little or no alkane is produced. Similarly, oxidative decarboxylation of primary

$$Pb^{IV}(O_2CR)_4 \rightarrow Pb^{II}(O_2CR)_2 + RCO_2R + CO_2 \quad (Eq.\ 14)$$

$$Pb^{IV}(O_2CR)_4 \rightarrow Pb^{II}(O_2CR)_2 + R(-H) + RCO_2H + CO_2 \quad (Eq.\ 15)$$

---

[61] J. K. Kochi *J. Amer. Chem. Soc.*, **87**, 2500 (1965).

[62] J. K. Kochi, *J. Org. Chem.*, **30**, 3265 (1965).

[63] Reference 25 gives a discussion of electron- and ligand-transfer oxidations of alkyl radicals. See also C. L. Jenkins and J. K. Kochi, *J. Org. Chem.*, **36**, 3095 (1971).

[64] P. D. Bartlett, G. N. Fickes, F. C. Haupt, and R. Helgeson, *Accounts Chem. Res.*, **3**, 177 (1970).

and secondary carboxylic acids in the presence of cupric acetate is in the same category.[58-60] On the other hand, formation of alkane formally does not consume lead tetraacetate, although the presence of the latter is important since acids are not otherwise decarboxylated. Alkanes are derived from alkyl radicals via hydrogen transfer with solvent (mainly from the α-C—H bond of carboxylic acids when they are employed as solvents). The requirements of lead tetraacetate in this process depend on the fate of the solvent-derived radical S·. If S· disappears via self-dimerization and disproportionation, the overall stoichiometry is given by Eq. 16. If S· is oxidized by lead tetraacetate, a 1:1 Pb^{IV} to CO_2 stoichiometry is obtained (Eq. 17).

$$Pb^{IV}(O_2CR)_4 + 2\ SH \rightarrow Pb^{II}(O_2CR)_2 + 2\ RH + 2\ CO_2 + S\text{—}S \qquad \text{(Eq. 16)}$$
$$Pb^{IV}(O_2CR)_4 + SH \rightarrow Pb^{II}(O_2CR)_2 + RH + CO_2 + RCO_2S \qquad \text{(Eq. 17)}$$

The data presented in the tabular survey should be examined with these values in mind. The complexities lend ambiguity to a valid universal assignment of the stoichiometry for lead tetraacetate decarboxylation. The lead tetraacetate requirement may formally vary from 1 to 2 because of differences in rates of oxidation of alkyl radicals.

Solvent Effects

The effect of solvent on oxidative decarboxylations with lead tetraacetate alone and in the presence of cupric acetate or pyridine has been studied in detail for the homoallylic series (allylacetic, cyclopropylacetic, and cyclobutanecarboxylic acids).[59] The homoallylic system was chosen for detailed study because it presented optimum opportunities to examine the effect of solvent on a number of competing processes, *viz.*, oxidative elimination, oxidative substitution, rearrangement, hydrogen transfer (reduction to alkanes), and aromatic substitution (in aromatic solvents). Solvents can have a profound effect on the *rate* as well as the products of oxidation. In solvents such as hexamethylphosphoramide, dimethylformamide, and tetrahydrofuran, the decarboxylation of cyclobutanecarboxylic acid occurs spontaneously and exothermically at room temperature. By comparison, reactions under the same conditions in benzene, acetonitrile, or ethyl acetate are too slow to measure.[59]

Stereochemistry of Oxidative Decarboxylation

The stereochemical aspects of oxidative decarboxylation with lead tetraacetate have been discussed in a recent paper.[65] The oxidative decarboxylation of the four stereoisomers of acetylglycyrrhetic acid

---

[65] L. Canonica, B. Danieli, P. Manitto, and G. Russo, *Gazz. Chim. Ital.*, **98**, 696 (1968).

(1–4) was examined, both uncatalyzed (in benzene) and catalyzed (in benzene containing pyridine or cupric acetate). In every case a mixture of the two diacetoxyenones **5a** and **5b** and the three acetoxydienones **6** was obtained. The rates of uncatalyzed decompositions were influenced by the

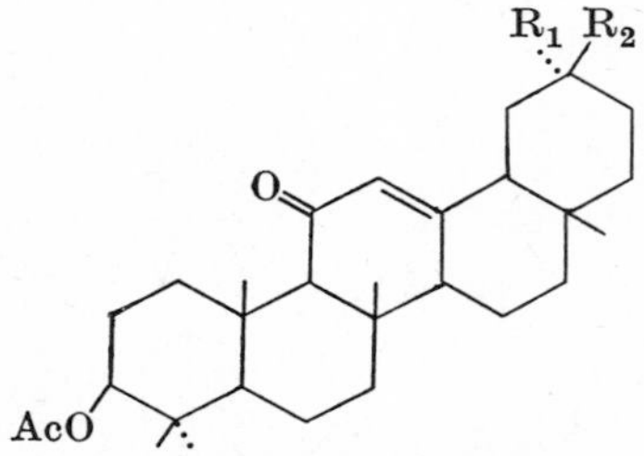

1 $R_1 = CH_3, R_2 = CO_2H$
2 $R_1 = CO_2H, R_2 = CH_3$

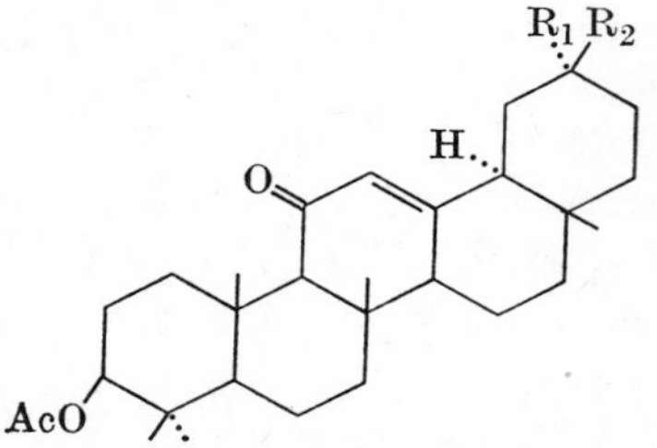

3 $R_1 = CH_3, R_2 = CO_2H$
4 $R_1 = CO_2H, R_2 = CH_3$

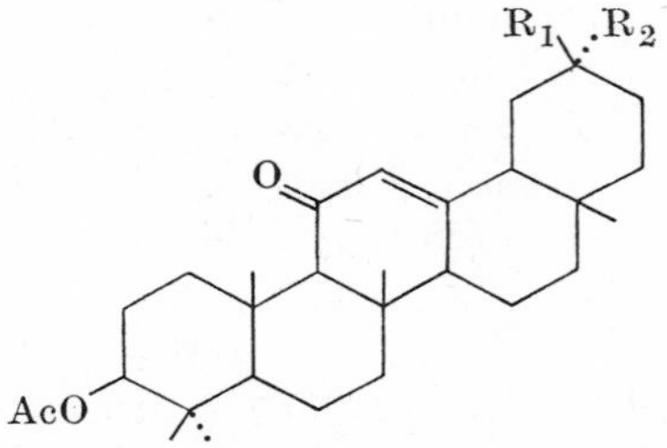

5a $R_1 = CH_3, R_2 = OAc$
5b $R_1 = OAc, R_2 = CH_3$

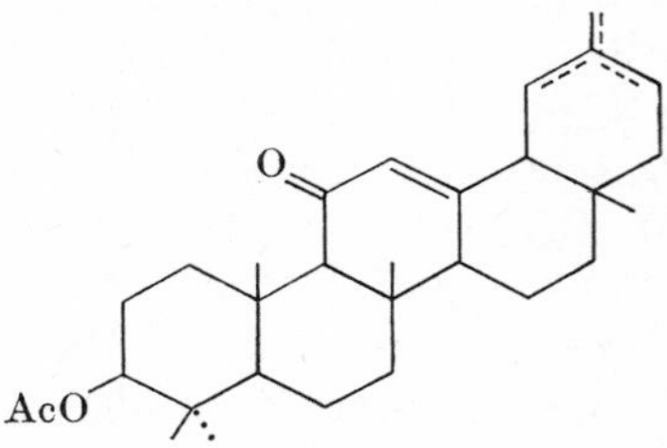

6

orientation of the carboxyl group, but the product distribution was unaffected. The orientation of the carboxyl group did not affect the relative amounts of the products in reactions catalyzed by pyridine or Cu^{II}. The products were considered to arise from the oxidation of alkyl radicals to carbonium ions. The acetoxydienone with an exocyclic double bond is formed in the larger amount, suggesting that approach of the metal complex oxidant to the radical **7** from an equatorial direction is preferred.

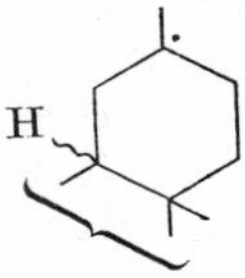

7

The oxidative decarboxylation of optically active *exo-* or *endo-*norbornane-2-carboxylic acid with lead tetraacetate affords mainly *exo-*2-norbornyl acetate.[26] Approximately the same degree of retention

[Scheme: norbornane-2-carboxylic acid (endo and exo CO2H isomers) → Pb(OAc)4, C6H6 → norbornyl acetate (40% retention of optical activity)]

of optical activity is obtained in both reactions.* Similarly, the oxidative decarboxylation of *cis*- and *trans*-2-phenylcyclopropanecarboxylic acid with the Pb^{IV}/Cu^{II} reagent produces the same mixture of acetates resulting from the opening of a 2-phenylcyclopropyl cation.[66] The absence of stereospecificity, particularly in the formation of esters 8 and 9, is in accord with the oxidation of the 2-phenylcyclopropyl radical and differs from the results of anodic decarboxylation.[67]

[Scheme: cis- and trans-2-phenylcyclopropanecarboxylic acid + Pb(IV)/Cu(II) → $C_6H_5CHCH=CH_2$ with OAc, plus alkenes 8 (C_6H_5 and H cis, H and CH_2OAc) and 9 (H and H, C_6H_5 and CH_2OAc)]

Decarboxylation of Formic Acid, α-Hydroxy Acids, α-Amino Acids, and α-Keto Acids

Although there is compelling evidence, as discussed above, for a homolytic mechanism for the oxidative decarboxylation by lead tetraacetate, in some instances a heterolytic or two-equivalent oxidation may be operative.[68] Formic acid is rapidly decarboxylated by lead tetraacetate at

---

* This observation is open to several interpretations and is consistent with radical as well as carbonium ion intermediates.[26,28b]

[66] T. Shono, I. Nishiguchi, and R. Oda, *Tetrahedron Lett.*, **1970**, 373.

[67] T. Shono, I. Nishiguchi, S. Yamane, and R. Oda, *Tetrahedron Lett.*, **1969**, 1965.

[68] The two-equivalent heterolytic mechanism originally proposed by Mosher and Kehr[24] has been recently reinvoked to account for the facile decarboxylation of triphenylacetic acid (W. A. Mosher and T. C. Mayberry, private communication).

room temperature to give a quantitative yield of lead diacetate and carbon dioxide. The reaction has been used for the rapid and quantitative determination of formic acid.[69] This reaction may well proceed via a direct two-electron oxidation.[24, 69, 70]

$$HCO_2H + Pb^{IV} \rightarrow Pb^{II} + 2 H^+ + CO_2$$

Decarboxylation of α-hydroxy acids by lead tetraacetate is also a very facile process and can be used for the quantitative analysis of α-hydroxy acids.[71] A concerted, heterolytic decomposition is usually postulated for the reaction, which is analogous to the well-known cleavage of 1,2-diols[72] by Pb^{IV} discovered by Criegee.[73]

The acetyl derivatives of α-hydroxy acids are also quite readily decarboxylated by lead tetraacetate. In some cases the *gem*-diacetate is obtained, as in the oxidative decarboxylation of acetylmandelic acid to benzylidene diacetate.[58] More commonly, the carbonyl compound is isolated as illustrated in the first equation on p. 297.[74]

It has been suggested that oxidation of α-keto acids, which takes place readily only in protic solvents such as water or ethanol, proceeds via the *gem*-diol as shown in Eq. 18.[75]

---

[69] A. S. Perlin, *Anal. Chem.*, **26**, 1053 (1954). Also see J. Zyka, *Pure Appl. Chem.*, **13**, 569 (1966).

[70] J. M. Grosheintz, *J. Amer. Chem. Soc.*, **61**, 3379 (1939).

[71] I. Nemec, A. Berka, and J. Zyka, *Mikrochem. J*, **6**, 525 (1962).

[72] See C. A. Bunton in *Oxidation in Organic Chemistry*, K. Wiberg, Ed., Chapter VI, Academic Press, New York, 1965.

[73] R. Criegee and E. Büchner, *Ber.*, **73**, 563 (1940).

[74] M. Tanabe and D. F. Crowe, *J. Org. Chem.*, **30**, 2776 (1965).

[75] E. Baer, *J. Amer. Chem. Soc.*, **62**, 1597 (1940).

$$H_2O + RCOCO_2H \rightleftharpoons R-\underset{\underset{OH}{|}}{\overset{\overset{OH}{|}}{C}}-CO_2H \xrightarrow{Pb(OAc)_4}$$

$$RCO_2H + CO_2 + HOAc + Pb(OAc)_2$$

(Eq. 18)

By analogy with α-hydroxy acids, α-amino acids might be expected to decompose by a concerted process. Indeed, N-acetylamino acids react with

$$\xrightarrow{} \underset{R}{\overset{R}{>}}C=NH + CO_2 + HOAc + Pb(OAc)_2$$

lead tetraacetate in dimethylformamide at 25° to produce acetamide, carbon dioxide, and the corresponding aldehyde.[76] The aldehyde can be accounted for by an initial concerted decomposition, followed by hydrolysis of the intermediate imide.

$$\xrightarrow{} \underset{H}{\overset{R}{>}}C=NCOCH_3 + CO_2 + Pb(OAc)_2 + HOAc$$

$$\underset{H}{\overset{R}{>}}C=NCOCH_3 \xrightarrow{H_2O} RCHO + CH_3CONH_2$$

[76] H. L. Needles and K. Ivanetich, *Chem. Ind.* (London), **1967**, 581.

However, it is reported that both N-benzyl- and N-benzoyl-glycine react with lead tetraacetate to give the corresponding acetates.[18]

$$C_6H_5CH_2NHCH_2CO_2H \xrightarrow{Pb(OAc)_4} C_6H_5CH_2NHCH_2OAc$$

A change in solvent can cause acetoxylation to accompany decarboxylation.[77]

Carboxyamidines undergo oxidative decarboxylation to aminonitriles.[78] The reaction was observed during degradative studies on kasugamycin, and its generality should be explored.

$$\underset{\underset{NH}{\|}}{RNHCCO_2H} \xrightarrow{Pb(OAc)_4} RNHCN$$

Bisdecarboxylation

1,2-Dicarboxylic acids are oxidatively bisdecarboxylated to alkenes by lead tetraacetate. Grob and co-workers[79–81] suggested a concerted elimination, but it is unlikely that elimination involves the cyclic

[77] H. L. Slates, D. Taub, C. H. Kuo, and N. L. Wendler, *J. Org. Chem.*, **29**, 1424 (1964).
[78] Y. Suhara, K. Maeda, and H. Umezawa, *Tetrahedron Lett.*, **1966**, 1239.
[79] C. A. Grob, M. Ohta, and A. Weiss, *Angew. Chem.*, **70**, 343 (1958).
[80] C. A. Grob, M. Ohta, E. Renk, and A. Weiss, *Helv. Chim. Acta*, **41**, 1191 (1958).
[81] C. A. Grob and A. Weiss, *Helv. Chim. Acta*, **43**, 1390 (1960).

intermediate **10** in view of its unfavorable geometry.

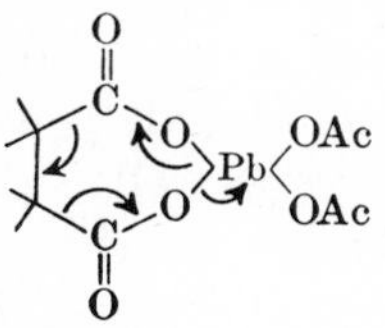

(structure 10)

Under carefully controlled conditions, oxygen inhibits the bisdecarboxylation of 1-carbethoxybicyclo[2,2,2]octane-2,3-dicarboxylic acid at 60° in benzene.[82] At 80°, the inhibition period is quite short and may not be noticeable if the reaction is carried out in the usual manner.[83] However, the yield of alkene is actually increased in the presence of oxygen in contrast to results obtained with monobasic acids. A homolytic mechanism cannot be completely ruled out because a stepwise decarboxylation of the bislead^{IV} complex **11** would lead to the alkyl radical **12**, which could undergo an easy intramolecular oxidation by Pb^{IV}.

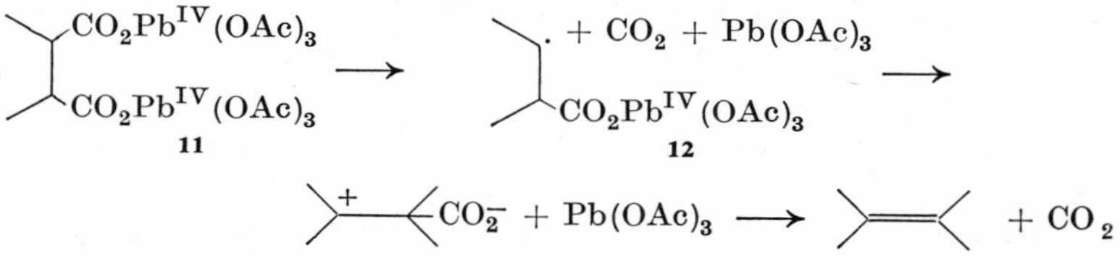

(reaction scheme showing 11 → 12 → alkene + CO_2)

Bisdecarboxylation of *meso-* or *dl*-1,2-diphenylsuccinic acid[26] produces approximately the same yield of *trans*-stilbene uncontaminated by the *cis* isomer. Electrochemical bisdecarboxylation gives the same results. On the basis of these results it was suggested that the reaction is not concerted and that a labile monocarboxylic intermediate is involved. Recent extensive product studies[82] of bisdecarboxylations also show that the reaction proceeds through free radicals and labile monocarboxylic intermediates (see pp. 342–344).

SCOPE AND LIMITATIONS

Primary and Secondary Carboxylic Acids (Tables IV and V)

Decarboxylation of primary and secondary carboxylic acids can lead to a variety of products which may be attributed to several factors: (a)

---

[82] H. P. Wagner, Ph.D. Thesis, University of Basel, 1968; C. A. Grob, private communication.

[83] C. M. Cimarusti and J. Wolinsky, *J. Amer. Chem. Soc.*, **90**, 113 (1968); see also J. Wolinsky and R. B. Login, *J. Org. Chem.*, **37**, 121 (1972).

oxidation of primary and secondary alkyl radicals by lead tetraacetate is a relatively slow process, and alkane formation *via* hydrogen transfer with solvent is competitive; (b) oxidation can give both alkenes and esters; (c) oxidation of primary and secondary alkyl radicals often leads to products from both a rearranged as well as the unrearranged alkyl group. Decarboxylation of the acetate associated with lead(IV) becomes competitive when primary acids are decarboxylated.

For the reaction to be of synthetic utility it is essential to divert the alkyl radical intermediates to one major product in high yield. There are several possible ways of achieving this goal.

Alkane Formation. Although alkanes are usually the major product from the decarboxylation of primary acids, they are not formed in synthetically useful amounts or in high purity. In principle, diversion of alkyl radicals to alkanes may be achieved by employing a solvent (SH) in which hydrogen transfer to alkyl radicals is rapid and able to compete effectively with the relatively slow oxidation of the radicals by Pb^{IV}. However, alkyl radicals may also be oxidized by Pb^{III} in a cage process, and they may therefore be difficult to divert.

$$R\cdot + Pb^{IV} \rightarrow [R^+] + Pb^{III}$$
$$R\cdot + Pb^{III} \rightarrow [R^+] + Pb^{II}$$
$$R\cdot + SH \rightarrow RH$$

Cage reactions involving the solvent would be expected to decrease with increasing temperature, whereas oxidation of alkyl radicals by Pb^{IV} increases with temperature.[44] Hence, depending on the relative amounts of oxidation by Pb^{IV} or Pb^{III}, a photochemical reaction at room temperature or a thermal reaction at elevated temperatures may be optimum. Good yields of alkanes (70–80%, based on carboxylic acid) have been obtained from the photochemical decarboxylation of primary carboxylic acids by lead tetraacetate in chloroform at room temperature.[44, 59] Decarboxylation of secondary carboxylic acids under the same conditions afforded alkanes in moderate yields (50–60%). Conditions have not been optimized. Under appropriate conditions of solvent, temperature, and concentration (combined with the recycling of unreacted carboxylic acid), this may represent a convenient method for converting primary and secondary carboxylic acids to alkanes in a single step.

There are only a few methods available for effecting this conversion. One method involves the homolytic decomposition of a perester. Thus the secondary carboxylic acid **13** was converted, in 45% overall yield, to the corresponding hydrocarbon by way of the perester.[84] A similar sequence

[84] (a) K. B. Wiberg, B. R. Lowry, and T. H. Colby, *J. Amer. Chem. Soc.*, **83**, 3998 (1961).
(b) P. E. Eaton and T. W. Cole, *J. Amer. Chem. Soc.*, **86**, 3157 (1964).

of reactions has been applied to the synthesis of cubane. As *pseudo*-longifolic acid (14) was recovered unchanged on treatment with Pb^{IV} in solution, conversion to the hydrocarbon was made through the perester.[85]

Alkene versus Ester. Attempts have been made to enhance the yield of one of the oxidation products from a carboxylic acid.[86] Thus the oxidation of cyclohexanecarboxylic acid with lead tetraacetate affords a mixture of cyclohexyl acetate and cyclohexene. The yield of ester can be increased by using a large excess of potassium acetate in acetic acid solution as shown in Table II (p. 302). By contrast, cyclohexene is the major product when the decarboxylation of the acid by lead tetraacetate is carried out in dimethylformamide solutions. It appears worthwhile to test the generality of these observations under controlled conditions, especially with respect to oxygen.

Some acetate esters are formed in reasonable yields by the decarboxylation of secondary acids in benzene. Thus the pentacylic carboxylic acid 15 affords the acetate in 60% yield at room temperature.[87] The acetate is one of the most reactive secondary esters in solvolysis.

[85] J. Lhomme and G. Ourisson, *Tetrahedron*, **24**, 3167 (1968).
[86] J. T. Marvel, Ph.D. Thesis, Massachusetts Institute of Technology, 1964. Kindly provided by Professor George Büchi.
[87] L. Birladeanu, T. Hanafusa, and S. Winstein, *J. Amer. Chem. Soc.*, **88**, 2315 (1966).

TABLE II. Esters and Alkenes from Oxidative Decarboxylation of Acids by Lead Tetraacetate

Acid	Solvent	Conditions	Products (% Yield)
$(CH_3)_2CHCO_2H$	HOAc/KOAc	80°	$(CH_3)_2CHOAc$ (52), $CH_3CH=CH_2$ (<1)
cyclohexyl-CO_2H	C_6H_6	80°, 16 hr	cyclohexyl–OAc (13), cyclohexene (47)
	HOAc/KOAc	60°, 0.3 hr	,, (77)
	$HCON(CH_3)_2$	Room temp, 3 hr	,, (8), cyclohexene (73)
	$CH_3CON(CH_3)_2$	110°, 8 hr	,, (41)
	None	63°, 39 hr	,, (42)
cyclooctyl-CO_2H	HOAc/KOAc	60°, 1.5 hr	Cyclooctyl acetate (50), cyclooctene (22)
	$CH_3CON(CH_3)_2$	Room temp, 0.8 hr	Cyclooctyl acetate (40), cyclooctene (41)
4-cyclooctenyl-CO_2H	HOAc/KOAc	60°, 1 hr	4-Cycloocten-1-yl acetate + 2-bicyclo[3,3,0]octyl acetates (83, total)
	$CH_3CON(CH_3)_2$	Room temp, 1.8 hr	4-Cycloocten-1-yl acetate + 2-bicyclo[3,3,0]octyl acetates (39, total), 2-bicyclo[3,3,0]octene (30)

As noted on p. 295, both *exo-* and *endo-*norbornane-2-carboxylic acid give the *exo* acetate (24–67 %) with lead tetraacetate-pyridine in benzene. When similarly treated, both bornane-2-carboxylic acids produce the complex mixture of six products shown below, together with three other acetates formed by further oxidation.[33]

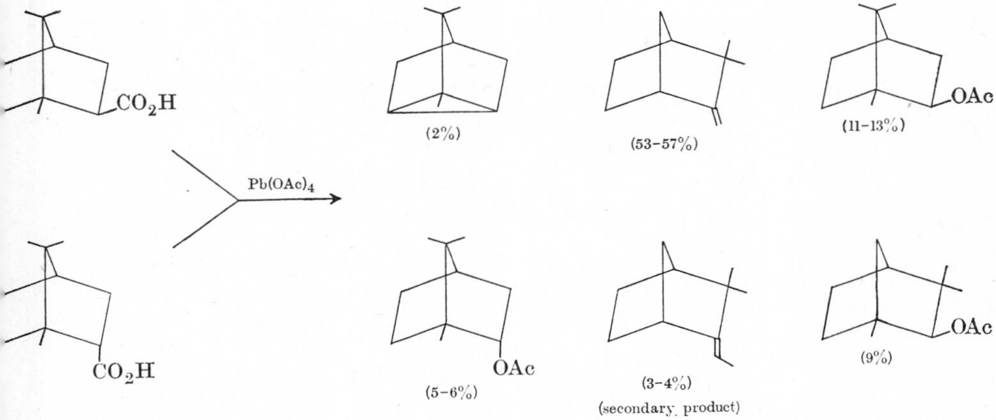

Alkene Formation through Use of Lead Tetraacetate/Cupric Acetate (Table X)

Another possible way to optimize the diversion of alkyl radicals to one major product is to facilitate the oxidative elimination (alkene formation) of the radicals. Electron-transfer oxidation of alkyl radicals by Cu^{II} is much more facile than oxidation by Pb^{IV}.[44, 51, 52] Cu^{II} alone does not effect decarboxylation under mild conditions, but the combination of lead tetraacetate and cupric acetate effects the smooth conversion of carboxylic acids to alkenes.[60]

$$\underset{H\ \ CO_2H}{\overset{}{\diagdown C - C \diagup}} \xrightarrow{Pb^{IV}/Cu^{II}} \diagdown C = C \diagup + 2\,H^+ + CO_2 \qquad \text{(Eq. 19)}$$

Catalytic amounts of cupric acetate can be used because it is regenerated in the reaction (see Eqs. 20–22 and p. 291).

$$RCO_2H + Pb^{IV} \rightarrow R\cdot + CO_2 + H^+ + Pb^{III} \qquad \text{(Eq. 20)}$$
$$R\cdot + Cu^{II} \rightarrow R(-H) + H^+ + Cu^{I} \qquad \text{(Eq. 21)}$$
$$Cu^{I} + Pb^{IV} \rightarrow Cu^{II} + Pb^{III},\ \ etc. \qquad \text{(Eq. 22)}$$

Oxidative decarboxylation of carboxylic acids with the Pb^{IV}/Cu^{II} reagent given in Eq. 19 has been used for the preparation of alkenes, particularly where R is a primary or secondary alkyl or a cycloalkyl[60] group. The reaction can be carried out equally well by thermal or photochemical initiation, and concentrations of cupric acetate as low as 0.005 M may be used. With primary acids, decarboxylation of acetate derived from lead tetraacetate is competitive but does not represent a serious limitation since excess oxidant may be used. For unsymmetrical secondary alkyl radicals, oxidative elimination can take place in more than one direction. For example, decarboxylation of 3-cyclohexene-1-carboxylic acid gives both possible isomeric dienes.[60] The oxidative decarboxylation of acetoxy

$$\text{cyclohexenyl-CO}_2\text{H} \xrightarrow{\text{Pb}^{IV}/\text{Cu}^{II}} \text{cyclohexadiene} + \text{cyclohexadiene}$$

acids with α-alkyl groups also affords a mixture of acetoxyalkenes with the Pb^{IV}/Cu^{II} reagent.[88]

$$\text{AcOCH}_2\text{CH}_2\text{CHCO}_2\text{H} \xrightarrow{\text{Pb}^{IV}/\text{Cu}^{II}}$$
$$\overset{|}{\text{CH}_2\text{CH}_3}$$

$\text{AcOCH}_2\text{CH}=\text{CHCH}_2\text{CH}_3$ + $\text{AcOCH}_2\text{CH}_2\text{CH}=\text{CHCH}_3$
(25%) (75%)

$$\text{AcOCH}_2\text{CH}_2\text{CH}_2\text{CHCO}_2\text{H} \xrightarrow{\text{Pb}^{IV}/\text{Cu}^{II}}$$
$$\overset{|}{\text{CH}_2\text{CH}_3}$$

$\text{AcOCH}_2\text{CH}_2\text{CH}=\text{CHCH}_2\text{CH}_3$ + $\text{AcCH}_2\text{CH}_2\text{CH}_2\text{CH}=\text{CHCH}_3$
(40%) (60%)

$$\text{AcO(CH}_2)_{11}\text{CHCO}_2\text{H} \xrightarrow{\text{Pb}^{IV}/\text{Cu}^{II}}$$
$$\overset{|}{\text{CH}_2\text{CH}_3}$$

$\text{AcO(CH}_2)_{10}\text{CH}=\text{CHCH}_2\text{CH}_3$ + $\text{AcO(CH}_2)_{11}\text{CH}=\text{CHCH}_3$
(50%) (50%)

Primary carboxylic acids give exclusively terminal alkenes. Nonanoic acid affords 1-octene in 87% yield and no 2-octene.[60] Cyclobutanecarboxylic acid in benzene gives cyclobutene in 78% yield, and the oxidation

$$n\text{-}C_8H_{17}CO_2H \xrightarrow{\text{Pb(IV)/Cu(II)}} 1\text{-}C_8H_{16}$$

$$\text{cyclobutyl-CO}_2\text{H} \xrightarrow{\text{Pb(IV)/Cu(II)}} \text{cyclobutene}$$

$$\text{cyclohexyl-CH}_2\text{CO}_2\text{H} \xrightarrow{\text{Pb(IV)/Cu(II)}} \text{cyclohexylidene=CH}_2$$

[88] Yu. N. Ogibin and G. I. Nikishin, private communication, 1971.

is a convenient synthesis of the cycloalkene.[60] Similarly, cyclohexylacetic acid produces methylenecyclohexane in 84% yield.[60]

The examples above illustrate that oxidative elimination of primary and secondary alkyl radicals by cupric acetate under these conditions leads specifically to alkene by loss of a β-hydrogen atom. A further example shown in the accompanying equation is the oxidative decarboxylation of the spiro acid to spiro-decene in high yield by the Pb^{IV}/Cu^{II} reagent.[89] The low yields of octalins indicate that the carbonium ion cannot be an important intermediate.

Solvent can also play an important role because relative rates of oxidative elimination (alkene) and oxidative substitution (ester) are influenced by the solvent.[59] Thus decarboxylation of cyclobutanecarboxylic acid by Pb^{IV}/Cu^{II} in acetonitrile-acetic acid gave a mixture of the three isomeric homoallylic acetates in 58% yield, with only 3% of cyclobutene as compared with the 78% in benzene. It is thus sometimes possible by an appropriate choice of solvent to convert carboxylic acids to either alkenes or esters.

The Pb^{IV}/Cu^{II} reagent has been successfully used for the conversion of cholanic acid (16) to Δ^{22}-24-norcholene.[90] Similarly, acetylglycyrrhetic acids gave a mixture of the three isomeric dienones 6 in 88% yield.[65]

An improved procedure for the oxidative decarboxylation of acids to alkenes using the Pb^{IV}/Cu^{II} reagent has been suggested.[88] Lead tetraacetate is added portionwise to a refluxing solution of benzene containing

---

[89] (a) R. I. Cargill and A. M. Foster, *J. Org. Chem.*, **35**, 1971 (1970); (b) D. L. Struble, A. L. J. Beckwith, and G. E. Gream, *Tetrahedron Lett.*, **1970**, 4795.
[90] A. S. Vaidya, S. M. Dixit, and A. S. Rao, *Tetrahedron Lett.*, **1968**, 5173.

the carboxylic acid, pyridine, and cupric acetate in the absence of air. For example, a mixture of 5-acetoxy-2-pentene and 5-acetoxy-1-pentene in a 40 : 60 molar ratio was formed from the oxidative decarboxylation of

$$\text{AcOCH}_2\text{CH}_2\text{CH}_2\text{CH(CH}_3\text{)CO}_2\text{H} \xrightarrow{\text{Pb}^{IV}/\text{Cu}^{II}}$$
$$\text{AcOCH}_2\text{CH}_2\text{CH=CHCH}_3 + \text{AcOCH}_2\text{CH}_2\text{CH}_2\text{CH=CH}_2$$
$$(40\%) \qquad\qquad (60\%)$$

2-methyl-5-acetoxypentanoic acid in yields of 110% based on lead tetraacetate and 80% based on the acid charged.

Aromatic Substitution. Benzene and other arenes can be used as solvents for oxidative decarboxylations with lead tetraacetate. Homolytic aromatic substitution,[28, 91] however, can be an important side reaction, especially with acids that yield primary alkyl radicals or other alkyl

$$\text{RCO}_2\text{H} + \text{ArH} \xrightarrow{\text{Pb(OAc)}_4} \text{RAr}$$

radicals not readily oxidized by Pb^{IV}. As shown in Table III, tertiary alkyl-, aralkyl-, and allyl-containing acids give little or no phenylated product. However, aromatic substitution is an important process during decarboxylation of cyclopropane acids and of tertiary acids which generate bridgehead radicals.

[91] D. R. Harvey and R. O. C. Norman, *J. Chem. Soc.*, **1964**, 4860.

TABLE III. AROMATIC SUBSTITUTION DURING OXIDATIVE DECARBOXYLATION BY LEAD TETRAACETATE

Acid	Alkylbenzene (Yield %)	Refs.
Primary		
n-Butyric	22	28b
Isovaleric	11	28b
Hydrocinnamic	23	58
Phenylacetic	1	58
Secondary		
Isobutyric	5	24
2-Methylbutyric	10	92
Cyclohexanecarboxylic	10	29
2-Phenylcyclopropanecarboxylic	27	66, 67
Tertiary		
Pivalic	0.4	29
α,α-Dimethylbutyric	0.3	29
Apocamphane-1-carboxylic	56	31
Adamantane-1-carboxylic	>20	51, 52
Norbornane-1-carboxylic	10	93

Intramolecular homolytic substitutions yielding tetralin (**17**) and 1,1-diphenylhydrindene (**18**) are the principal reactions during oxidative decarboxylation of 5-phenylvaleric[31, 32] acid and 4,4,4-triphenylbutyric acid,[27b] respectively.

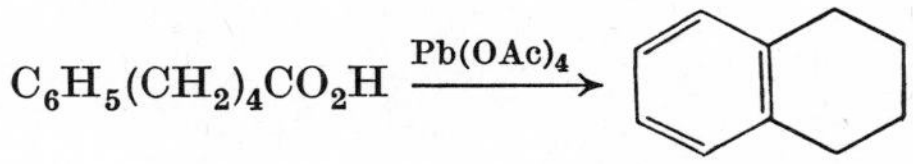

$$C_6H_5(CH_2)_4CO_2H \xrightarrow{Pb(OAc)_4}$$

17 (54%)

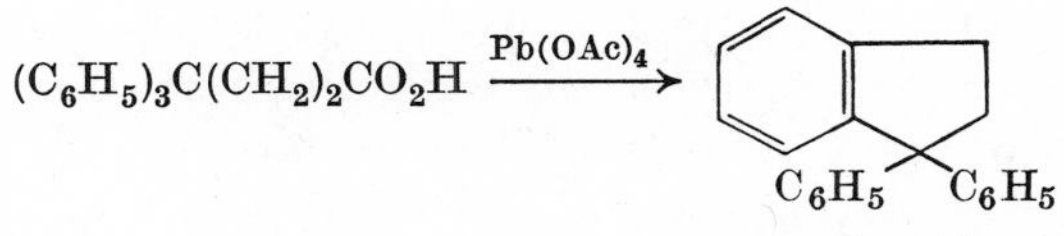

$$(C_6H_5)_3C(CH_2)_2CO_2H \xrightarrow{Pb(OAc)_4}$$

$C_6H_5 \quad C_6H_5$
18 (~40%)

Quinones are also readily alkylated with Pb^{IV}. The procedure is synthetically useful.[94] However, electron-rich aromatics such as anisole

---

[92] W. A. Ayer and C. E. McDonald, *Can. J. Chem.*, **43**, 1429 (1965); see also W. A. Ayer, C. E. McDonald, and J. B. Stothers, *ibid.*, **41**, 1113 (1963).
[93] W. J. Kissel, Ph.D. Thesis, State University of New York at Buffalo, 1968.
[94] L. F. Fieser and F. C. Chang, *J. Amer. Chem. Soc.*, **64**, 2043 (1942).

$$\text{[naphthoquinone]} \xrightarrow{\text{Pb(OAc)}_4} \text{[2-methyl-1,4-naphthoquinone]}$$

are readily acetoxylated by lead tetraacetate and are not usefully alkylated by this reagent.[91]

Alkyl Halides. The diversion of alkyl radicals to alkyl halides is discussed under Halodecarboxylation (p. 326).

Tertiary Carboxylic Acids (Table VI)

Oxidation of tertiary alkyl radicals by lead tetraacetate occurs readily. Thus decarboxylation of tertiary carboxylic acids gives alkenes as the main products, together with smaller amounts of esters. Addition of cupric acetate has little or no effect on the reaction, in contrast to the behavior of primary and secondary acids.

Decarboxylations may be carried out thermally at 80° or photochemically (350 nm) at room temperature.[29, 34] Better yields of alkenes are obtained photochemically because further oxidation of the alkenes by lead tetraacetate is less serious at lower temperatures. Oxidation of alkenes by lead tetraacetate is well known[1-8, 95] and can lead to a number of products, *e.g.*, esters of 1,2-glycols or of allylic alcohols. These com-

$$\text{C=C} + \text{Pb(OAc)}_4 \xrightarrow{\text{RCO}_2\text{H}} \text{RCO}_2-\text{C}-\text{C}-\text{OCOR}$$

$$\text{HC}-\text{C=C} + \text{Pb(OAc)}_4 \xrightarrow{\text{RCO}_2\text{H}} \text{C=C}-\text{C}-\text{OCOR}$$

peting side reactions, which result in reduced yields of alkene, occur in all oxidative decarboxylations with lead tetraacetate but may be especially serious with the highly substituted alkenes formed from tertiary carboxylic acids. For example, pivalic acid reacts with lead tetraacetate in benzene photochemically (350 nm, 10 minutes at room temperature) to give isobutylene in 74% yield.[29] The same reaction carried out thermally (30–90 minutes at 81°) produces only 48% of isobutylene. Similar results are obtained in carboxylic acid media.[34] The latter studies also showed that further oxidation of the alkene can be retarded by bases and promoted by acids.

---

[95] R. O. C. Norman and C. B. Thomas, *J. Chem. Soc.*, B, **1967**, 771; **1968**, 994.

The tertiary acid **19** with lead tetraacetate in refluxing benzene produces a mixture of alkenes in 56% yield.[96]

Oxidative decarboxylation with lead tetraacetate has been effectively employed in the synthesis of compounds related to podocarpic acid (**20**) and in synthetic approaches to the diterpenoid alkaloid, atisine [97–100] (see p. 341). Thus 12-methoxypodocarpa-8,11,13-trien-19-oic acid (**21**) gives the methoxyalkenes, the tertiary acetate, and the lactone.[97, 98] It was

[96] G. Büchi, R. E. Erickson, and N. Wakabayashi, *J. Amer. Chem. Soc.*, **83**, 927 (1961).
[97] C. R. Bennett and R. C. Cambie, *Tetrahedron*, **23**, 927 (1967).
[98] C. R. Bennett, R. C. Cambie, and W. A. Denny, *Aust. J. Chem.*, **22**, 1069 (1969).
[99] J. W. Huffman and P. G. Arapakos, *J. Org. Chem.*, **30**, 1604 (1965).
[100] L. H. Zalkow and N. N. Girotra, *Chem. & Ind.* (London), **1964**, 704.

suggested that the most likely mechanism for formation of the lactone **22** is initial acetoxylation at the benzylic position by lead tetraacetate,[98] a well-documented reaction of this reagent.[1-8] Loss of acetic acid, which is known to occur readily from these compounds,[98] gives the unsaturated acid **23**. Lead tetraacetate oxidation of this acid to yield the lactone **22** is analogous to the oxidative decarboxylation of other γ,δ-unsaturated acids (see p. 316).

The unsaturated acid 23 was synthesized by another route and shown to give the lactone 22 in 71% yield with lead tetraacetate. Moreover, the keto acid 24, which cannot be acetoxylated in the benzylic position does not give any of the lactone corresponding to 22 (p. 310).

It was originally reported that dehydroabietic acid (25) and lead tetraacetate-pyridine in refluxing benzene gave the alkene 26 in 80% yield.[99] However, repetition of this work showed that all three possible alkenes and the tertiary acetate were formed.[65] Similar situations are encountered in the oxidative decarboxylation of tetrahydroabietic acid and the synthesis of fichtelite from abietic acid.[101]

As mentioned earlier, tertiary carboxylic acids are decarboxylated by lead tetraacetate approximately 100 times faster than primary acids and 5 times faster than secondary acids. Advantage has been taken of this to decarboxylate selectively compounds containing two kinds of carboxyl groups. For example, camphoric acid has been converted to campholytolactone.[102] Another example of the selective removal of a tertiary carboxyl group is furnished by the glutaric acid derivative on p. 351.

Apocamphane-1-carboxylic acid (27) is unusual for a tertiary carboxylic acid in that the decarboxylation carried out in refluxing benzene gives

[101] N. P. Jensen and W. S. Johnson, *J. Org. Chem.*, **32**, 2045 (1967).
[102] L. L. McCoy and A. Zagalo, *J. Org. Chem.*, **25**, 824 (1960).

1-phenylapocamphane in 56% yield.[31] Oxidation of the bridgehead radical to the carbonium ion is expected to be much less favorable than for other tertiary alkyl radicals. Similarly, photolysis of the lead(IV) salt of 1-adamantanecarboxylic acid (28) in benzene produces phenyladamantane, but no products expected from the 1-adamantyl cation.[10] However, oxidative decarboxylation of norbornane-1-carboxylic acid (29) in acetic

acid affords 1-norbornyl acetate in addition to 2-norbornyl acetate.[93] The formation of the latter is strong support for a bridgehead cation as an intermediate. No 2-norbornyl acetate was found when the reaction was carried out in benzene, but 1-phenylnorbornane (10%) and norbornane (8%) were isolated.

α-Aralkanecarboxylic Acids (Table VII)

α-Aralkanecarboxylic acids react with lead tetraacetate in acetic acid or benzene-acetic acid solution, thermally at 80° or photochemically at room temperature, to give the corresponding acetates in high yield.[29, 58, 59, 103] When oxidative elimination (alkene formation) is also possible, acetates

$$\text{Ar}-\underset{\underset{R'}{|}}{\overset{\overset{R}{|}}{C}}-CO_2H \xrightarrow{Pb(OAc)_4} \text{Ar}-\underset{\underset{R'}{|}}{\overset{\overset{R}{|}}{C}}-OAc$$

are still the major products. For example, α,α-dimethylphenylacetic acid affords α-methylstyrene (14%) and α,α-dimethylphenyl acetate (74%) in both thermal and photochemical reactions.[29, 58] The product distribution is little changed by the presence of cupric acetate. As discussed earlier

$$C_6H_5-\underset{\underset{CH_3}{|}}{\overset{\overset{CH_3}{|}}{C}}-CO_2H \xrightarrow{Pb(OAc)_4} C_6H_5-\overset{\overset{CH_3}{|}}{C}=CH_2 + C_6H_5-\underset{\underset{CH_3}{|}}{\overset{\overset{CH_3}{|}}{C}}-OAc$$

[103] D. I. Davies and C. Waring, *J. Chem. Soc., C*, **1968**, 2332.

(p. 289), the relative importance of oxidative elimination and oxidative substitution is related to the stability of the carbonium ion [R+] and the nature of the solvent. Thermal and photochemical decomposition of lead(IV) tetrakis(α,α-dimethylphenylacetate) in cyclohexane gave α-methylstyrene in 55% and 100% yields, and no ester was found.[10]

Yields of benzylic acetates are dependent on the nature of the substituents in the aromatic ring. Electron-releasing groups capable of stabilizing the benzylic cation favor the reaction. Poor yields of ester are obtained when electron-withdrawing groups are present. Thus both Cu^{II} and Pb^{IV} oxidize benzyl radicals slowly, and benzyl acetate in only 62% yield, together with small amounts of toluene and dibenzyl, was obtained from phenylacetic acid.[58] In contrast, p-methoxyphenylacetic acid was decarboxylated smoothly to give p-methoxybenzyl acetate in 99% yield, whereas p-nitrophenylacetic acid gave only 1% of p-nitrobenzyl acetate.[58] Oxidative decarboxylation by lead tetraacetate, therefore, can be used for the conversion of benzylic carboxyl groups to the corresponding acetates (and subsequently to the alcohols by saponification). However, except in special cases, it is doubtful if the reaction is of much synthetic value.

Decarboxylation of α-aralkyl acids in the presence of added lithium chloride affords the corresponding aralkyl chloride in high yield.[58] (See Table XIII, and "Halodecarboxylation," pp. 326–330.)

Unsaturated Acids (Table VIII)

α,β-Unsaturated Acids. The oxidative decarboxylation of either cis- or trans-cinnamic acid produces a mixture of two stereoisomeric enol acetates and propenylbenzenes.[104] Equilibration of the vinylic radicals is indicated.

[104] B. Danieli, P. Manitto, and G. Russo, *Chim. Ind.* (Milan), **51(6)**, 609 (1969).

The yields of products from the decarboxylation of α,β-unsaturated acids are generally poor because of the high reactivity of the vinyl radicals. The oxidative decarboxylation of these acids presents a situation similar to that of aromatic acids (see p. 322). Hence competitive decarboxylation of acetate would be serious. Moreover, oxidation of the double bond by lead tetraacetate will compete with decarboxylation, perhaps to its exclusion. However, a photochemical reaction carried out at room temperature in the absence of acetate may be feasible, for relatively thermostable lead(IV) salts such as lead(IV) tetra-1-adamantanecarboxylate (which is stable up to 280°) can be readily decomposed photochemically.[10]

β,γ-Unsaturated Acids. Decarboxylation of β,γ-unsaturated acids by lead tetraacetate in mixtures of benzene and acetic acid usually affords the two isomeric allylic acetates. Thus 3-octenoic acid gives a mixture of 1-acetoxy-2-heptene and 3-acetoxy-1-heptene.[58] Yields are improved and the rate enhanced by cupric acetate (see Table X). Added cupric acetate also influences the ratio of acetates, the terminal alkene being favored in

$$n\text{-}C_4H_9\diagdown\diagup H$$
$$\phantom{n\text{-}C_4H_9}C{=}C$$
$$\diagup\diagdown$$
$$HCH_2CO_2H$$

$\longrightarrow$

$$n\text{-}C_4H_9\diagdown\diagup H$$
$$\phantom{n\text{-}C_4H_9}C{=}C \qquad + \; n\text{-}C_4H_9CHCH{=}CH_2$$
$$\diagup\diagdown \qquad\qquad\qquad\qquad |$$
$$HCH_2OAc \qquad\qquad\;\; OAc$$

$\quad$ 30% [Pb(OAc)$_4$]$\qquad\qquad$ 21% [Pb(OAc)$_4$]
$\quad$ 13% (Pb^{IV}/Cu^{II})$\qquad\quad\;\,$ 68% (Pb^{IV}/Cu^{II})

the presence of Cu^{II}. The same distribution of isomeric heptenyl acetates is observed in the decarboxylation of 3-octenoic acid by the Pb^{IV}/Cu^{II} reagent as in other allylic systems. Butyl radicals, derived from the thermolysis of valeryl peroxide, can be trapped by 1,3-butadiene.[105] Oxidation of the resulting octenyl radicals with Cu^{II} in benzene-acetic

$$n\text{-}C_4H_9\cdot \; + \; CH_2{=}CHCH{=}CH_2 \; \longrightarrow \; n\text{-}C_5H_{11}CH\overset{\displaystyle CH}{\diagdown}CH_2, \quad etc.$$

acid produces an 85:15 mixture of 3-acetoxy-1-octene and 1-acetoxy-2-octene. The distribution of heptenyl acetates from the oxidative decarboxylation of 3-octenoic acid is also affected by added pyridine, which enhances the rate of decarboxylation. The distribution of allylic acetates derived by oxidation of allylic radicals by metal oxidants has been discussed at length.[105]

---

[105] J. Kochi and H. Mains, *J. Org. Chem.*, **30**, 1862 (1965); also see refs. 51, 52, 59.

Vinylacetic acid reacts with lead tetraacetate in benzene-acetic acid at 80° to produce allyl acetate (70%) and allyl vinylacetate (6%).[58] The Cu^{II}-catalyzed decarboxylation gives allyl acetate in 87% yield.

$$CH_2{=}CHCH_2CO_2H \xrightarrow{Pb(OAc)_4} CH_2{=}CHCH_2OAc + CH_2{=}CHCH_2CO_2CH_2CH{=}CH_2$$

A study of β,γ-unsaturated acids showed that some furnish a mixture of allyl acetates and others furnish a single isomer.[106] Thus decarboxylation of the arylated acid **30** by lead tetraacetate in acetic acid affords a mixture of two isomeric acetates. Decarboxylation of the homologous acid **31** under identical conditions forms only one acetate.

$$\underset{\mathbf{30}}{CH_2{=}C\genfrac{}{}{0pt}{}{C(CH_3)_2CO_2H}{Ar}} \xrightarrow{Pb(OAc)_4} CH_2{=}C\genfrac{}{}{0pt}{}{C(CH_3)_2OAc}{Ar} + AcOCH_2C\genfrac{}{}{0pt}{}{C(CH_3)_2}{Ar}$$

$$\underset{\mathbf{31}}{\genfrac{}{}{0pt}{}{CH_3}{H}C{=}C\genfrac{}{}{0pt}{}{C(CH_3)_2CO_2H}{Ar}} \xrightarrow{Pb(OAc)_4} \genfrac{}{}{0pt}{}{CH_3}{H}AcO{-}C{-}C\genfrac{}{}{0pt}{}{C(CH_3)_2}{Ar}$$

Hence, on decarboxylation, these β,γ-unsaturated acids generate intermediate allylic radicals which are subsequently oxidized by lead tetraacetate to allylic esters.[28b]

Oxidative decarboxylation of the cyclopropanecarboxylic acid **32** gave only the lactone **33**.[107] Attempted iododecarboxylation of the same acid

[106] J. Jacques, C. Weidmann, and A. Horeau, *Bull. Soc. Chim. Fr.*, **1959**, 424.
[107] J. Meinwald, J. W. Wheeler, A. A. Nimetz, and J. S. Liu, *J. Org. Chem.*, **30**, 1038 (1965).

with Pb^{IV}/I_2 gave only an iodolactone. Formation of lactones is also common in the reaction of γ,δ-unsaturated acids with lead tetraacetate.

An oxidative aromatization during decarboxylation with lead tetraacetate has been observed with dihydroaromatic acids.[108]

γ,δ-Unsaturated Acids. In either benzene or acetonitrile, allylacetic acid is decarboxylated by lead tetraacetate to give a mixture of homoallylic acetates in poor yield (<10%). The principal products are 1-butene (resulting from the reaction of allylcarbinyl radical with the solvent) and allylcarbinylbenzene (when the reaction is run in benzene).[59] (The latter result is in agreement with the inefficient oxidation of other primary alkyl radicals by Pb^{IV}.) The poor overall material balance of the

$$CH_2{=}CHCH_2CH_2CO_2H \xrightarrow[C_6H_6]{Pb(OAc)_4} CH_2{=}CHCH_2CH_3$$
(9%)

$$+ \ CH_2{=}CHCH_2CH_2C_6H_5$$
(6%)

$$+ \ CH_2{=}CHCH_2CH_2OAc \ + \ \square\!\!-\!OAc \ + \ \triangleright\!\!-\!CH_2OAc$$
(5%, total)

C_4H_7-alkyl fragments may have been due to competitive decarboxylation of acetic acid. Decarboxylation in the presence of Cu^{II} afforded two new products, 1,3-butadiene and a mixture of octadienyl acetates, together with a reduced yield of 1-butene and allylcarbinylbenzene (Eq. 23).

Thus allylcarbinyl radicals, like other primary alkyl radicals, show great selectivity toward oxidation by Pb^{IV} and Cu^{II}. Octadienyl acetates are formed by way of addition of allylcarbinyl radical to 1,3-butadiene followed by Cu^{II} oxidation of the resultant allylic radical. (Compare the trapping of butyl radicals, p. 314.) The yields of octadienyl acetates are increased significantly by the prior addition of butadiene.

[108] A. J. Birch, *J. Chem. Soc.*, **1950**, 1551.

$$CH_2=CHCH_2CH_2CO_2H \xrightarrow{Pb(IV)/Cu(II)} CH_2=CHCH_2CH_3 + CH_2=CHCH=CH_2$$
$$(1\%) \qquad\qquad (24\%)$$

$$+ CH_2=CHCH_2CH_2OAc + \underbrace{\square\!\!-\!OAc + \triangleright\!-\!CH_2OAc +}_{(3\%,\ total)}$$

$$\underbrace{CH_2=CHCH_2CH_2CH_2CH(OAc)CH=CH_2 +}_{CH_2=CHCH_2CH_2CH_2CH=CHCH_2OAc}$$
$$(10\%,\ total)$$

(Eq. 23)

Essentially the same results are obtained from the decarboxylation of cyclopropylacetic acid.[59] All the products can be considered to arise from the allylcarbinyl radical. The first-order isomerization of cyclopropylcarbinyl radical must have been too rapid compared with other second-order processes such as hydrogen transfer, oxidation, addition, and substitution reactions which could also have occurred.[109–111]

$$\triangleright\!-\!CH_2CO_2H \xrightarrow[-CO_2]{Pb(OAc)_4} \triangleright\!-\!CH_2\cdot \longrightarrow$$

$$\begin{array}{c} CH_2 \\ | \quad \diagdown \\ \quad\ \ CH=CH_2 \longrightarrow Products \\ | \\ CH_2 \end{array}$$

Decarboxylation of 3-cyclohexene-1-carboxylic acid with Pb^{IV}/Cu^{II} yields a mixture of the two isomeric cyclohexadienes together with benzene, an oxidation product of 1,3-cyclohexadiene.[60]

$$\bigcirc\!\!-\!CO_2H \xrightarrow{Pb^{IV}/Cu^{II}} \bigcirc + \bigcirc + \bigcirc$$
$$\qquad\qquad\qquad (38\%) \quad (8\%) \quad (18\%)$$

A number of interesting products have been obtained from lead tetraacetate oxidation of some γ,δ-unsaturated acids in which the double bond is held in close proximity to the carboxyl group. Oxidation of *endo*-5-carboxybicyclo[2,2,1]hept-2-ene (**34**) with lead tetraacetate forms the acetoxy lactone **35** in 80% yield.[112] An analogous reaction is the conversion of the tricyclic acid **23** to a tetracyclic lactone **22** (p. 310). (There are

---

[109] D. J. Carlsson and K. U. Ingold, *J. Amer. Chem. Soc.*, **90**, 7047 (1968).
[110] J. K. Kochi and P. J. Krusic, *J. Amer. Chem. Soc.*, **90**, 7157 (1968).
[111] L. K. Montgomery and J. W. Matt, *J. Amer. Chem. Soc.*, **89**, 6556 (1967).
[112] R. M. Moriarty, H. G. Walsh, and H. Gopal, *Tetrahedron Lett.*, **1966**, 4363.

similar reactions in the bisdecarboxylation of unsaturated dicarboxylic acids; see pp. 346–348.)

The corresponding *exo* acid **36** gives an isomeric acetoxy lactone under identical conditions. Similarly, 2-cyclopentenylacetic acid produces an acetoxy lactone in 70% yield.

The authors considered initial electrophilic addition of lead tetraacetate to the double bond in the endo acid **34**, followed by intramolecular nucleophilic opening by the carboxylate function, as the most likely mechanism.[112] Evidence in support of this mechanism is furnished by the observation that the same acetoxy lactone is formed from the corresponding methyl ester or the amide, presumably by the process shown in Eq. 24.

Attempted halodecarboxylation of the *endo* acid **34** (but not its methyl ester or amide) with Pb(OAc)$_4$-LiCl gives the acetoxy lactone **37** in 70% yield.[113] This lactone may be formally regarded as the product of a Wagner-Meerwein rearrangement of the lactone **35**. The failure of the methyl ester or amide to form the acetoxy lactone **37** suggested that the acyl hypochlorite is an intermediate.

endo-5-Carboxybicyclo[2,2,2]oct-2-ene (**38**) gives substantial decarboxylation with lead tetraacetate, and the yield of the acetoxy lactone is only 15%.[112] Oxidative decarboxylation of the *exo* isomer **39** is reported

---

[113] R. M. Moriarty, H. Gopal, and H. G. Walsh, *Tetrahedron Lett.*, **1966**, 4369; see also R. M. Moriarty, C. R. Romain, and T. O. Lovett, *J. Amer. Chem. Soc.*, **89**, 3927 (1967).

to give the complex mixture of acetates (total yield 30–38%) shown in the accompanying scheme[114] with the relative distribution of products indicated.

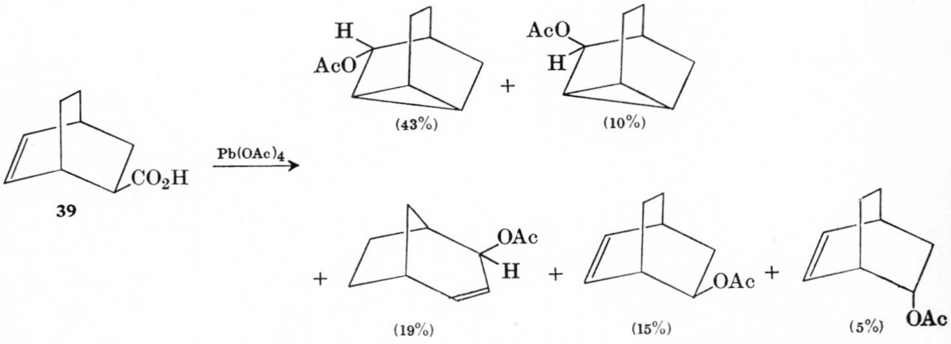

From the preceding discussion it is apparent that decarboxylation of acids in the bicyclo[2,2,1]heptene and bicyclo[2,2,2]octene series is extremely complex because of the easy rearrangement of these compounds. Similar difficulties have been encountered in bisdecarboxylation of analogous compounds (see pp. 346–348).

Whether oxidation of unsaturated acids by lead tetraacetate proceeds by attack at the double bond followed by participation of the carboxylic function to form lactones or by oxidative decarboxylation depends on the alkene. The rate of electrophilic addition by metal ions is increased in strained bicyclic systems[115] as well as in alkenes which are extensively alkylated.

Effects due to neighboring groups are not restricted to alkenes, since lactone formation presumably involving the enol is observed with the keto acid **40**.[116a] The reaction is especially noteworthy since lactone formation appears to take precedence over the ready decarboxylation of the α-hydroxy acid function.

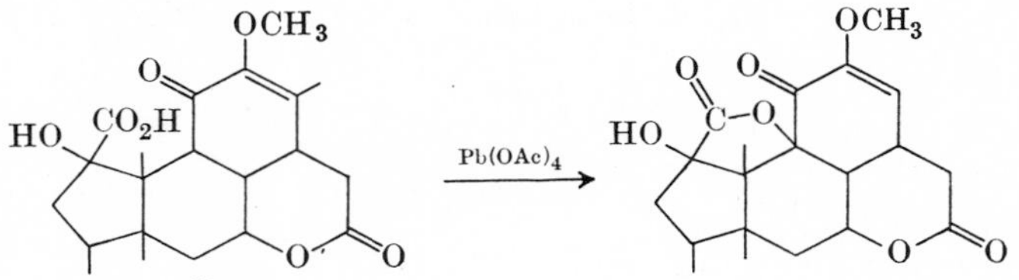

---

[114] N. A. LeBel and J. E. Huber, *J. Amer. Chem. Soc.*, **85**, 3193 (1963).

[115] T. G. Traylor, *Accts. Chem. Res.*, **2**, 152 (1969).

[116] (a) Z. Valenta, A. H. Gray, S. Papadopoulos, and C. Podesva, *Tetrahedron Lett.* No. **20**, 25 (1960); (b) A. C. Cope, C. H. Park, and P. Scheiner, *J. Amer. Chem. Soc.*, **84**, 4862 (1962).

An interesting example of participation by an aromatic ring is given by the phenyl migration during oxidation of β,β,β-triphenylpropionic acid.[27a]

$$(C_6H_5)_3CCH_2CO_2H \xrightarrow{Pb(OAc)_4} (C_6H_5)_2C=CHCO_2Ph$$

Miscellaneous Unsaturated Acids. 4-Cycloheptene-1-carboxylic acid with lead tetraacetate in acetic acid at 70° gives 4-cyclohepten-1-yl acetate in 70% yield.[116b]

Oxidative decarboxylation of the higher homolog, 4-cyclooctene-1-carboxylic acid, affords the expected 4-cycloocten-1-yl acetate as well as products which could be accounted for by a transannular ring closure of the 4-cyclooctenyl cation.[86]

Oxidative decarboxylation of 2,4,6-cycloheptatriene-1-carboxylic acid (**41**) has been used to generate the tropylium ion.[117] However, the *benzo*

analog **42** on oxidation with lead tetraacetate does not produce the benzotropylium ion but is oxidized further to 2H-benz[cd]azulene.[118] The dibenzo analog **43** undergoes an interesting transformation to an unsaturated ketone (p. 322).[119]

[117] M. J. S. Dewar, C. R. Ganellin, and R. Pettit, *J. Chem. Soc.*, **1958**, 55; also see C. R. Ganellin and R. Pettit, *J. Amer. Chem. Soc.*, **79**, 1767 (1957).
[118] V. Boekelheide and C. D. Smith, *J. Amer. Chem. Soc.*, **88**, 3950 (1966).
[119] R. Munday and I. O. Sutherland, *J. Chem. Soc., C*, **1969**, 1427.

Aromatic Carboxylic Acids (Table IX)

Decarboxylation of aromatic carboxylic acids has a high activation energy and is not a facile process, presumably because of the strong aryl-carboxyl bond and the instability of the incipient aryl radicals. Consequently, competitive decarboxylation of acetate is common. Lead tetrabenzoate has been used as a phenylating agent.[120] Its decomposition in chlorobenzene at 130° afforded a mixture of isomeric chlorobiphenyls. Similarly, pyridine is phenylated by lead tetrabenzoate in pyridine at 105°. The ratio of isomers in both of these reactions is in good agreement with those obtained from homolytic phenylations using benzoyl peroxide.

Davies and Waring studied the reaction of aromatic carboxylic acids with lead tetraacetate.[31, 42, 121] 2′-Substituted-2-carboxybiphenyls give 3,4-benzocoumarins. By analogy with the cyclization of biphenyl-2-carboxylic acid to 3,4-benzocoumarin induced by other homolytic oxidants,[122] the homolytic mechanism shown in Eq. 25 was suggested. Only a trace of benzocoumarin is obtained with X = Me, which reflects

$X = H, NO_2, Cl, OCH_3, CO_2CH_3, CO_2H$

both the difficulty of displacement of a methyl group and the ease of oxidation of the substrate by a different process. 2-Phenyl-2-carboxybiphenyl (**44**) likewise affords no benzocoumarin but does form triphenylene

---

[120] D. H. Hey, C. J. M. Stirling, and G. H. Williams, *J. Chem. Soc.*, **1954**, 2747; see also D. H. Hey, C. J. M. Stirling, and G. H. Williams, *J. Chem. Soc.*, **1955**, 3963; N. A. Maier, L. I. Guseva, and Yu. A. Ol'dekop, *J. Gen. Chem. USSR*, **37**, 2528 (1967).

[121] D. I. Davies and C. Waring, *J. Chem. Soc.*, C, **1968**, 2337.

[122] G. W. Kenner, M. A. Murray, and C. M. B. Taylor, *Tetrahedron*, **1**, 259 (1957).

in quantitative yield, presumably as shown in the accompanying scheme.

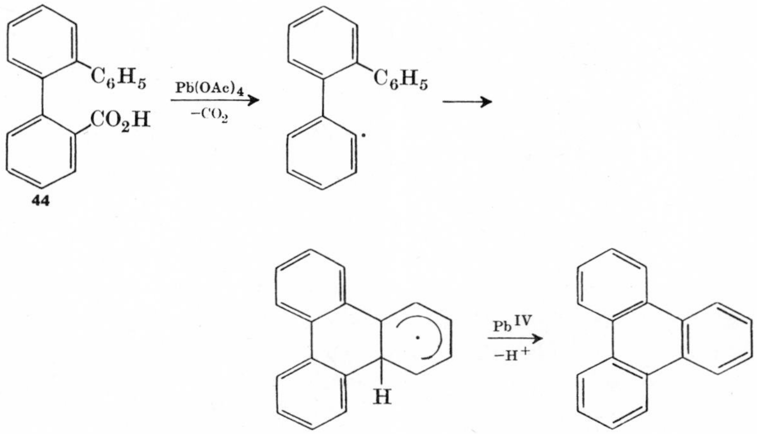

Other authors consider an ionic mechanism plausible for the formation of benzocoumarin from biphenyl-2,2'-dicarboxylic acid.[123]

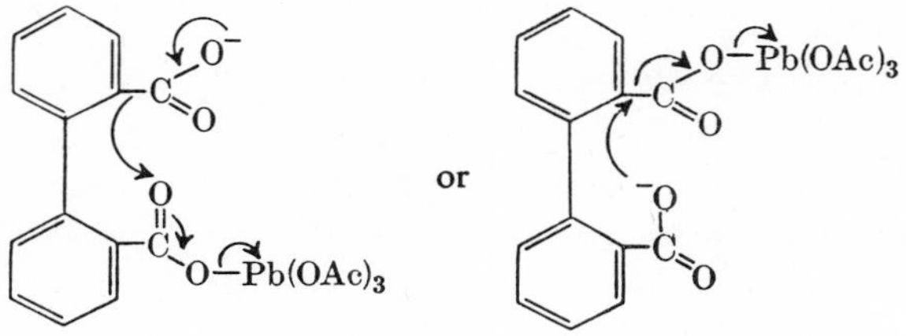

β-Keto Acids, β-Cyano Acids, and Malonic Half-Esters
(Table XI)

Half-esters of substituted malonic acids are decarboxylated by lead tetraacetate to the corresponding acetates in moderate yield.[124-126] Similarly, β-keto acids and β-cyano acids give the corresponding acetates.[127] To carry out the reaction the acid is heated with lead tetraacetate

---

[123] W. R. Moore and H. Arzoumanian, *J. Org. Chem.*, **27**, 4667 (1962).
[124] M. Vilkas and M. Rouhi-Laridjani, *Compt. Rend.*, **251**, 2544 (1960).
[125] M. Rouhi-Laridjani and M. Vilkas, *Compt. Rend.*, **254**, 1090 (1962).
[126] M. Rouhi-Laridjani, *Publ. Sci. Tech. Min. Air* (France) *Notes Tech.*, **1964**, No. 141 [C.A., **63**, 8188g (1965)].
[127] J. C. Chottard, M. Julia, and J. M. Salard, *Tetrahedron*, **25**, 4967 (1969).

$$\underset{H}{\overset{R}{>}}C\underset{CO_2Et}{\overset{CO_2H}{<}} \xrightarrow{Pb(OAc)_4} \underset{H}{\overset{R}{>}}C\underset{CO_2Et}{\overset{OAc}{<}}$$

$$\underset{H}{\overset{R}{>}}C\underset{COR'}{\overset{CO_2H}{<}} \xrightarrow{Pb(OAc)_4} \underset{H}{\overset{R}{>}}C\underset{COR'}{\overset{OAc}{<}}$$

$$\underset{H}{\overset{R}{>}}C\underset{CN}{\overset{CO_2H}{<}} \xrightarrow{Pb(OAc)_4} \underset{H}{\overset{R}{>}}C\underset{CN}{\overset{OAc}{<}}$$

in benzene at 20–80°, followed by decarboxylation at a higher temperature, usually by distillation in the presence of copper powder. It has been suggested that the reaction proceeds via initial acetoxylation to an acetoxy acid which is decarboxylated at higher temperatures as indicated in Eq. 26. Acetoxylation of activated C—H bonds by lead tetraacetate is a well-

$$\underset{H}{\overset{R}{>}}C\underset{CO_2Et}{\overset{CO_2H}{<}} \xrightarrow{Pb(OAc)_4} \left[\underset{AcO}{\overset{R}{>}}C\underset{CO_2Et}{\overset{CO_2H}{<}}\right] \xrightarrow{200°} \underset{AcO}{\overset{R}{>}}C\underset{CO_2Et}{\overset{H}{<}} \quad \text{(Eq. 26)}$$

established process.[1–8, 128] The plausibility of initial acetoxylation has been supported recently by isolation of the acetoxy acid **45** from the oxidation of the half-ester of γ-phenylpropylmalonic acid.[127] The tetralin derivatives no doubt arose from decarboxylation of the starting material

$$C_6H_5CH_2CH_2CH_2\underset{CO_2C_2H_5}{\overset{|}{C}HCO_2H} \xrightarrow{Pb(OAc)_4} C_6H_5CH_2CH_2CH_2\underset{CO_2C_2H_5}{\overset{OAc}{\underset{|}{C}CO_2H}} +$$

45 (33%)

$$C_6H_5CH_2CH_2CH_2\underset{}{\overset{OAc}{\underset{|}{C}HCO_2C_2H_5}} +$$
(10%)

[tetralin derivative with CO₂C₂H₅] (5%) + [tetralin derivative with AcO and CO₂C₂H₅] (7%)

---

[128] J. J. Riehl and A. Fougerousse, *Bull. Soc. Chim. Fr.*, **1968**, 4083.

and the intermediate acetate, respectively, followed by homolytic cyclization. Initial acetoxylation of the half ester, however, is not a necessary condition for decarboxylation, since the α-methyl homolog, which cannot undergo prior acetoxylation, is oxidatively decarboxylated by lead tetraacetate to give a tetralin ester in high yield.[127]

$$C_6H_5CH_2CH_2CH_2\underset{CO_2C_2H_5}{\underset{|}{\overset{CH_3}{\overset{|}{C}}}}CO_2H \xrightarrow{Pb(OAc)_4}$$ [tetralin with CH₃ and CO₂C₂H₅]
(66–80%)

It is interesting that no α,β-unsaturated esters were formed under these conditions even in the presence of Cu^{II}, although by analogy with alkene formation from other acids they would have been expected. However, the

$$\underset{H}{\overset{RCH_2}{\diagdown}}\underset{CO_2Et}{\overset{CO_2H}{\diagup}}C \xrightarrow{Pb(OAc)_4} \!\!\!\!/\!\!\!\!/\!\!\!\! \longrightarrow RCH=CHCO_2Et$$

yields of acetates were only moderate, and α,β-unsaturated esters, if formed, may have undergone further oxidation.

γ-Keto Acids (Table XII)

The oxidative decarboxylation of γ-keto acids can be effectively carried out with lead dioxide. The reaction is usually performed heterogeneously in a hot tube with powdered glass or in a high-boiling solvent at 135° or higher. In the absence of an acetate ligand, α,β-unsaturated ketones are formed in yields as high as 85%. Nepetonic acid was converted to methyl (−)-3-methyl-1-cyclopentenyl ketone.[129a] The reaction is general,

[cyclopentane with CH₃, CO₂H, COCH₃] $\xrightarrow{PbO_2}$ [cyclopentene with CH₃, COCH₃]

but the yields vary with substituents located on the α- and β-carbon atoms.[129b] Some representative examples are shown on p. 326.

---

[129] (a) S. M. McElvain and E. J. Eisenbraun, *J. Amer. Chem. Soc.*, **77**, 1599 (1955). (b) D. V. Hertzler, J. M. Berdahl, and E. J. Eisenbraun, *J. Org. Chem.*, **33**, 2008 (1968).

$$\text{CH}_3\text{-}\underset{\text{O}}{\overset{\text{O}}{\bigcirc}}\text{-}\underset{\text{CH}_3}{\overset{\text{CH}_3}{\text{C}}}\text{-CO}_2\text{H} \xrightarrow{\text{PbO}_2} \text{CH}_3\text{-}\underset{\text{O}}{\overset{\text{O}}{\bigcirc}}\text{=}\underset{\text{CH}_3}{\overset{\text{CH}_3}{\text{C}}}$$

(76%)

$$\text{C}_6\text{H}_5\text{COCH}_2\overset{\text{C}_6\text{H}_5}{\underset{|}{\text{CH}}}\text{CO}_2\text{H} \xrightarrow{\text{PbO}_2} \text{C}_6\text{H}_5\text{COCH=CHC}_6\text{H}_5$$

(84%)

[cyclohexanone-CO₂H] —PbO₂→ [cyclohexenone] (92%)

[cyclohexane with CO₂H and COCH₃] —PbO₂→ [cyclohexene-COCH₃] (22%)

In order to effect these conversions under milder conditions, more extensive use of lead tetraacetate should be considered. A recently reported example is the decarboxylation of the spiro acid shown in the accompanying equation.[130] Rearrangement accompanying the oxidative decarboxylation is consistent with carbonium ion intermediates.

[spiro diketone acid] $\xrightarrow[\text{C}_6\text{H}_6,\text{reflux}]{\text{Pb(OAc)}_4}$ [bicyclic enone]

Halodecarboxylation (Tables XIII, XIV)

The Hunsdiecker method for the decarboxylation of an acid to an alkyl halide by the action of halogen on the dry silver carboxylate is well known.[131, 132] It is normally used for preparing alkyl bromides but can also be used for the preparation of alkyl chlorides.[131, 132] The Hunsdiecker method suffers from the difficulty and expense of preparing dry silver salts. A modified procedure using mercuric salts has certain advantages in

$$\text{RCO}_2\text{Ag} \xrightarrow{\text{Br}_2} \text{RBr} + \text{CO}_2 + \text{AgBr}$$

[130] R. L. Cargill and A. M. Foster, *J. Org. Chem.*, **35**, 1971 (1970).
[131] R. G. Johnson and R. K. Ingham, *Chem. Rev.*, **56**, 219 (1956).
[132] C. V. Wilson, *Org. Reactions*, **9**, 332 (1957).

that the metal carboxylate is not isolated.[133] Other modifications have also been tried.[134, 135] Both the silver and the mercury methods give good yields with primary acids, moderate yields with secondary acids, and poor yields with tertiary acids and dicarboxylic acids.[131–133] Various halodecarboxylation procedures are compared in Table XIV.

Metal Halide. Consideration of the homolytic mechanism for Pb^{IV} decarboxylation and the ligand transfer oxidation of alkyl radicals[25] led to the conclusion that decarboxylation of acids by Pb^{IV}/halide species should afford alkyl halides.[61, 62] Indeed, decarboxylation of acids by lead tetraacetate in the presence of metal halides such as lithium, potassium, or calcium chloride gives alkyl chlorides in high yield; the reagent will be referred to as Pb^{IV}/LiCl. Similarly, in the presence of lithium bromide or iodide, the corresponding bromide or iodide is formed, although in diminished yields.

$$RCO_2H + Pb(OAc)_4 + LiCl \rightarrow RCl + CO_2 + LiOAc + Pb(OAc)_2 + HOAc$$

The reaction is carried out by adding 1 molar equivalent of the halide to a solution of 1 molar equivalent of the acid and of lead tetraacetate in benzene, and heating under reflux. Reactions are complete in a few minutes, in contrast to the much longer times in the absence of added halides. The reaction is strongly inhibited by oxygen and can actually be interrupted by the deliberate addition of oxygen. Alkyl chloride yields decrease with increasing amounts of added halide above one equivalent. When 4 equivalents of lithium chloride are used, no n-butyl chloride is formed from valeric acid, and copious amounts of chlorine are produced.[61, 62]

Halodecarboxylation with lead tetraacetate affords no products characteristic of the rearrangement of the carbonium ion intermediates, in contrast to oxidative decarboxylation. Thus t-butylacetic acid affords only neopentyl chloride and no t-amyl chloride. Similarly, cyclobutanecarboxylic acid forms cyclobutyl chloride in 98% yield. This should be

$$(CH_3)_3CCH_2CO_2H \xrightarrow{Pb^{IV}/LiCl} (CH_3)_3CCH_2Cl$$
(92%)

$$\text{cyclobutyl-}CO_2H \xrightarrow{Pb^{IV}/LiCl} \text{cyclobutyl-}Cl$$
(98%)

---

[133] S. J. Cristol and W. C. Firth, *J. Org. Chem.*, **26**, 280 (1961); S. J. Cristol, J. R. Douglass, W. C. Firth, and R. E. Krall, *ibid.*, **27**, 2711 (1962); J. S. Meek and D. T. Osuga, *Org. Syntheses*, **43**, 9 (1963); see also refs. 134 and 135 for modifications of the Hunsdiecker reaction.

[134] J. A. Davis, J. Herynk, S. Carroll, J. Bunds, and D. Johnson, *J. Org. Chem.*, **30**, 415, (1965).

[135] A. McKillop, D. Bromley, and E. C. Taylor, *J. Org. Chem.*, **34**, 1172 (1969).

compared with the oxidative decarboxylation of cyclobutanecarboxylic acid by lead tetraacetate alone[59] (Table V).

The chlorodecarboxylation of *cis*- and *trans*-4-*t*-butylcyclohexanecarboxylic acids separately with 1 molar equivalent of lead tetraacetate and lithium chloride gave the same mixture of *cis*- and *trans*-4-*t*-butylcyclohexyl chlorides as obtained from the homolytic decomposition of the

$(CH_3)_3C$-cyclohexane-CO_2H $\xrightarrow{Pb^{IV}/LiCl}$ $(CH_3)C$-cyclohexane-Cl + $(CH_3)_3C$-cyclohexane-Cl

$(CH_3)_3C$-cyclohexane-CO_2H $\xrightarrow{Pb^{IV}/LiCl}$

2.05 : 1.0

tertiary hypochlorite.[136–138] The results, in contrast to oxidative decarboxylation with lead tetraacetate itself, are consistent with a homolytic transfer of chlorine to the alkyl radical.[138] For example, the oxidative decarboxylation of the 4-*t*-butylcyclohexanecarboxylic acids in acetic acid led to a mixture of *cis*- and *trans*-4-*t*-butylcyclohexyl acetates in the molar ratio 1.0:0.9 which is typical of a carbonium ion process.[136]

Replacement of the carboxyl group by chlorine during chlorodecarboxylation occurs with loss of asymmetry. Thus optically active ($[\alpha]D = 11.72°$) 2-benzyl-2-cyanopropanoic acid afforded only racemic 1-phenyl-2-chloro-2-cyanopropane.[139] Racemization probably occurs in the free-radical intermediate.

$$C_6H_5CH_2\underset{CN}{\overset{CH_3}{\underset{|}{\overset{|}{C}}}}-CO_2H \xrightarrow{Pb^{IV}/LiCl} C_6H_5CH_2\underset{CN}{\overset{CH_3}{\underset{|}{\overset{|}{C}}}}-Cl$$

$[\alpha]D = 11.72°$ $\qquad\qquad [\alpha]D = 0.000 \pm 0.002°$

Primary and secondary carboxylic acids give almost quantitative yields of alkyl chlorides (Table XIII). Lead tetraacetate itself under these

---

[136] S. D. Elakovich and J. G. Traynham, *Tetrahedron Lett.*, **1971**, 1435.

[137] F. D. Greene, C. Chu, and J. Walia, *J. Amer. Chem. Soc.*, **84**, 2463 (1962); *J. Org. Chem.*, **29**, 1285 (1964); F. R. Jensen, L. H. Gale, and J. E. Rodgers, *J. Amer. Chem. Soc.*, **90**, 5793 (1968).

[138] R. D. Stolow and T. W. Giants, *Tetrahedron Lett.*, **1971**, 695; *J. Amer. Chem. Soc.*, **93**, 3536 (1971).

[139] J. Casanova and L. E. Eberson, private communication, 1971.

conditions produces methyl chloride (40–60%). Tertiary acids also give alkyl chlorides in good yield, together with small amounts of alkenes. This method thus represents a superior method for the halodecarboxylation of acids, especially tertiary acids.

$$(CH_3)_3CCO_2H \xrightarrow{Pb^{IV}/LiCl} \underset{(72\%)}{(CH_3)_3CCl} + \underset{(9\%)}{(CH_3)_2C=CH_2}$$

$$C_2H_5C(CH_3)_2CO_2H \xrightarrow{Pb^{IV}/LiCl}$$

$$\underset{(91\%)}{C_2H_5C(CH_3)_2Cl} + \underset{(4\%)}{C_2H_5C(CH_3)=CH_2} + \underset{(4\%)}{CH_3CH=C(CH_3)_2}$$

A slightly modified procedure involving the use of excess lithium chloride (2.5 molar eq.) in an acetic acid-benzene solution led to excellent yields of α-deuteriocyclooctyl chloride from α-deuteriocyclooctanecarboxylic acid.[140]

Benzoic acid gives a poor yield of chlorobenzene due to competitive halodecarboxylation of acetate.[61, 62] Removal of acetic acid prior to addition of halide should increase the yield of aryl chlorides. Aralkyl acids produce good yields of benzyl chlorides.[58] Interestingly, the succinic acid derivative **46** on treatment with lead tetraacetate and lithium chloride affords the vicinal dichloride.[141] Oxidative elimination does not appear to be a complicating factor in this case, nor in the chlorodecarboxylation of the β-chloro acid **47** (p. 330).[142]

[140] J. G. Traynham and R. L. Frye, private communication, 1970.
[141] H. J. Reich and D. J. Cram, *J. Amer. Chem. Soc.*, **91**, 3517 (1969).
[142] E. N. Cain, *Tetrahedron Lett.*, **1971**, 1865.

The inhibition by oxygen and lack of skeletal rearrangement in the neopentyl and cyclobutyl systems is taken as evidence for a homolytic mechanism (see pp. 287, 327).

Brominative decarboxylation of the tricyclic acid **49** with Pb(OAc)$_4$/LiBr affords the corresponding bromide.[143] This reaction was used to confirm the structure of the tetracyclic ketone **48** by reducing the bromide with sodium in liquid ammonia to the known hydrocarbon **50**. Decarboxylation of the tricyclic acid by the Hunsdiecker[131, 132] or Cristol-Firth[133] procedure forms the rearranged bromide **51**. This is one of the few cases of brominative decarboxylation reported using the Pb^{IV}/LiBr reagent. Further possibilities should be examined for the generality of the procedure.

Pseudohalides, such as thiocyanate and cyanide, also initiate rapid decomposition of Pb^{IV} carboxylates.[61, 62] Hence it should be possible to extend the halodecarboxylation reaction to the preparation of thiocyanates, azides, *etc.*, although there remains the possibility that thiocyanate and azide may be oxidized by lead tetraacetate.

$$\text{RCO}_2\text{H} \xrightarrow[\text{SCN}^-]{\text{Pb}^{IV}} \text{RSCN}$$

$$\text{RCO}_2\text{H} \xrightarrow[\text{N}_3^-]{\text{Pb}^{IV}} \text{RN}_3$$

---

[143] R. R. Sauers, K. Kelly, and B. Sickles, Abstr. 158th A.C.S. National Meeting, New York, Sept. 1969, Org. Section, Paper 74.

Iodine (Photochemical). Iododecarboxylation of carboxylic acids has been effected using a combination of lead tetraacetate with iodine.[144, 145] The reaction is carried out by adding 1 equivalent of iodine to 1 equivalent of lead tetraacetate and 1 equivalent of carboxylic acid in refluxing carbon tetrachloride. The solution is illuminated with a tungsten lamp until the iodine color disappears. In the dark the reaction proceeds more slowly and in lower yield. Good yields of alkyl iodides are obtained from primary and secondary acids (Table XIII). Pivalic acid gives t-butyl iodide in only a 10% yield. Dicarboxylic acids afford low yields of diiodides. Benzoic acid produces iodobenzene in moderate yield.

$$C_6H_{11}\text{-}CO_2H \xrightarrow{Pb^{IV}/I_2} C_6H_{11}\text{-}I \quad (91\%)$$

$$(CH_3)_3CCO_2H \xrightarrow{Pb^{IV}/I_2} (CH_3)_3CI + (CH_3)_2C=CH_2$$
$$(10\%) \quad (20\%)$$

$$HO_2C(CH_2)_4CO_2H \xrightarrow{Pb^{IV}/I_2} I(CH_2)_4I \quad (33\%)$$

$$C_6H_5CO_2H \xrightarrow{Pb^{IV}/I_2} C_6H_5I \quad (56\%)$$

3β-Acetoxy-11-oxobisnorallocholanic acid **(52)** forms approximately a 1:1 mixture of 3β-acetoxy-20α-iodoallopregnan-11-one and its 20-β isomer in 85% yield.[145]

[144] D. H. R. Barton and E. P. Serebryakov, *Proc. Chem. Soc.*, **1962**, 309.
[145] D. H. R. Barton, H. P. Faro, E. P. Serebryakov, and N. F. Woolsey, *J. Chem. Soc.*, **1965**, 2438.

The mixture undergoes replacement of iodine by hydrogen on reaction with zinc and acetic acid. Reduction of alkyl iodides formed in iododecarboxylations represents a convenient two-step synthesis of hydrocarbons from acids.

$$RCO_2H \xrightarrow{Pb^{IV}/I_2} RI \xrightarrow[Li/t-C_4H_9OH]{Zn/HOAc \text{ or}} RH$$

By analogy with the formation of hypoiodites from alcohols and lead tetraacetate[146–148] it was postulated that iododecarboxylation proceeds

$$ROH \xrightarrow{Pb^{IV}/I_2} ROI$$

via an acyl hypoiodite. However, there is no experimental evidence

$$RCO_2H \xrightarrow{Pb^{IV}/I_2} RCO_2I \longrightarrow RI + CO_2$$

against an initial decarboxylation to alkyl radicals which are then converted to alkyl iodide by reaction with iodine. The presence of alkyl radicals as intermediates is consistent with the absence of stereospecificity in the iododecarboxylation of cyclopropane acids.[149, 150]

Since acyl hypoiodites are possible intermediates, the reaction of carboxylic acids with t-butyl hypoiodite was studied.[145] Carboxylic acids are decarboxylated by t-butyl hypoiodite in benzene, presumably by the sequence shown.

$$t\text{-}C_4H_9OI + RCO_2H \rightleftharpoons RCO_2I + t\text{-}C_4H_9OH$$
$$RCO_2I \rightarrow RI + CO_2$$

Moderate yields of diiodides are obtained from dicarboxylic acids. t-Butyl hypoiodite is the preferred reagent for effecting this reaction. Pb^{IV}/I_2 gives only poor yields of diiodides.

The Hunsdiecker reaction has also been assumed to proceed via the acyl hypohalite, with subsequent homolytic decomposition.[131, 132] Similarly, a free-radical decomposition of hypohalite was favored for the

---

[146] K. Heusler and J. Kalvoda, *Ang. Chem. Int. Ed.*, **3**, 525 (1964).

[147] C. Meystre, K. Heusler, J. Kalvoda, P. Wieland, G. Anner, and A. Wettstein, *Experientia*, **17**, 475 (1961).

[148] R. H. Hesse in *Advances in Free Radical Chemistry*, G. H. Williams, Ed., Vol. III, Logos Press, London, 1969.

[149] H. M. Walborsky and J. C. Chen, *J. Amer. Chem. Soc.*, **93**, 671 (1971).

[150] H. M. Walborsky, C. J. Chen, and J. L. Webb, *Tetrahedron Lett.*, **1964**, 3551.

modified Cristol-Firth procedure.[133] Acyl hypobromites were recently prepared by reaction of Br_2O (a possible intermediate in the modified Hunsdiecker reaction) with carboxylic acids and were shown to give alkyl bromides by loss of carbon dioxide.[151] The overall reaction scheme is as shown.

$$HgO + Br_2 \rightarrow HgBr_2 + [Br_2O]$$
$$[Br_2O] + RCO_2H \rightarrow [RCO_2Br] + HOBr$$
$$[RCO_2Br] \rightarrow RBr + CO_2$$

If acyl hypohalites are intermediates in halodecarboxylations with lead tetraacetate, it should be possible to trap them by addition of alkenes (the Prevost reaction,[152] Eq. 27). This has not been attempted.

$$RCO_2X + \begin{matrix} \diagdown \\ \diagup \end{matrix} C{=}C \begin{matrix} \diagup \\ \diagdown \end{matrix} \longrightarrow \begin{matrix} \diagdown \\ \diagup \end{matrix} \underset{XO_2CR}{C{-}C} \begin{matrix} \diagup \\ \diagdown \end{matrix} \qquad \text{(Eq. 27)}$$

Iododecarboxylation with lead tetraacetate and iodine has been used in the synthesis of 3,1,1-bicycloheptane (53) shown in Eq. 28.[153, 154]

[structure: CO₂H-bicycloheptane] $\xrightarrow[74\%]{Pb(OAc)_4/I_2}$ [structure: I-bicycloheptane] $\xrightarrow[45\%]{Li\text{-}t\text{-}C_4H_9OH}$ [structure: bicycloheptane] (Eq. 28)

53

Iodine (Thermal). Decarboxylation of acids with Pb^{IV}/I_2 can also be used for the preparation of esters under slightly different conditions.[155] The reaction is carried out by removing the acetic acid, adding iodine, and heating in a suitable solvent, first at 100°, then at 150–180°. The formation of esters involves several steps. Initial decomposition gives ester, alkyl iodide, carbon dioxide, and Pb^{II} carboxylate (Eq. 29). Double decomposition of the latter with alkyl iodide, which requires a higher temperature, gives more ester. The overall stoichiometry is shown in Eq. 30.

$$2\ Pb(O_2CR)_4 + I_2 \xrightarrow{85\text{--}105°} RCO_2R + 2\ RI + 3\ CO_2 + 2\ Pb(O_2CR)_2 \quad \text{(Eq. 29)}$$

$$Pb(O_2CR)_2 + 2\ RI \xrightarrow{150\text{--}180°} 2\ RCO_2R + PbI_2$$

$$2\ Pb(OAc)_4 + 8\ RCO_2H + I_2 \longrightarrow 3\ RCO_2R + 3\ CO_2 + Pb(O_2CR)_2 + PbI_2 + 8\ HOAc \quad \text{(Eq. 30)}$$

---

[151] P. W. Jennings and T. D. Ziebarth, *J. Org. Chem.*, **34**, 3216 (1969).
[152] C. Prevost, *Compt. Rend.*, **196**, 1129 (1933).
[153] H. Musso and K. Naumann, *Angew. Chem.*, **78**, 116 (1966).
[154] H. Musso, K. Naumann, and K. Grychtol, *Chem. Ber.*, **100**, 3614 (1967).
[155] G. B. Bachman and J. W. Wittmann, *J. Org. Chem.*, **28**, 65 (1963).

It has been shown recently that the first part of the reaction goes via an iodine triacylate, which was independently synthesized from the acid anhydride, iodine, and ozone.[156] Thermal decomposition of iodine triacylates affords ester and alkyl iodide according to the stoichiometry below. A free-radical chain mechanism was proposed for the decomposition

$$(RCO_2)_3I \xrightarrow{120°} RCO_2R + RI + 2\ CO_2$$

of iodine acylates.

The thermal reaction of Pb^{IV}/I_2 is analogous to the formation of esters from silver carboxylates and iodine (2:1), which is the well-known Simonini reaction.[131, 132, 157] However, both methods give good yields of

$$2\ RCO_2Ag + I_2 \rightarrow RCO_2R + CO_2 + 2\ AgI$$

esters only with primary acids and are of very limited synthetic value.

The conversion of alcohols to alkyl iodides by a modification of these procedures is suggested by the halodecarboxylation of oxalic acid half-esters. Thus 1-apocamphanol is readily converted to 1-apocamphanyl hydrogen oxalate and is then treated with mercuric oxide and iodine. Photolysis affords 1-iodoapocamphane in 30% overall yield.[158]

Bisdecarboxylation of 1,2-Dicarboxylic Acids (Table XV)

In 1952 Doering and co-workers discovered that 1,2-dicarboxylic acids and their anhydrides can be converted to alkenes by the action of lead dioxide.[159] The reactions are carried out under vigorous conditions

---

[156] G. B. Bachman, G. F. Kite, S. Tuccarbasu, and G. M. Tullman, *J. Org. Chem.*, **35**, 3167 (1970).
[157] A. Simonini, *Monatsh.*, **13**, 320 (1892); **14**, 81 (1893).
[158] A. Goosen, *Chem. Commun.*, **1969**, 145.
[159] W. von E. Doering, M. Farber, and A. Sayigh, *J. Amer. Chem. Soc.*, **74**, 4370 (1952).

(~200°) and yields are low, probably owing to further oxidation of alkenes by Pb^{IV}. A later report indicated that yields are not reproducible and are very much dependent on the method of preparation of the lead dioxide.[160] Despite these limitations, the reaction has been applied successfully to the synthesis of some bicyclic alkenes. Bicyclo[2,2,2]octane-2,3-dicarboxylic anhydride **(54)** gives bicyclo[2,2,2]oct-2-ene in 20% yield. Yields as high as 60% were obtained by bisdecarboxylating the diacid with lead tetraacetate in the presence of oxygen.[161]

Many of the dicarboxylic acids studied so far are readily available from the addition of maleic anhydride to cyclic dienes. Thus a combination of a Diels-Alder reaction using maleic anhydride followed by bisdecarboxylation with lead tetraacetate represents a convenient synthesis of a wide variety of bicyclic alkenes.

Alkene Synthesis by Lead Dioxide or Lead Tetraacetate. Bisdecarboxylation was also used in the syntheses of 2-methylenenorbornane (Eq. 31)[162] and aposantene (Eq. 32).[163]

In 1958, Grob showed that the yields in oxidative bisdecarboxylation could be greatly improved (to 50–70%) by the use of lead tetraacetate in

---

[160] W. von E. Doering and M. Finkelstein, *J. Org. Chem.*, **23** 141 (1958).
[161] J. K. Crandall, L. C. Crowley, D. B. Banks, and L. C. Lin, to be published.
[162] R. R. Sauers, *J. Amer. Chem. Soc.*, **81** 4873 (1959).
[163] S. Beckmann and R. Schaber, *Ann.*, **585**, 154 (1954).

benzene or acetonitrile, at reflux temperatures, in the presence of an organic base such as pyridine.[79-81] A further modification involved the use of pyridine as solvent.[164]

There are now many examples of the use of the bisdecarboxylation reaction for the preparation of alkenes. With lead tetraacetate-pyridine in acetonitrile or benzene, Grob carried out the conversions of the bicyclic dicarboxylic acids **55a** and **55b** shown in Eqs. 33 and 34.[79-81]

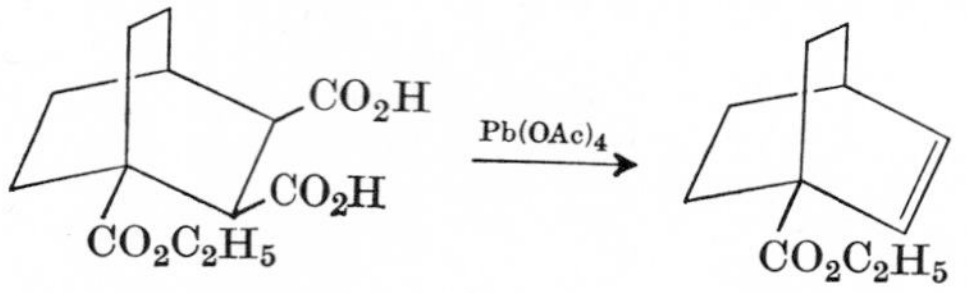

(Eq. 33)

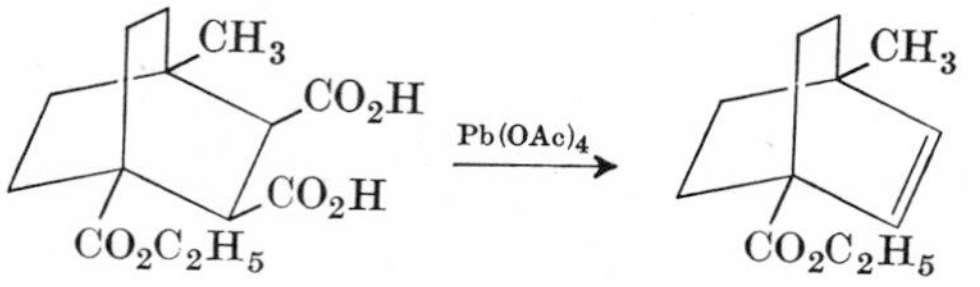

(Eq. 34)

The effect of solvent, temperature, and reaction time on the yields in Eqs. 33 and 35 has been studied.[165] The best yields (45%) of the dimethyl ester **56** are obtained in dimethyl sulfoxide or dioxane in the presence of

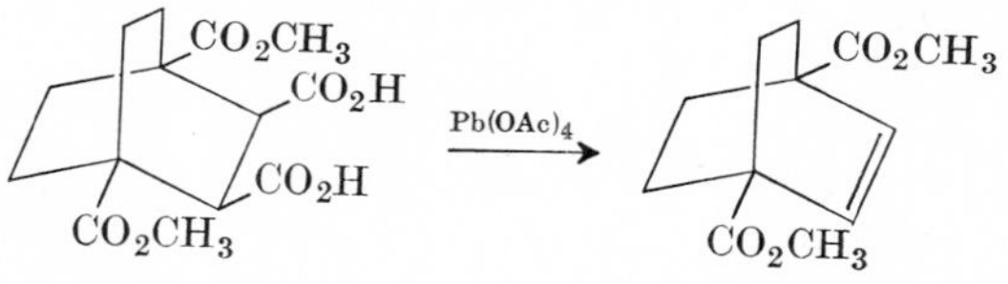

(Eq. 35)

pyridine at room temperature with reaction times of 3–4 hours and 2 days, respectively. Initial attempts to prepare the ester **56** by the Grob procedure failed.[166, 167] However, it was later reported that it could be prepared in 45% yield by this reaction.[168, 169] The same authors found that it could also be prepared from dimethyl dihydroterephthalate in one step in 80% yield by the Diels-Alder addition of ethylene under high pressure.

[164] E. Grovenstein, D. V. Rao, and J. W. Taylor, *J. Amer. Chem. Soc.*, **83**, 1705 (1961).
[165] N. B. Chapman, S. Sotheeswaran, and K. J. Toyne, *Chem. Commun.*, **1965**, 214; *J. Org. Chem.*, **35**, 917 (1970).
[166] G. Smith, C. L. Warren, and W. R. Vaughan, *J. Org. Chem.*, **28**, 3323 (1963).
[167] L. G. Humber, G. Myers, L. Hawkins, C. Schmidt, and M. Boulerice, *Can. J. Chem.*, **42**, 2852 (1964).
[168] J. C. Kauer, R. E. Benson, and G. W. Parshall, *J. Org. Chem.*, **30**, 1431 (1965).
[169] J. Colonge and R. Vuillemet, *Bull. Soc. Chim. Fr.*, **1961**, 2235.

Cimarusti and Wolinsky found that yields of alkenes were increased in the presence of added oxygen.[83] Better yields were also obtained with increasing amounts of lead tetraacetate and shorter reaction times. As a consequence, 4-cyclohexene-1,2-dicarboxylic acid can be converted to 1,4-cyclohexadiene in excellent yields. A number of other 1,2-diacids also give improved yields of alkene by this procedure.

Grob and co-workers reported that the *cis*-dicarboxylic acid **57** affords a slightly higher yield of olefin than does the *trans* isomer.[79-81] Under slightly different conditions, a higher yield is obtained from the *trans* isomer.[83] Hence it may be concluded that the stereochemical relationship of the two carboxyl groups has no dramatic effect on the reaction; this suggests that a concerted mechanism is not operative (compare p. 344).

Cimarusti and Wolinsky prepared oxygenated bicyclic alkenes by addition of maleic anhydride to cyclic 2-acetoxy-1,3-dienes (formed *in situ* from unsaturated cyclic ketones and isopropenyl acetate) followed by hydrolysis and bisdecarboxylation of the adducts with lead tetraacetate.[83] Two examples are shown.

Decarboxylation of the cyclohexadienedicarboxylic acid **58** affords phenylacetic acid, but in only 12% yield.[170] The expected product of bisdecarboxylation is the valence tautomer of toluene, **59**, which would certainly have rearranged to toluene.

Oxidative bisdecarboxylation was used in the final step in the first synthesis of Dewar benzene.[171] It was later reported that electrochemical

bisdecarboxylation of the corresponding dibasic acid gives a better

[170] H. Plieninger and G. Ege, *Chem. Ber.*, **94**, 2095 (1961).
[171] E. E. van Tamelen and S. P. Pappas, *J. Amer. Chem. Soc.*, **85**, 3297 (1963).

yield.[172, 173] Anodic oxidation,[174–176] like oxidation with lead tetraacetate, can be used for both the decarboxylation of monocarboxylic acids and the bisdecarboxylation[177] of dicarboxylic acids. In many cases, both methods give similar results.[26, 177, 178]

Oxidative bisdecarboxylation of the dibasic acid **60** with lead tetraacetate forms bicyclohexene in 30–38% yield.[179] Similarly, basketene (**61**) can be prepared by oxidative decarboxylation of either the *cis* or *trans* diacid.[180–183] Likewise, the alkene **62** is formed by oxidative decarboxylation of the either the *trans* or *cis* diacids.[183,184]

[172] T. Whitesides, unpublished results, quoted in E. E. van Tamelen and D. Carty, *J. Amer. Chem. Soc.*, **89**, 3922 (1967); see also *Ann. Rep. Prog. Chem.*, **64B**, 207 (1967) and ref. 173.

[173] P. Radlick, R. Klem, S. Spurlock, J. J. Sims, E. E. Van Tamelen, and T. Whitesides, *Tetrahedron Lett.*, **1968**, 5117.

[174] For a recent review of electrochemical decarboxylations of both mono- and di-carboxylic acids, see L. Eberson in *The Chemistry of Carboxylic Acids and Esters*, S. Patai, Ed., Chapter 6, Interscience, New York, 1969. For general reviews see also refs. 175 and 176; A. K. Vijh and B. E. Conway, *Chem. Rev.*, **67**, 623 (1967); C. K. Mann and K. K. Barnes, *Electrochemical Reactions in Nonaqueous Systems*, M. Dekker, New York, 1970.

[175] B. C. L. Weedon, *Quart. Rev.* (London), **6**, 380 (1952).

[176] B. C. L. Weedon in *Advances in Organic Chemistry*, R. A. Raphael, E. C. Taylor, and H. Wynberg, Eds., Vol. 1, p. 1, Interscience, New York, 1960.

[177] H. Plieninger and W. Lehnert, *Chem. Ber.*, **100**, 2427 (1967).

[178] W. W. Paudler, R. E. Herbener, and A. G. Zeiler, *Chem. Ind.* (London), **1965**, 1909.

[179] R. N. McDonald and C. E. Reineke, *J. Amer. Chem. Soc.*, **87**, 3020 (1965); *J. Org. Chem.*, **32**, 1878 (1967).

[180] S. Masamune, M. Cuts, and M. G. Hogben, *Tetrahedron Lett.*, **1966**, 1017.

[181] W. G. Dauben and D. L. Whalen, *Tetrahedron Lett.*, **1966**, 3743.

[182] R. Furtoss and J. M. Lehn, *Bull. Soc. Chim. Fr.*, **1966**, 2497.

[183] W. G. Dauben, C. H. Schallhorn, and D. L. Whalen, *J. Amer. Chem. Soc.*, **93**, 1446 (1971).

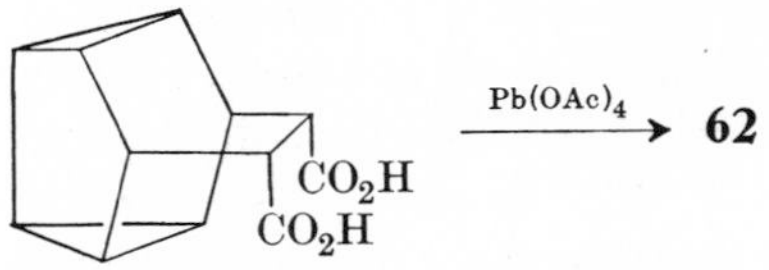

Bisdecarboxylation was a key step in the synthesis of benzobarrelene (Eq. 36).[83, 185, 186] Electrochemical bisdecarboxylation, however, gave

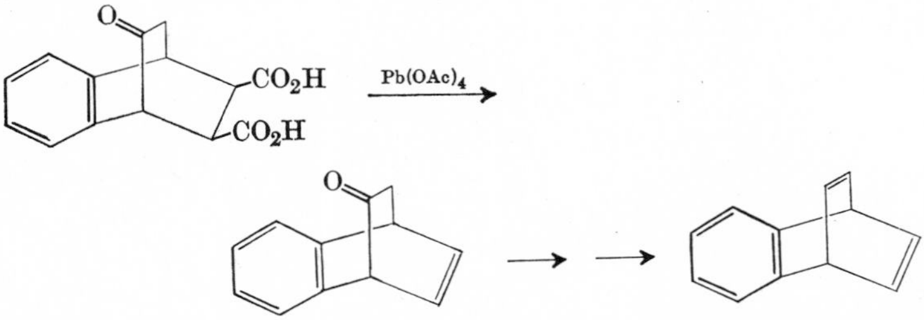

(Eq. 36)

better yields.[177] The lower yield of alkene with lead tetraacetate is the result of competitive formation of the lactone **63**. Formation of the ketal prior to decarboxylation improved the yield of alkene somewhat.[177]

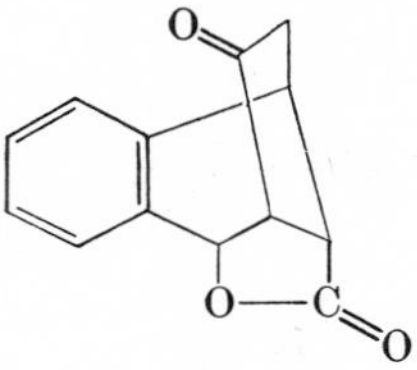

63

The first reported synthesis of dibenzocyclooctatetraene (**64**) also involved bisdecarboxylation with lead tetraacetate.[187] The same authors

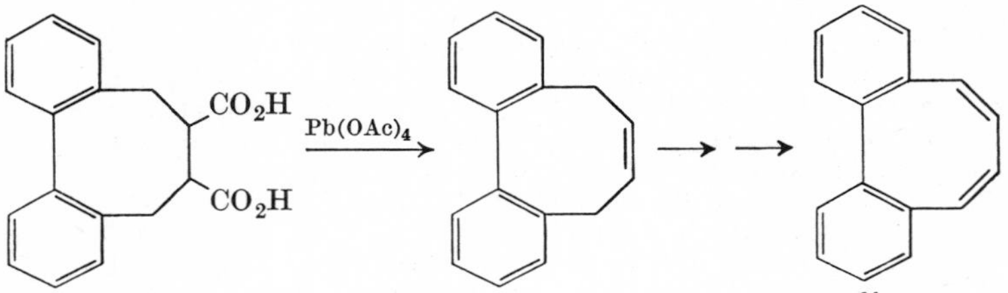

[184] L. A. Paquette and J. C. Stowell, *J. Amer. Chem. Soc.*, **92**, 2584 (1970), and private communication.

[185] K. Kitahonoki and Y. Takano, *Tetrahedron Lett.*, **1963**, 1597.

[186] K. Tori, Y. Takano, and K. Kitahonoki, *Chem. Ber.*, **97**, 2798 (1964).

[187] E. Vogel, W. Frass, and J. Wolpers, *Angew. Chem.*, **75**, 979 (1963); *ibid.*, *Int. Ed.*, **2**, 625 (1963).

prepared the valence tautomer **65** by bisdecarboxylation with lead tetraacetate and also photochemically.

Bisdecarboxylation with lead tetraacetate was used in the structure elucidation of the photochemical adduct of benzene with 2 moles of maleic anhydride (Eq. 37).[164] The product was the well-characterized maleic anhydride adduct of cyclooctatetraene.

(Eq. 37)

It is reported that the interesting triene **66** can be prepared by reaction of an appropriate diacid with lead tetraacetate.[188] The triene and basketene (**61**) have also been prepared by the decomposition of the peresters of their acid precursors.[189]

Bisdecarboxylation with lead tetraacetate has been used as a synthetic tool with many complex systems (see Table XV). For example, a tetracyclic dicarboxylic acid is converted to an alkene (Eq. 38) which is considered a possible intermediate in a route to the diterpenoid alkaloid atisine.[100]

[188] C. D. Nenitzescu, M. Avram, J. J. Pogany, C. D. Mateescu, and M. Farcasin, *Acad. Rep. Populare Romine, Studii Cercetari Chem.*, **11**, 7 (1963); quoted by R. Criegee in ref. 1. Other work, however, has shown that the diacid has the opposite configuration: R. C. Cookson, J. Hudek, and J. Marsden, *Chem. Ind.* (London), **1961**, 21.

[189] E. N. Cain, R. Vukov, and S. Masamune, *Chem. Commun.*, **1969**, 98.

[Structure with CO$_2$H groups and AcOCH$_2$] $\xrightarrow{Pb(OAc)_4}$ [Alkene product with AcOCH$_2$] (Eq. 38)

An interesting synthesis of 1,1-diaryllallenes is illustrated by the bisdecarboxylation of an unsaturated derivative of succinic acid.[190]

$$(C_6H_5)_2C=CCH_2CO_2H \underset{CO_2H}{} \xrightarrow{Pb(OAc)_4} (C_6H_5)_2C=C=CH_2$$
(22%)

Rearrangements and Side Reactions Accompanying Bisdecarboxylation. One general limitation of bisdecarboxylation is the tendency of the product alkene to be further oxidized by lead tetraacetate. Destruction of product by this route may be minimized by using mild conditions and dilute solutions. The vigorous conditions, sometimes with no diluent, used in the original lead dioxide method probably resulted in extensive decomposition of the product.[159, 160] Rather surprisingly, the use of a photochemical method at room temperature has not been reported, although monocarboxylic acids have given improved yields of alkenes by such a procedure.[29, 60]

Recent intensive product studies of the bisdecarboxylation of the bicyclic acid **55** and the *cis*- and *trans*-tetracyclic acids **57** and **68** have provided an insight into the mechanism of bisdecarboxylation and the many side reactions that are possible.[82] Results (products, rates, oxygen inhibition, solvent trapping) indicate that bisdecarboxylation involves both radical and carbonium ion intermediates analogous to the decarboxylation of monocarboxylic acids. The second carboxyl group is lost from a carbonium ion as shown. Therefore a dicarboxylic acid capable of generating a carbonium ion which is susceptible to rearrangement may be expected to give a low yield of alkene.

$$-\overset{+}{C}-\overset{|}{C}-CO_2H \longrightarrow C=C + CO_2 + H^+$$

Although decarboxylation of the bicyclic acid **55** affords the alkene **67** in 88% yield, a careful examination revealed that many side products were formed in small yields (see Table XV). Products were explained by Scheme 1.[82]

[190] J. K. Crandall, R. A. Colyer, and D. C. Hampton, unpublished results.

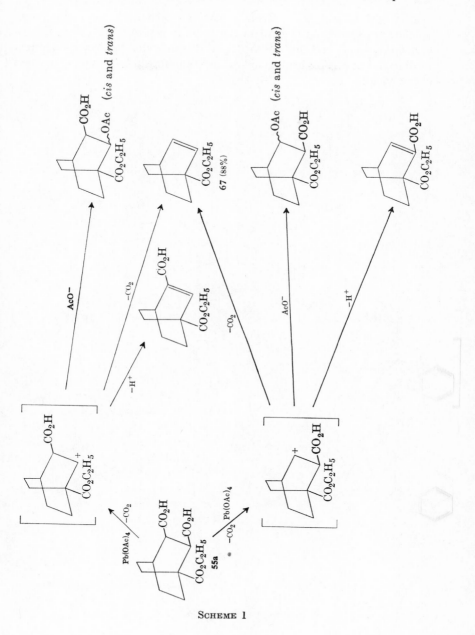

SCHEME 1

Similarly, Scheme 2 involving a common carbonium ion as an intermediate was proposed to explain the formation of products from the *cis* and *trans* isomers **57** and **68** in dimethyl sulfoxide. The common intermediate is prone to rearrangement, and therefore the alkene is obtained in low yield.

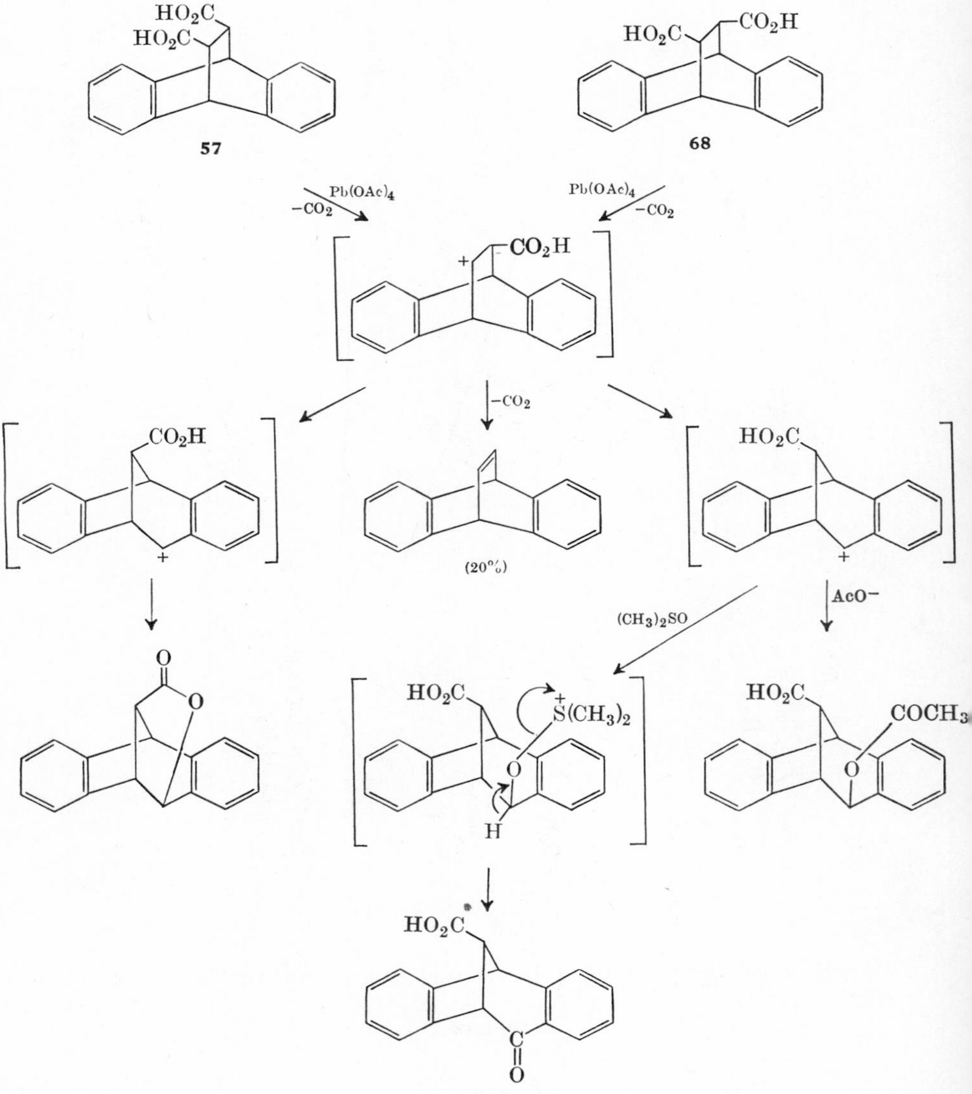

SCHEME 2

Many anomalous products formed in bisdecarboxylation reactions may be readily explained within the framework outlined for the two reactions in Schemes 1 and 2. Thus β-truxinic acid reacts with lead tetraacetate to give a lactone[178] whose formation can be readily explained by rearrangement of the carbonium ion 69. Anodic oxidation of β-truxinic acid forms the same lactone. 1,4-Diphenyl-1,4-diacetoxy-2-butene, the other decarboxylation product, may also arise from the same intermediates.

Although no cyclobutene derivatives are found in the preceding reaction, decarboxylation of cyclobutane-1,2-dicarboxylic acid is reported to give cyclobutene in 8.5% yield.[191] Similarly, decarboxylation of the corresponding tetradeuterio derivative can be used to prepare cyclobutene-1,2-d_2.[192]

The oxidative decarboxylation of tetradeuteriocyclohexanecarboxylic acid in xylene yields the deuterated cyclohexene.[193]

Another side reaction in bisdecarboxylations is anhydride formation. Bisdecarboxylation of anhydrides must proceed by opening of the anhydride to form the lead(IV) carboxylate. When anhydride formation is

---

[191] M. Seki, quoted in ref. 1.
[192] S. Borcic and J. D. Roberts, *J. Amer. Chem. Soc.*, **87**, 1056 (1965).
[193] H. Bodot, R. Lauricella, and L. Pizzala, *Bull. Soc. Chim. Fr.*, **1968**, 984.

especially favorable, bisdecarboxylation may not occur. Thus anhydride formation from the *cis* acid is facile, and the isolated carboxylic acid group is decarboxylated preferentially to give an anhydride product (Eq. 39).

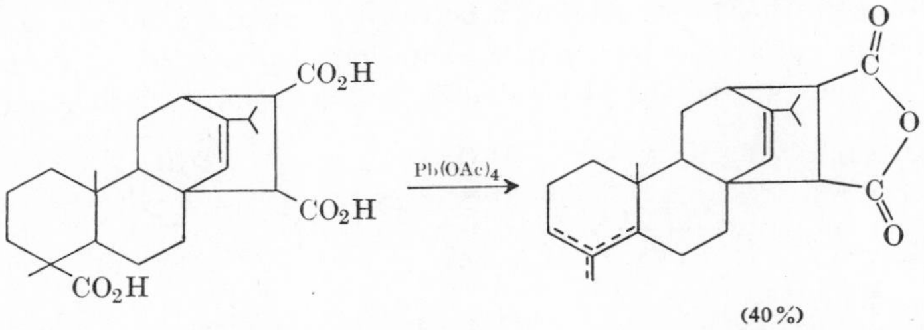

(Eq. 39)

(40%)

Anhydride formation is also observed in the bisdecarboxylation of the bicyclic acid **55a** (p. 336) and the *cis* and *trans* acids of Schemes 1 and 2.[82, 165]

As with monocarboxylic acids, many competing side reactions have been observed when a double bond is in close proximity to the carboxyl groups (see Table XV). The only product isolated from the reaction of the bicyclic acid **70** and lead tetraacetate is a dilactone.[194] Similarly, the dibasic acid **71** gives a dilactone in small yield, and none of the expected diene is obtained.[83] Analogously, the dibasic acid **72** yields a dilactone.[195] It should be noted, however, that the triene **66** was prepared by the oxidative decarboxylation of a similar diacid (p. 341).

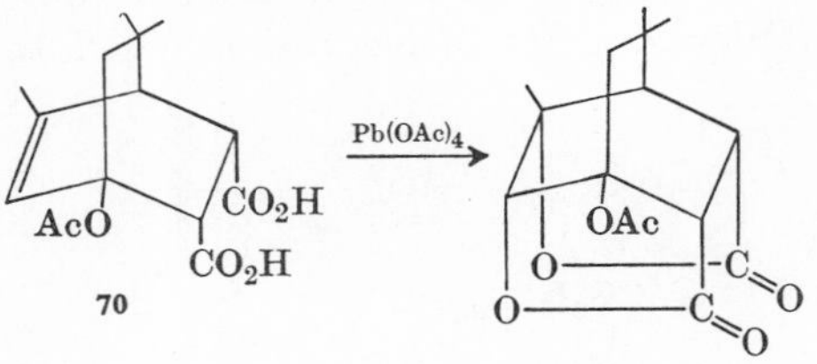

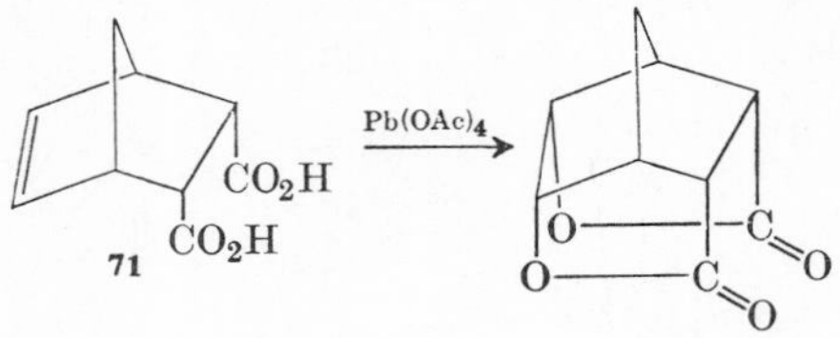

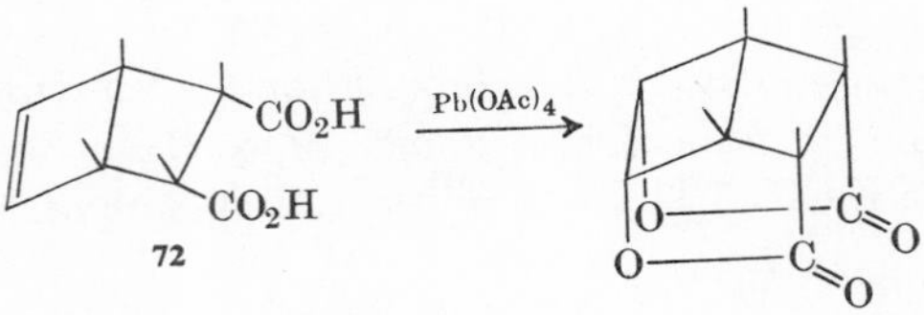

[194] K. Alder and S. Schneider, *Ann.*, **524**, 189 (1936).
[195] R. Criegee, H. Kristinsson, D. Seebach, and F. Zanker, *Chem. Ber.*, **98**, 2331 (1965).

The necessity for close proximity of the strained double bond to the carboxyl groups is demonstrated by the fact that Δ^4-tetrahydrophthalic acid gave 1,4-cyclohexadiene in 76% yield.[83] The exo diacid **73** is reported

to furnish none of the diene when treated with lead dioxide.[196] However, the products of the reaction were not reported, and the reactions should be repeated using the improved procedure with lead tetraacetate.

A second type of lactone formation involves rearrangement and is analogous to the lactone formation from unsaturated acids **57** and **68** discussed earlier (Scheme 2). The *trans* diacid **74** was reported to form the lactone as the major identifiable product.[197] Formation of this lactone may be readily accounted for by the intermediate carbonium ion as shown in

Eq. 40, and falls within the general framework of the schemes outlined previously. However, a reinvestigation of the decarboxylation resulted in

(Eq. 40)

the isolation of three products (p. 348) in addition to the lactone.[92]

[196] N. S. Zefirov, R. A. Ivanova, R. S. Filatova, and Yu. K. Yur'ev, *J. Gen. Chem. USSR*, **37**, 1893 (1966).
[197] L. H. Zalkow and D. R. Brannon, *J. Chem. Soc.*, **1964**, 5497.

(10%)   75

Thus the tetracyclic diacid ester **74** is another example of an acid which gives only a low yield of alkene in the bisdecarboxylation reaction as a consequence of extensive rearrangement of the intermediate β-carboxycarbonium ion. The common carbonium ion intermediate (Eq. 40) is the probable precursor of all the products except the olefinic lactone **75**. Its formation does not involve decarboxylation and is readily explained by intramolecular attack of the double bond on the lead(IV) carboxylate.

A similar steric example was given earlier for the rearranged lactone **63** on p. 340.[185]

In summary, there are numerous possibilities for side reactions in the bisdecarboxylation with lead tetraacetate, and, in comparison with monocarboxylic acids, the reaction presents many more complications than were originally recognized. Bisdecarboxylation undoubtedly generates both free-radical and carbonium ion intermediates. When complications due to lactone formation arise from the presence of double bonds nearby, the electrochemical technique for bisdecarboxylation appears to be preferable.[173, 198] The electrochemical method also enjoys certain advantages in the workup, although scale-up presents some limitations.

Decarboxylation of Malonic Acids (Table XI) and Glutaric and Adipic Acids (Table XV)

The reaction of malonic acid and lead tetraacetate in the presence of pyridine, studied as a possible method for generating methylene by bisdecarboxylation, gave traces of methylene diacetate and isomeric

$$CH_2(CO_2H)_2 \xrightarrow{Pb(OAc)_4} \!\!\!\!\!\!\!/\!\!\!/ \; CH_2 :$$

picolines; methylene was not detected.[198a]

[198] H. H. Westberg and H. J. Dauben, Jr., *Tetrahedron Lett.*, **1968**, 5123.
[198a] T. D. Walsh and H. Bradley, *J. Org. Chem.*, **33**, 1276 (1968).

Disubstituted malonic acids are decarboxylated to ketones by lead tetraacetate.[199, 200] Workup of the reaction under neutral conditions[199] gives products whose spectral characteristics are those of *gem* diacetates and which on hydrolysis in acidic or basic media afford the corresponding ketones (Eq. 41).

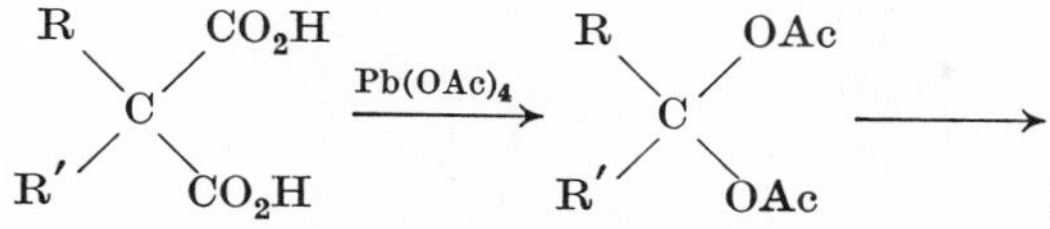

$$\begin{array}{c} R \\ \diagdown \\ R' \end{array} C \begin{array}{c} CO_2H \\ \diagup \\ CO_2H \end{array} \xrightarrow{Pb(OAc)_4} \begin{array}{c} R \\ \diagdown \\ R' \end{array} C \begin{array}{c} OAc \\ \diagup \\ OAc \end{array} \longrightarrow \begin{array}{c} R \\ \diagdown \\ R' \end{array} C{=}O \quad (Eq.\ 41)$$

The alternative method most commonly used for this conversion is the accompanying five-step sequence.

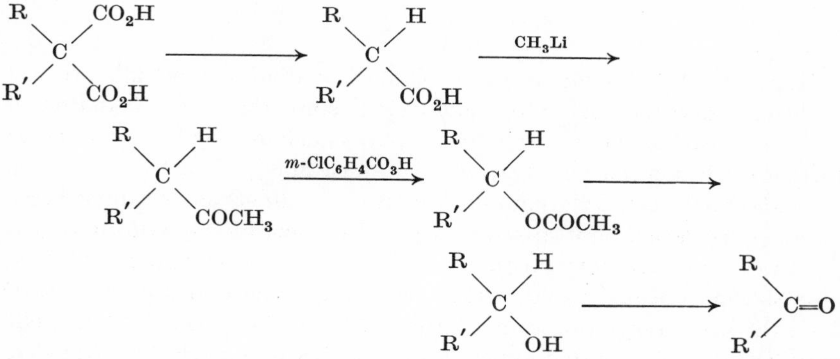

$$\begin{array}{c} R \\ \diagdown \\ R' \end{array} C \begin{array}{c} CO_2H \\ \diagup \\ CO_2H \end{array} \longrightarrow \begin{array}{c} R \\ \diagdown \\ R' \end{array} C \begin{array}{c} H \\ \diagup \\ CO_2H \end{array} \xrightarrow{CH_3Li}$$

$$\begin{array}{c} R \\ \diagdown \\ R' \end{array} C \begin{array}{c} H \\ \diagup \\ COCH_3 \end{array} \xrightarrow{m\text{-}ClC_6H_4CO_3H} \begin{array}{c} R \\ \diagdown \\ R' \end{array} C \begin{array}{c} H \\ \diagup \\ OCOCH_3 \end{array} \longrightarrow$$

$$\begin{array}{c} R \\ \diagdown \\ R' \end{array} C \begin{array}{c} H \\ \diagup \\ OH \end{array} \longrightarrow \begin{array}{c} R \\ \diagdown \\ R' \end{array} C{=}O$$

A bicyclic malonic acid derivative forms the ketone in 16% yield.[154] (Compare results with the corresponding monoacid, Eq. 28, p. 333.)

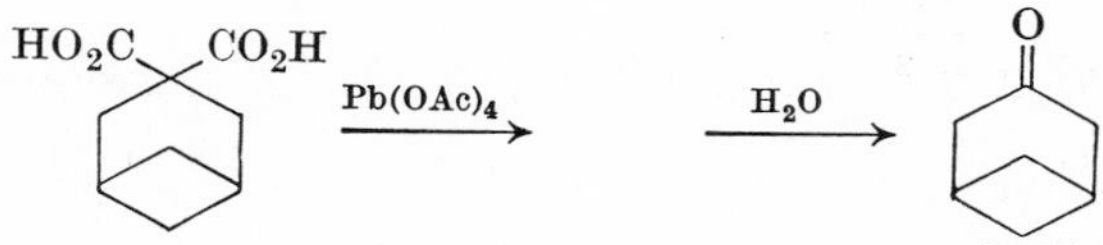

However, the closely related acid **76** produces only a trace of ketone by this method;[200] for this transformation the five-step sequence shown above

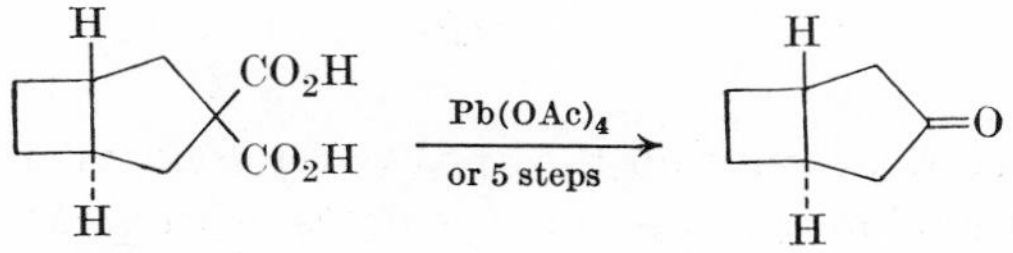

---

[199] J. J. Tufariello and W. J. Kissel, *Tetrahedron Lett.*, **1966**, 6145.
[200] J. Meinwald, J. J. Tufariello, and J. J. Hurst, *J. Org. Chem.*, **29**, 2914 (1964).

proved superior. The failure of the procedure using lead tetraacetate is curious since cyclopentane-1,1-dicarboxylic acid is converted to cyclopentanone in 53% yield in this way. Moreover, a similar transformation has recently been successfully carried out with lead tetraacetate.[201]

[Reaction: cyclopentane-1,1-dicarboxylic acid → Pb(OAc)$_4$ → H$_2$O → cyclopentanone]

[Reaction: methyl-substituted bicyclic dicarboxylic acid → Pb(OAc)$_4$ → diacetoxy product (60%)]

Monosubstituted malonic acids, like the parent compound, are not effectively decarboxylated to aldehydes despite the facile oxidation of α-acetoxy carboxylic acids.[58] From phenylmalonic acid a product of further oxidation (benzoic acid) was obtained in 8% yield.[93]

Oxidative decarboxylation of disubstituted malonic acids cannot begin with acetoxylation as it appears to do with some of the half-esters (see p. 324). Steps involving successive decarboxylations are preferred; and, recently, an α-acetoxy acid, a presumed intermediate, has been isolated from the decarboxylation of norbornane-2,2-dicarboxylic acid.[93] A cyclic lead(IV) intermediate that undergoes two successive one-electron transfers

[Reaction scheme: norbornane-2,2-dicarboxylic acid → 1. Pb(OAc)$_4$, 2. H$_2$O → norcamphor + norbornyl-OAc with CO$_2$H + norbornyl-OAc with CO$_2$H + nortricyclyl-CO$_2$H]

has been suggested to account for the apparent non-chain character of the reaction. Carbonium ions are also intermediates, because as much as 40% of cationic rearrangement products (especially in acetic acid) have been

[201] K. Shirahata, T. Kato, Y. Kitahara, and N. Abe, *Tetrahedron*, **25**, 3179 (1969).

[Scheme showing oxidative decarboxylation mechanism with Pb(OAc)$_4$]

observed in norbornyl systems.

[Scheme showing norbornyl dicarboxylic acid ester reacting with Pb(OAc)$_4$ to give OAc products]

The diarylglutaric acid **77** with lead tetraacetate gives a lactone, presumably by way of a carbonium ion.[102] Exclusive formation of the

[Scheme showing compound 77 (diarylglutaric acid) reacting with Pb(OAc)$_4$ to give a carbonium ion intermediate and then a diphenyl butyrolactone]

secondary rather than the isomeric primary carbonium ion is reasonable

[Structure of primary carbonium ion isomer: C$_6$H$_5$-CH(-CH$_2^+$)-CH(C$_6$H$_5$)-CO$_2$H]

considering the relative stabilities of primary and secondary radicals and carbonium ions. Glutaric acid under the same conditions gives no reaction, in agreement with the much slower rate of decarboxylation of primary acids.[102]

Anodic oxidation of alicyclic 1,3-dicarboxylic acids such as **78** and **79** has been used to prepare strained-ring systems related to bicyclobutane and bicyclopentane, although in low yields.[202] Presumably the products are formed from diradical intermediates and not carbonium ions.

[Structure 78: cyclobutane with CH₃O₂C, CO₂H, HO₂C, CO₂CH₃ substituents] →(Electrolysis)→ [cyclobutane with diagonal, CH₃O₂C, CO₂CH₃]

[Structure 79: cyclopentane with two CO₂H groups] →(Electrolysis)→ [bicyclopentane]

Adipic acids may produce lactones with Pb^{IV}.[203]

[benzene with two C(CH₃)₂CO₂H groups] →(Pb(OAc)₄)→ [isobenzofuranone-type lactone]

Decarboxylation of α-Amino and α-Hydroxy Acids (Table XVI) and α-Keto Acids (Table XVII)

Decarboxylation of a wide variety of α-amino,[16–18, 76] α-hydroxy,[9–11, 71, 73] and α-keto acids[75] by lead tetraacetate has been studied (see pp. 295–298). The reactions usually proceed under very mild conditions and often give quantitative yields of carbon dioxide. Some have been used for quantitative analysis.[71] The reactions are probably of little use in synthetic organic chemistry.

An interesting series of reactions provided by jaconecic acid and lead tetraacetate gives the keto acid shown.[204] The sequence of reactions in Eq. 42 is suggested.

[Jaconecic acid structure] →(Pb(OAc)₄)→ [lactone intermediate] →(Pb(OAc)₄)→ [keto acid: HO₂C—CH₂—CH(CH₃)—C(=O)—CH₃] (Eq. 42)

---

[202] A. F. Vellturo and G. W. Griffin, *J. Amer. Chem. Soc.*, **87**, 3021 (1965); T. Campbell, A. Vellturo, and G. W. Griffin, *Chem. Ind.* (London), **1969**, 1235.

[203] L. R. C. Barclay, C. E. Milligan, and N. D. Hall, *Can. J. Chem.*, **40**, 1664 (1962).

[204] R. B. Bradbury and S. Masamune, *J. Amer. Chem. Soc.*, **81**, 5201 (1959).

Summary and Comparison with Other Methods of Decarboxylation

Lead tetraacetate can be used for the conversion of a variety of carboxylic acids to a spectrum of decarboxylated products under relatively mild conditions. Experimental evidence indicates that both free radicals and carbonium ions are intermediates in most, if not all, of these reactions. Side reactions result from the many possible reaction pathways open to these reactive intermediates. However, despite the numerous possible pathways available for decomposition, conversion of the acids to a principal product (alkene, ester, alkyl halide, *etc.*) is often possible, and the reactions can be of synthetic value. Summarized below are some of the reactions which can be effected.

$RCO_2H \rightarrow RH$. Lead tetraacetate in chloroform gives moderate yields (50–60%) of the alkane from secondary acids and good yields (70–80%) from primary acids. Oxidative decarboxylation to olefin is the main reaction with tertiary acids.

The two-step conversion, halodecarboxylation followed by reduction, is also successful with primary and secondary acids, but not with tertiary acids. Alternatively, alkyl chlorides and bromides prepared by the Pb^{IV}-metal halide procedure can be selectively reduced by some recently reported methods.[205, 206]

Probably the best alternative for most acids is homolytic decomposition of the perester or diacyl peroxide.

$$RCO_2H \rightarrow RCO_3C_4H_9\text{-}t \rightarrow RH$$
$$RCO_2H \rightarrow (RCO_2)_2 \rightarrow RH$$

Decarboxylation with lead tetraacetate is usually more convenient, and the yields of alkane are probably at least as good as, if not better than, those obtained by the peroxide method.

Decarboxylation with copper powder in quinoline is not generally applicable. Its use is ordinarily restricted to α,β-unsaturated or aromatic acids and generally requires vigorous conditions that may result in low yields and complex mixtures.

$RCO_2H \rightarrow R(-H)$. The reagent of choice for this conversion is the $Pb(OAc)_4$-$Cu(OAc)_2$ combination. It is applicable to a wide variety of carboxylic acids and furnishes alkenes in very high yields.

$RCO_2H \rightarrow ROAc$. This reaction competes with alkene formation in oxidative decarboxylations with lead tetraacetate. Although there is no

---

[205] R. O. Hutchings, D. Hoke, J. Keogh, and D. Koharski, *Tetrahedron Lett.*, **1969**, 3495.
[206] S. Matsumura and N. Tokura, *Tetrahedron Lett.*, **1969**, 363.

special reagent for effecting this conversion, the use of potassium acetate in acetic acid appears to enhance the yields of acetates. Only in special cases, e.g., α-aralkyl and β,γ-unsaturated acids, does this reaction proceed in high yield.

$RCO_2H \rightarrow RX$ (X = Cl, Br, I). A variety of reagents can be used to effect this conversion, e.g., $Pb(OAc)_4$-MCl or -MBr, $Pb(OAc)_4 + I_2$. The reactions are analogous to the Hunsdiecker reaction and its various modifications. The method using lead tetraacetate is generally at least as good as other methods and is often preferred for yield and simplicity of operation.

$HO_2CC-CCO_2H \rightarrow C=C$. The bisdecarboxylation reaction is a method of broad synthetic application and has been used in the synthesis of many complex alkenes, some unobtainable by other simple procedures. Anodic oxidation can also be used for the same purpose and sometimes may be superior because it is less prone to side reactions involving rearrangements.

Decarboxylation by Other Metal Salts. Several other oxidizing metal ions such as Ce^{IV}, Mn^{III}, Co^{III}, Tl^{III} and Ag^I can effect the decarboxylation of acids.[45–50] Although each of these metal ions exhibits its own peculiarities in mechanism and product formation, by and large they do not differ significantly from Pb^{IV} in synthetic application. Since lead tetraacetate is more readily available than the other metal acetates, it will probably be the reagent of choice.

EXPERIMENTAL CONDITIONS

Preparation and Properties of Lead(IV) Salts. Lead tetraacetate is available commercially and can be prepared easily from red lead oxide and acetic acid.[1, 207] Other lead(IV) carboxylates can be prepared by metathesis of the acetate with the corresponding acid. The acetic acid is removed by distillation* or by azeotropic distillation with chlorobenzene, the latter being used as solvent for the reaction. Many lead(IV) carboxylates have been prepared in this manner.[208, 209]

---
* In general, the acetic acid is not removed for decarboxylation reactions. The lead(IV) carboxylate is generated *in situ* by metathesis and usually decarboxylates much faster than does the acetate.

[207] J. C. Bailar, *Inorg. Syn.*, **1**, 47 (1939).
[208] C. D. Hurd and P. R. Austin, *J. Amer. Chem. Soc.*, **53**, 1543 (1931).
[209] Y. Yukawa and M. Sakai, *Bull. Chem. Soc. Jap.*, **36**, 761 (1963).

Lead tetraacetate crystallizes from hot glacial acetic acid containing 1% of acetic anhydride in colorless monoclinic crystals which are best stored moist with a trace of acetic acid. Dry crystals rapidly turn brown on exposure to moist air.

Solvents. There are many inert solvents for carrying out decarboxylations with lead tetraacetate. Benzene, chlorobenzene, chloroform, hexamethylphosphoramide, dimethylformamide, dimethylacetamide, tetrahydrofuran, acetonitrile, pyridine, dioxane, and dimethyl sulfoxide have been used. The reaction is usually more rapid in coordinating solvents such as amines, nitriles, and ethers. To retard hydrolysis, the presence of acetic acid is desirable. Decarboxylations are often carried out in mixtures of acetic acid and an inert solvent. The solubility of lead tetraacetate is very low in ether, petroleum ether, and carbon tetrachloride. Alcohols cannot be used as solvents since they react readily with lead(IV) salts.

Reaction Temperature and Time. The temperature and time of reaction for decarboxylation with lead tetraacetate are dependent on the structure of the acid, on the presence of added catalysts (see below), and on the solvent. The relative rates of decarboxylation of acids increase in the order $1° < 2° < 3°$. Lead tetraacetate is stable indefinitely if stored in a cold, dark place with exclusion of moisture. Decomposition with gas evolution starts only at 140°. Lead tetraacetate undergoes first-order decomposition with a half-life of 40 minutes at 80° in pivalic acid, 200 minutes at 80° in isobutyric acid, and 900 minutes at 100° in n-butyric acid.[34] Thus, at least for acids other than primary, 80° is a convenient temperature for reaction. In polar coordinating solvents such as dimethylformamide, dimethyl sulfoxide, acetonitrile, and pyridine the reaction may be much faster. Some reactions in these solvents have been carried out at room temperature. The lowest temperature possible should be used in order to avoid further oxidation of products, such as alkenes, by lead tetraacetate. Reactions may also be carried out photochemically at room temperature.

Catalysts and Inhibitors. Added bases such as pyridine accelerate the decarboxylation.[28] Added metal acetates such as lithium or sodium acetate also accelerate the reaction.[28, 72] Decarboxylations with lead tetraacetate are also accelerated by cupric salts or metal halides, and these reactions give mainly alkenes or alkyl halides, respectively. Reactions are often inhibited by oxygen or other radical inhibitors such as phenols. Consequently, reaction mixtures should be degassed with nitrogen, and pure solvents, free of inhibitors, should be employed.

EXPERIMENTAL PROCEDURES

The progress of a decarboxylation can be followed by measuring the disappearance of lead tetraacetate by iodometric titration. Reactions can also be followed by measuring the evolution of carbon dioxide. Excess lead tetraacetate can be destroyed by adding a few drops of ethylene glycol or glycerol.

The workup depends on the nature of the reaction product. The usual procedure is to pour the reaction mixture into water and extract with ether. The ether solution is washed with aqueous sodium bicarbonate to remove acetic acid. When an anhydrous workup is required, the acetic acid is removed in vacuum, and the residue is extracted with ether to remove the product from the lead(II) salts.

1-Hexene from n-Heptanoic Acid (Use of Pb^{IV}/Cu^{II} Reagent).[60]

In a 100-ml volumetric flask is placed $Cu(OAc)_2 \cdot H_2O$ (0.107 g, 0.535 mmol), pyridine (0.110 g, 1.39 mmols), heptanoic acid (2.75 g, 2.12 mmols), and 10–20 ml of benzene or chlorobenzene. The mixture is stirred magnetically for 30–40 minutes. To the resulting homogeneous solution is added 97% lead tetraacetate (4.50 g, 9.86 mmols) and the remainder of the solvent (77–87 ml). The resulting mixture is stirred in the dark for 1 hour to effect solution and metathesis of the carboxylic acid with lead tetraacetate.

Aliquots of this solution are purged of oxygen by sweeping with a stream of nitrogen and heated at 80° in a thermostated oil bath, until the reaction, followed volumetrically by gas evolution, is complete.

Alternatively, aliquots can be decomposed photochemically at room temperature in a Rayonet photochemical reactor (So. New England Ultraviolet Co.) with 3500 Å region lamps.

The reaction mixture is poured into water and washed with aqueous sodium bicarbonate. Yields of 1-hexene are determined by quantitative gas chromatographic analysis using an internal standard and are found to be 70 and 72%, respectively, for the photochemical and thermal reactions. Identification is confirmed by hydrogenation to n-hexane.

Cyclobutene from Cyclobutanecarboxylic Acid (Use of Pb^{IV}/Cu^{II} Reagent).[59]

(a) A solution of $Cu(OAc)_2 \cdot H_2O$ (0.525 g, 0.253 mmol), cyclobutanecarboxylic acid (2.13 g, 21.3 mmols), glacial acetic acid (5 ml), and lead tetraacetate (97%, 2.284 g, 5.0 mmols) in benzene (43 ml) is prepared by agitation in a 50-ml volumetric flask. Degassed aliquots are then decomposed thermally (81°) or photochemically (3500 Å). Cyclobutene is formed in 78% yield (estimated by quantitative gas chromatography using an internal standard). (See p. 304 for equation.)

(b) A mixture of $Cu(OAc)_2 \cdot H_2O$ (0.239 g, 0.120 mmol), cyclobutanecarboxylic acid (0.674 g, 6.74 mmols), lead tetraacetate (97%, 0.689 g, 1.51 mmols), and glacial acetic acid (1.5 ml) is prepared and degassed with argon in a flask equipped with a rubber septum. Dimethylacetamide (13 ml), previously flushed with argon, is added with a syringe. (The reaction can also be run in dimethylformamide, dimethyl sulfoxide, tetrahydrofuran, or pyridine.) Some heat is evolved on mixing. The resulting homogeneous solution is then stirred in the dark at 28° for 2 hours. The yield of cyclobutene analyzed by quantitative gas chromatography is 80%.

6-Heptenoic Acid from Suberic Acid (Use of Pb^{IV}/Cu^{II} Reagent).[59]

Suberic acid is converted to ethyl hydrogen suberate by equilibrating a mixture of the acid with diethyl suberate. A mixture of ethyl hydrogen suberate (121 g, 0.60 mol), pyridine (55 g, 0.70 mol), $Cu(OAc)_2 \cdot H_2O$ (21 g, 0.11 mol), and lead tetraacetate (95%, 421 g, 0.95 mol) is stirred with 2 l of benzene. The mixture is flushed with nitrogen and carefully warmed to avoid rapid and uncontrollable liberation of carbon dioxide. It is then refluxed for 2 hours.

Excess ethylene glycol is added, and the mixture is diluted with water and washed several times with dilute nitric or perchloric acid to remove all salts. The benzene is removed after drying. Distillation yields ethyl 6-heptenoate (33 g, 35%), bp 88–88.5° (23 mm). Hydrolysis of 30 g of the ester with potassium hydroxide gives 6-heptenoic acid (23 g, 30%), bp 82–84° (1.5 mm), uncontaminated by heptanoic acid.

Δ^{22}-24-Norcholene from Cholanic Acid (Use of Pb^{IV}/Cu^{II} Reagent).[90]

A mixture of cholanic acid (3.0 mmols), $Cu(OAc)_2 \cdot H_2O$ (0.67 mmol) pyridine (1.1 mmols) and lead tetraacetate (5.45 mmols) in benzene (20 ml) is heated on a steam bath under an atmosphere of nitrogen for 1 hour. The product (60% yield) is Δ^{22}-24-norcholene, mp 74°, $[\alpha]^{29}D + 16°$. (See p. 306 for equation.)

5-Acetoxy-2-pentene and 5-Acetoxy-1-pentene from 2-Methyl-5-acetoxypentanoic Acid.[88]

In a 250-ml 3-necked flask fitted with a stirrer, thermometer, reflux condenser, inlet for the introduction of nitrogen, tube for the addition of crystalline lead tetraacetate, and gas buret is placed 0.5 mmol of $Cu(OAc)_2 \cdot H_2O$, 5 ml of pyridine, 80 mmol of 2-methyl-5-acetoxypentanoic acid and 100 ml of benzene. Lead tetraacetate (60 mmols) is placed in the tube, and the system is swept with nitrogen for ½ hour. The reaction is heated to 78–80° and maintained at this temperature while the lead tetraacetate is added portionwise over a 2–3 hour period. Each successive portion of lead tetraacetate is added only after the evolution of carbon dioxide from the previous addition has ceased. The mixture is finally heated with stirring for an additional half-hour at 78–80°. After cooling, the mixture is filtered and the filtrate is distilled directly without prior workup. The benzene is removed by distillation at atmospheric pressure, and the residue is distilled at 20 mm to afford 0.5 g of a mixture containing 40% 5-acetoxy-2-pentene and 60% 5-acetoxy-1-pentene. The combined yields represent 110% based on lead tetraacetate and 80% based on the starting acid charged (not recovered). (See p. 306 for equation.)

Triphenylcarbinol from Triphenylacetic Acid (Acetoxydecarboxylation).[29] A solution of triphenylacetic acid (1.00 g, 3.45 mmols) and lead tetraacetate (96%, 1.18 g, 2.52 mmols) in a mixture of glacial acetic acid (15 ml) and benzene (15 ml) is heated at 81° with stirring. When gas evolution ceases, the reaction is cooled and the solvent is removed by rotary evaporation. Soluble Pb(II) salts are removed by washing with water, and the insoluble material is stirred with 2% aqueous sodium bicarbonate. Extraction with benzene followed by drying and evaporation gives a residue which, when washed with petroleum ether, affords 0.583 g (89%) of white, crystalline triphenylcarbinol, mp 160.5–161.5°.

exo-Norbornyl Acetate from exo-Norbornane-2-carboxylic Acid (Acetoxydecarboxylation).[26] To a solution of exo-norbornane-2-carboxylic acid (2.884 g, 20.0 mmols) in 40 ml of anhydrous benzene and 2.37 g (30.0 mmols) of dry pyridine, under nitrogen, is added lead tetraacetate (7.59 g, 40.0 mmols). The mixture is stirred and heated under gentle reflux for 400 minutes. The crude reaction mixture is chromatographed on alumina. The product is eluted with ether and fractionally distilled to give exo-norbornyl acetate, bp 137–140° (0.991 g, 32%), n^{25}D 1.4605.

5-Nonanone from Di-n-butylmalonic Acid.[199] Di-n-butylmalonic acid (6.31 g, 0.0292 mol) is dissolved in a solution of benzene (40 ml) and pyridine (6 ml, ~0.07 mol). Lead tetraacetate (29.8 g, 0.0642 mol) is added, and the mixture is gently warmed until carbon dioxide evolution commences. The heat is removed after carbon dioxide evolution begins. After it ceases, the mixture is heated under reflux for 3 hours.

Ether (100 ml) is added to the cooled mixture, and lead(II) salts are removed by filtration and washed with more ether. The combined ether solutions are washed successively with cold $2\,N$ hydrochloric acid, aqueous sodium bicarbonate, and aqueous sodium chloride and dried over anhydrous magnesium sulfate. Ether is removed and the residue, the geminal diacetate $(n\text{-}C_4H_9)_2C(OCOCH_3)_2$, is hydrolyzed by heating it with a solution of 7 g of potassium hydroxide in 15 ml of water and 15 ml of methanol on a steam bath for 30 minutes. Ether extraction of the cooled reaction mixture affords 5-nonanone (2.97 g, 70%).

Ethyl 2-Acetoxyhexanoate from Ethyl Hydrogen n-Butylmalonate (Acetoxydecarboxylation).[124] n-Butylmalonic acid monoethyl ester (15 g) and lead tetraacetate (39 g) in benzene (300 ml) are stirred at 50° for 1 hour. A mixture of ether and dilute hydrochloric acid is added. The ether extract is dried (sodium sulfate), and the ether is removed. The residue is heated at 200° until decarboxylation is complete. The mixture is distilled to give ethyl 2-acetoxyhexanoate (9.2 g, 60%), bp 100–104° (12 mm), n^{19}D 1.421. Hydrolysis gives 2-hydroxyhexanoic acid, mp 60°.

General Procedure for Iododecarboxylation with $Pb(OAc)_4$-I_2.[145] The acid (2.5–10 mmols) and lead tetraacetate (slight excess of equimolar amount) are heated under reflux in an inert solvent (75–100 ml) with stirring and illumination with a 300 W tungsten lamp for 5–20 minutes. Iodine is added in small portions until the color persists. A stream of dry, oxygen-free nitrogen is passed through the solution during the experiment, and the evolved carbon dioxide is collected in saturated barium hydroxide (gravimetric determination). A suitable solvent is carbon tetrachloride (redistilled and dried over calcium chloride) or benzene (redistilled and dried over sodium wire). When the reaction is complete, the lead(II) salts are removed by filtration and the filtrate is decolorized by washing with aqueous sodium thiosulfate. Extraction with 1 N sodium hydroxide removes unchanged acid. Removal of solvent gives the product.

11-Acetoxy-1-iodoheptadecane from 12-Acetoxystearic Acid (Iododecarboxylation).[145] A mixture of 12-acetoxystearic acid (6.55 g, 19.2 mmols) and lead tetraacetate (8.58 g, 19.4 mmols) in carbon tetrachloride (25 ml) is irradiated with a 300 W tungsten lamp and heated under reflux in a slow stream of dry oxygen-free nitrogen. A saturated solution of iodine (5 g) in carbon tetrachloride is added until the iodine color is no longer discharged. The lead(II) diacetate is removed by filtration, and the solution is washed with aqueous sodium thiosulfate and with water. Chromatography of the organic layer over alumina gives, on elution with light petroleum (40–60°) and with benzene-light petroleum (40–60°, 1:1), 11-acetoxy-1-iodoheptadecane (5.95 g, 82%), bp 173–176° (1.5 mm), n^{24}D 1.4782.

General Procedure for Chlorodecarboxylation with $Pb(OAc)_4$-LiCl.[61,62] A weighed amount of lead tetraacetate (2.0 g, 4.5 meq) is added to a solution of carboxylic acid in benzene (10 ml). The mixture is stirred at room temperature until homogeneous. If it is anhydrous, the resulting solution is colorless. A weighed amount of lithium chloride is added, and the mixture is immediately flushed with argon or nitrogen. (Optimum conditions for chlorodecarboxylation require 1 eq of lithium chloride per mole of lead(IV) tetraacetate.) For smooth decarboxylation it is necessary to flush the solution carefully to remove oxygen. The salt at room temperature rapidly turns yellow. The mixture is decomposed by heating to 81°, the rate of gas evolution being followed volumetrically.

When the reaction is complete, the reaction mixture often consists of a clear colorless solution and a colorless amorphous solid. The solution is extracted with dilute perchloric acid, then with aqueous sodium carbonate, and is dried with magnesium sulfate.

The clear colorless solution contains the alkyl chloride, which can be obtained by removal of solvent or by preparative gas chromatography.

cis,cis- and trans,trans-3,5-Dimethyl-1-chlorocyclohexane from cis-cis-3,5-Dimethylcyclohexanecarboxylic Acid (Chlorodecarboxylation).[138] To a solution of 2.15 g (13.8 mmols) of cis,cis-3,5-dimethylcyclohexanecarboxylic acid in 35 ml of dried benzene is added 5.7 g (12.9 mmols) of lead tetraacetate. The solution is stirred for 15 minutes until nearly homogeneous. To this is added 0.60 g (14.2 mmols) of anhydrous lithium chloride, and the solution is placed under a nitrogen atmosphere in a bath preheated to 75°. The solution is stirred at 75° for 30 minutes and for an additional 30 minutes while cooling to room temperature. The reaction mixture is diluted with 50 ml of benzene and washed twice with 50 ml of water. Removal of solvent under reduced pressure gives 3 g of yellow to red-brown liquid, which is chromatographed on an 8-cm column of alumina (neutral, activity I) 1.9 cm in diameter. Elution with hexane and removal of solvent (reduced pressure) gives 1.3 g (65%) of colorless liquid which, after distillation, bp 60° (16 mm), shows two peaks by gas chromatography corresponding to a 38:62 mixture of cis,cis and trans,trans stereoisomers.

1-Phenyl-2-chloro-2-cyanopropane from 2-Benzyl-2-cyanopropanoic Acid (Chlorodecarboxylation).[139] In a 200-ml three-necked flask equipped with a gas-inlet tube which extends to the bottom, a mechanical stirrer, and a reflux condenser were placed 2-benzyl-2-cyanopropanoic acid (1.89 g, 10 mmols) and glacial acetic acid (0.60 g, 10 mmols), dissolved in 60 ml of benzene. The flask was flushed with argon for several minutes with stirring, and then lead tetraacetate (4.5 g, 10 mmols) was added all at once. A nearly clear solution resulted after a short time. A gentle current of argon was maintained. After 10 minutes, lithium chloride (9.42 g, 10 mmols) was added, and the reaction flask was heated with an oil bath at 100°. A vigorous reaction with gas evolution commenced, and a white solid was deposited in the flask. After 20 minutes of reflux, formic acid was added to the reaction mixture dropwise through the reflux condenser until gas evolution ceased. The cooled reaction mixture was filtered into a separatory funnel, and the solid residue was washed several times with a few milliliters of fresh benzene. The combined benzene solutions were extracted with two 50-ml portions of 5% aqueous sodium bicarbonate and with water and were dried over anhydrous magnesium sulfate. Evaporation of the solvent using an aspirator left a colorless oil, approximately 1.8 g, which showed a single peak on glpc. Bulb-to-bulb distillation of this oil from a bath at 90–100° (2 mm) afforded 1.17 g (65%) of 1-phenyl-2-chloro-2-cyanopropane homogeneous by glpc. When the experiment above was conducted using (+)-2-benzyl-2-cyanopropanoic acid, $[\alpha]^{20}_D$ +11.72° (c = 1.4, $CHCl_3$), the resulting chloride showed no optical rotation (α_D 0.000 ± 0.002°).

General Procedure for Bisdecarboxylation.[83] Oxygen is bubbled for 15 minutes through stirred pyridine (ca. 10 ml/g of diacid; distilled from barium oxide). Weighed diacid (0.003–0.02 mol) and lead tetraacetate (50% excess, recrystallized from acetic acid and dried over potassium hydroxide under reduced pressure) are added, and the flask is immersed in an oil bath at 67 ± 2°. After several minutes, carbon dioxide evolution begins. After 8–10 minutes the reaction mixture is cooled, poured into excess dilute nitric acid, and extracted with ether. The ether solution is washed with aqueous sodium bicarbonate and saturated sodium chloride and dried. Removal of solvent usually gives the alkene in substantially pure form.

trans-Stilbene from d,l-2,3-Diphenylsuccinic Acid (Bisdecarboxylation).[26] A suspension of *d,l*-2,3-diphenylsuccinic acid (405 mg, 1.5 mmols) in pyridine (177 mg, 2.2 mmols) and dry benzene (3.0 ml) is cooled under nitrogen and treated with lead tetraacetate (700 mg, 1.9 mmols). The mixture is heated under reflux in nitrogen for 200 minutes during which a white precipitate forms. The mixture is cooled and filtered, and the solid is washed with 10 ml of benzene. The combined benzene solutions are washed successively with 20-ml portions of water, 1 *N* sodium hydroxide, water, 1 *N* hydrochloric acid, and water. After drying (magnesium sulfate), evaporation gives the crude product. This upon chromatography on alumina gives *trans*-stilbene, mp 122–122.5° (44% yield). Quantitative ultraviolet analysis of the crude product shows that no *cis*-stilbene is present.

Similarly, *meso*-2,3-diphenylsuccinic acid gives *trans*-stilbene in 41% yield.

1-Ethoxycarbonylbicyclo[2,2,2]oct-2-ene (Bisdecarboxylation).[82] A solution of 1-ethoxycarbonylbicyclo[2,2,2]octane-2,3-dicarboxylic acid (2.70 g, 10.0 mmols), pyridine (1.2 ml, 1.5 mmols), and lead tetraacetate (4.9 g, 11 mmols) in benzene (15 ml) is heated under reflux in a nitrogen atmosphere for 90 minutes. The lead diacetate is filtered and washed with benzene. The combined organic layers are washed successively with 2 N nitric acid, 10% potassium bicarbonate, and water and dried. Evaporation of solvent gives 1-ethoxycarbonylbicyclo[2,2,2]oct-2-ene (1.58 g, 88%). (See eq. 33, p. 336.)

Bicyclo[2,2,2]oct-2-en-5-one.[83] The general procedure with 601 mg (2.83 mmols) of bicyclo[2,2,2]octan-5-one-2,3-dicarboxylic acid gives 315 mg (84%) of crude bisdecarboxylation product. Sublimation at 50° (1 mm) affords 130 mg of bicyclo[2,2,2]oct-2-en-5-one; λ_{max} 5.78, 6.2, and 14.3 μ; 2,4-dinitrophenylhydrazone, mp 148.5–149.5° (lit. 150.5°). (See p. 338 for equation.)

TABULAR SURVEY

The following tables list oxidative decarboxylations of acids reported up to early 1970.

Within each table, compounds are arranged in order of increasing number of carbon atoms. Compounds with the same number of carbon atoms are arranged in order of increasing complexity.

Where more than one product is formed in a reaction, the products are arranged roughly in order of complexity, starting with alkanes and alkenes and going on to esters and other more complex compounds.

Yields are based on the limiting component of the reaction mixture, either the carboxylic acid or lead tetraacetate. Yields based on Pb^{IV} are calculated according to the stoichiometry 1 $Pb^{IV} \rightarrow$ 1 mole of product. Yields thus calculated may exceed 100% (e.g., propane from butyric acid, Table IV; the total of propane, propene, and ester from isobutyric acid and light, Table V) because 1 mole of Pb^{IV} can yield up to 2 moles of alkane (Eq. 16). The yields are often based on gas chromatography but may be based on the weight of isolated product. Configuration of products is not consistently given.

When light is needed, as in some iododecarboxylations, the wavelength of the light is generally in parentheses after the reaction temperature, e.g., 30° (3500 Å). In cases where the wavelength is not known, the use of light is indicated by $h\nu$, e.g., 77° ($h\nu$).

The following solvent abbreviation is used: AcOH, acetic acid.

A dash in the temperature or solvent column means that no information was given.

TABLE IV. Decarboxylation of Primary Carboxylic Acids by Lead Tetraacetate

	Reactant	Solvent	Temp, °C	Product(s) (% Yield)	Refs.
C_1	HCO_2H[a]	AcOH	Room temp	CO_2 (97)	24
C_2	CH_3CO_2H	AcOH	120	CH_4 (30), $CH_2(OAc)_2$ (6), $AcOCH_2CO_2H$ (40), $(CH_2CO_2H)_2$ (5)	20–23
C_4	$n\text{-}C_3H_7CO_2H$	$n\text{-}C_3H_7CO_2H$	30 (3500 Å)	Propane (116), propene (6), n-propyl butyrate (19), isopropyl butyrate (4)	34
C_5	$n\text{-}C_4H_9CO_2H$	C_6H_6	80	Butane (25), 1-butene (2), 2-butene (cis 0.4; trans 0.7), n-butylbenzene (20), s-butylbenzene (0.6), s-butyl butyrate (8), n-butyl butyrate (2), s-butyl acetate (2), n-butyl acetate (0.3)	28
	$i\text{-}C_4H_9CO_2H$	C_6H_6	80	Isobutane (—), butene (0.3), isobutylbenzene (11), isobutyl isovalerate (2), s-butyl isovalerate (0.3)	28
		$i\text{-}C_4H_9CO_2H$	155–160	Butane (8), isobutane (11), 1-butene (3), 2-butenes (6), isobutene (0.2), s-butyl isovalerate (9)	24
	▷–CH_2CO_2H	C_6H_6/AcOH	30 (3500 Å)	1-Butene (24), 4-phenyl-1-butene (12), mixture of cyclobutyl acetate, 3-butenyl acetate, and cyclopropylmethyl acetate (8)	59
C_8	[cyclohexane with OH and CH_2CO_2H substituents]	C_6H_6	80	[1-oxaspiro structure] (—)	26

Note: References 210–245 are on pp. 420–421.

[a] Formic acid is not a primary acid, but fits in this table better than in any other.

TABLE IV. Decarboxylations of Primary Carboxylic Acids by Lead Tetraacetate (*Continued*)

	Reactant	Solvent	Temp, °C	Products (% Yield)	Refs.
C$_9$	n-C$_8$H$_{17}$CO$_2$H	CHCl$_3$	30 (3500 Å)	n-C$_8$H$_{18}$ (72), 1-octene (1)	44
	C$_6$H$_5$CH$_2$CH$_2$CH$_2$CO$_2$H	C$_6$H$_6$	80	C$_6$H$_5$CH$_2$CH$_2$CH$_2$C$_6$H$_5$ (71)	103
		C$_6$H$_6$/AcOH	80	C$_6$H$_5$CH$_3$ (14), C$_6$H$_5$C$_2$H$_5$ (7), C$_6$H$_5$CH$_2$CH$_2$CH$_2$C$_6$H$_5$ (22), C$_6$H$_5$CH=CH$_2$ (0.3), C$_6$H$_5$CH$_2$CH$_2$OAc (0.9)	58
		,,	30 (3500 Å)	C$_6$H$_5$CH$_2$CH$_3$ (17), C$_6$H$_5$CH$_2$CH$_2$C$_6$H$_5$ (25), C$_6$H$_5$CH=CH$_2$ (5), C$_6$H$_5$CH$_2$CH$_2$OAc (—)	58
C$_{10}$	C$_6$H$_5$CH(CH$_3$)CH$_2$CO$_2$H	C$_6$H$_6$/AcOH	80	Isopropylbenzene (I, 6), C$_6$H$_5$CH(CH$_3$)CH$_2$C$_6$H$_5$ (II, 13), α-methylstyrene (III, 0.7), C$_6$H$_5$CH(CH$_3$)CH$_2$OAc (IV, 10)	58
		,,	30 (3500 Å)	I (16), II (13), III (0.3), IV (1)	58
C$_{11}$	C$_6$H$_5$(CH$_2$)$_4$CO$_2$H	C$_6$H$_6$	80	n-Butylbenzene (4), tetralin (54)	32
	p-O$_2$NC$_6$H$_4$(CH$_2$)$_4$CO$_2$H	C$_6$H$_6$	80	6-Nitro-1,2,3,4-tetrahydronaphthalene (44)	32
C$_{13}$	![biphenyl-CH$_2$CH$_2$CO$_2$H]	C$_6$H$_6$	80	(42), (trace)	103
C$_{15}$	2-C$_{10}$H$_7$(CH$_2$)$_4$CO$_2$H	C$_6$H$_6$	80	(11), (44)	32

C_{16}	CH₃O–(biphenyl)–CH₂CH₂CO₂H	C_6H_6	80	(methoxy phenanthrene structure, CH₃O)	(—) 103
C_{18}	CH₃O–C₆H₄–CH(C₆H₅)–(CH₂)₃CO₂H	C_6H_6	80	(4-methoxyphenyl tetralin) (22), (6-methoxy-1-phenyltetralin) (23)	32
C_{21}	$(C_6H_5)_3CCH_2CO_2H$	CH_3CN	80	$(C_6H_5)_2C{=}CHC_6H_5$ (6), $(C_6H_5)_2C{=}CHCO_2C_6H_5$ (68), benzophenone (2), (4,4-diphenylchroman-2-one, C_6H_5, C_6H_5) (7)	27

Note: References 210–245 are on pp. 420–421.

TABLE V. Decarboxylation of Secondary Carboxylic Acids by Lead Tetraacetate

	Reactant	Solvent	Temp, °C	Product(s) (% Yield)	Ref.
C_4	$i\text{-}C_3H_7CO_2H$	$i\text{-}C_3H_7CO_2H$	135–140	Propane (I, 24), propene (II, 9), isopropyl isobutyrate (III, 3)	24
		"	30 (3500 Å)	I (56), II (44), III (24)	34
C_5	$C_2H_5CH(CH_3)CO_2H$	C_6H_6	80	Butane (15), 1-butene (14), 2-butenes (cis 5; trans 11), s-butylbenzene (10), s-butyl acetate (20), $C_2H_5CH(CH_3)CO_2CH(CH_3)C_2H_5$ (35)	28
	⟨cyclobutyl⟩–CO_2H	C_6H_6/AcOH	30 (3500 Å)	Cyclobutane (6), cyclobutene (2), phenylcyclobutane (15), mixture of cyclobutyl, cyclopropylmethyl, and 3-butenyl acetates (4–9), cyclobutyl cyclobutanecarboxylate (13)	59
C_7	⟨cyclohexyl⟩–CO_2H	C_6H_6	80	Cyclohexane (I, 14), cyclohexene (II, 38), phenylcyclohexane (III, 10), cyclohexyl acetate (IV, 20), cyclohexyl cyclohexanecarboxylate (V, 13)	29
		"	30 (3500 Å)	I (22), II (28), III (2), IV (9), V (4)	29
		$CHCl_3$	30 (3500 Å)	I (52), II (5)	44
C_8	(3-methyl-cyclopentane-1,2-dicarboxylic acid derivative with CO_2H and $COCH_3$)	—	150	(3-methyl-1-acetylcyclopentene) (34)	129

C₉ [structure: CO₂H] or [structure: CO₂H]	C₆H₆/pyridine	80	[structure: OAc] (24–67) [structure: OAc] (24–67)	26
C₁₀ [structure: CO₂H]	C₆H₆	80	[structure: OAc] (75)	210
[structure: OH, CO₂H adamantane]	—	—	[structure: O adamantane] (—)	210a

Note: References 210–245 are on pp. 420–421.

TABLE V. Decarboxylation of Secondary Carboxylic Acids by Lead Tetraacetate (*Continued*)

Reactant	Solvent	Temp, °C	Product(s) (% Yield)	Ref.
C_{11} (decalin-CO₂H)	C_6H_6	80	decalin-OAc (66)	211
norbornane-CO₂H	C_6H_6/pyridine	80	(2), (53–57), (3–4)	33
or norbornane-CH-CO₂H			(9), (5–6), (11–13)	

C_{15}	[structure with CO$_2$H]	C_6H_6/pyridine	Room temp	[structure with OAc] (60)	87
C_{16}	[structure with CH(CH$_3$)CO$_2$H]	—	—	[structure with CH(CH$_3$)OAc] (—)	212

Note: References 210–245 are on pp. 420–421.

TABLE VI. Decarboxylation of Tertiary Carboxylic Acids by Lead Tetraacetate

	Reactant	Solvent	Temp, °C	Products (% Yield)	Ref.
C_5	$(CH_3)_3CCO_2H$	C_6H_6	80	$(CH_3)_3CC_6H_5$ (0.4), isobutene (I, 48), t-butyl acetate (II, 9), $(CH_3)_3CCO_2C(CH_3)_3$ (III, 8)	28
		"	30 (3500 Å)	Isobutane (0.5),	28
		$(CH_3)_3CCO_2H$	135–140	I (74), II (15), III (8)	24
		$(CH_3)_3CCO_2H/AcOH$	30 (3500 Å)	1-Butene (4), I (38), II (3), III (3)	34
C_6	$C_2H_5C(CH_3)_2CO_2H$	C_6H_6	80	I (92), II + III (6)	28
				t-Amylbenzene (0.3), $C_2H_5C(CH_3)=CH_2$ (I, 25), $CH_3CH=C(CH_3)_2$ (II, 25), $C_2H_5C(CH_3)_2CO_2C(CH_3)_2C_2H_5$ (III, 9)	
		"	30 (3500 Å)	Isopentane (0.5), I (44), II (36), III (4), t-amyl acetate (8)	28
C_7	$n\text{-}C_3H_7C(CH_3)_2CO_2H$	C_6H_6	80	$n\text{-}C_3H_7C(CH_3)=CH_2$ (I, 42), $C_2H_5CH=C(CH_3)_2$ (II, 23)	28
		"	30 (3500 Å)	Isohexane (0.5), I (54), II (28)	28
C_9	![CO2H structure]	C_6H_6	80	$C_6H_5CH_2CO_2H$ (12)	170
C_{10}	![bicyclic CO2H structure]	C_6H_6	80	![product structure] (56)	31

C_{11}	[structure: adamantane-CO_2H]	C_6H_6	30 (>2800 Å) 1-Phenyladamantane (—)	36
C_{14}	[structure with CO_2CH_3, CO_2H, H_3C]	C_6H_6	80	[structures with CO_2CH_3 + CO_2CH_3] (total 56)	96
C_{18}	[structure with OCH_3, CO_2H]	C_6H_6/pyridine	80	[structure with OCH_3] (70),	97, 98

Note: References 210–245 are on pp. 420–421.

TABLE VI. Decarboxylation of Tertiary Carboxylic Acids by Lead Tetraacetate (*Continued*)

Reactant	Solvent	Temp, °C	Products (% Yield)	Ref.
C₁₈ (Cont.)				
[structure: OCH₃-tricyclic ketone with CO₂H]	C₆H₆	80	[structure: OCH₃-tricyclic with OAc] (15), [structure: OCH₃-tetracyclic lactone with OAc] I (7), I (71)	98
	"	80	[structure: OCH₃-tricyclic enone] (—), [structure: OCH₃-tricyclic ketone with OAc] (—)	98

372

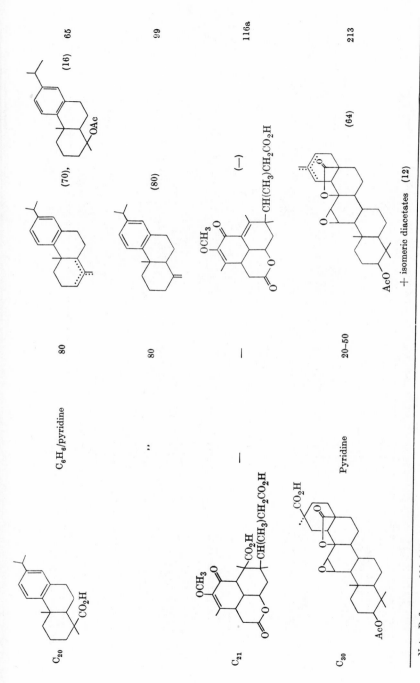

TABLE VII. DECARBOXYLATION OF α-ARYLALKANECARBOXYLIC ACIDS BY LEAD TETRAACETATE

	Reactant	Solvent	Temp, °C	Products (% Yield)	Ref.
C_8	$p\text{-}O_2NC_6H_4CH_2CO_2H$	$C_6H_6/AcOH$	80	$p\text{-}O_2NC_6H_4CH_2OAc$ (1)	58
	$C_6H_5CH_2CO_2H$	C_6H_6	80	$C_6H_5CH_2CH_2C_6H_5$ (I, 3), $(C_6H_5)_2CH_2$ (2), $C_6H_5CH_2OAc$ (II, 2), $(C_6H_5CH_2)_2CO$ (10)	103
		$C_6H_6/AcOH$	80	$C_6H_5CH_3$ (3), I (0.2), II (62)	58
C_9	$p\text{-}CH_3OC_6H_4CH_2CO_2H$	$C_6H_6/AcOH$	80	$p\text{-}CH_3OC_6H_4CH_2OAc$ (99)	58
C_{10}	$C_6H_5CH(C_2H_5)CO_2H$	$C_6H_6/AcOH$	80	$C_6H_5CH=CHCH_3$ (I, cis and trans, total 5), $C_6H_5CH(C_2H_5)OAc$ (II, 90)	58
		"	30 (3500 Å)	I (cis 1, trans 5), II (92)	58
	$C_6H_5C(CH_3)_2CO_2H$	$C_6H_6/AcOH$	80	$C_6H_5C(CH_3)=CH_2$ (I, 14), $C_6H_5C(CH_3)_2OAc$ (II, 74)	58
		"	30 (3500 Å)	I (11), II (78)	58
	$C_6H_5CH(OAc)CO_2H$	$C_6H_6/AcOH$	80	$C_6H_5CH(OAc)_2$ (—), C_6H_5CHO (—)	58
C_{12}	$1\text{-}C_{10}H_7CH_2CO_2H$	$C_6H_6/AcOH$	80	$1\text{-}C_{10}H_7CH_2OAc$ (71)	58
C_{14}	$o\text{-}C_6H_5C_6H_4CH_2CO_2H$	C_6H_6	80	$o\text{-}C_6H_5C_6H_4CH_2OAc$ (17)	103
	$(C_6H_5)_2CHCO_2H$	$C_6H_6/AcOH$	80	$(C_6H_5)_2CHOAc$ (85–90)	29
		"	30 (3500 Å)	$(C_6H_5)_2CH_2$ (0.5), $(C_6H_5)_2CHOAc$ (91)	29
C_{20}	$(C_6H_5)_3CCO_2H$	AcOH	100	$(C_6H_5)_3COAc$ (95)	24
		$C_6H_6/AcOH$	80	$(C_6H_5)_3COAc$ (84–89)	29
C_{21}	$(C_6H_5)_2C(OH)CH(C_6H_5)CO_2H$	C_6H_6	80	$C_6H_5COCH(C_6H_5)_2$ (71), $(C_6H_5)_2C(OH)CH(C_6H_5)OAc$ (14)	26
		CH_3CN	80	$C_6H_5COCH(C_6H_5)_2$ (76), $(C_6H_5)_2C(OH)CH(C_6H_5)OAc$ (8)	26

Note: References 210–245 are on pp. 420–421.

TABLE VIII. DECARBOXYLATION OF UNSATURATED CARBOXYLIC ACIDS BY LEAD TETRAACETATE

A. β,γ-Unsaturated Acids

	Reactant	Solvent	Temp, °C	Product(s) (% Yield)	Ref.
C_4	CH_2=$CHCH_2CO_2H$	C_6H_6/AcOH	80	CH_2=$CHCH_2C_6H_5$ (13), CH_2=$CHCH_2OAc$ (70), CH_2=$CHCH_2CO_2CH_2CH$=CH_2 (6)	58
		″	30 (3500 Å)	Biallyl (36), CH_2=$CHCH_2C_6H_5$ (3), CH_2=$CHCH_2OAc$ (11), CH_2=$CHCH_2CO_2CH_2CH$=CH_2 (1)	58
C_8	n-C_4H_9CH=$CHCH_2CO_2H$	C_6H_6/AcOH	80	n-C_4H_9CH=$CHCH_2OAc$ (30), $C_4H_9CH(OAc)CH$=CH_2 (21)	58
C_9	![cyclopropane with C(CH3)2, CO2H]	C_6H_6/pyridine	80	![butenolide with C(CH3)2 OAc] (35)	107
	![cyclohexadiene with CO2H, CH3, CH3]	AcOH	50	o-Xylene (50)	108
C_{11}	![cyclohexene with CH3 and C(CH3)2CO2H]	AcOH	40	![cyclohexane with OAc, =C(CH3)2, CH3] (—)	106

Note: References 210–245 are on pp. 420–421.

TABLE VIII. Decarboxylation of Unsaturated Carboxylic Acids by Lead Tetraacetate (Continued)

	Reactant	Solvent	Temp, °C	Product(s) (% Yield)	Ref.
				B. γ,δ-Unsaturated Acids (Continued)	
C_{12}	$CH_2=C(C_6H_5)C(CH_3)_2CO_2H$	AcOH	40	$CH_2=C(C_6H_5)C(CH_3)_2OAc$ (—), $AcOCH_2C(C_6H_5)=C(CH_3)_2$ (—)	71
	[structure: tetrahydronaphthalene with CH₃, CO₂H]	AcOH	50	1-Methylnaphthalene (—)	108
C_{17}	[structure: 6-methoxy-2-naphthyl-C(=CH₂)C(CH₃)₂CO₂H]	AcOH	40	$6\text{-}CH_3O\text{-}2\text{-}C_{10}H_6C(=CH_2)C(CH_3)_2OAc$ (—), $6\text{-}CH_3O\text{-}2\text{-}C_{10}H_6C(CH_2OAc)=C(CH_3)_2$ (—)	106
C_{18}	[structure: 6-methoxy-2-naphthyl-C(=CHCH₃)C(CH₃)₂CO₂H]	AcOH	40	$6\text{-}CH_3O\text{-}2\text{-}C_{10}H_6C[CH(CH_3)OAc]=C(CH_3)_2$ (—)	106
				B. γ,δ-Unsaturated Acids	
C_5	$CH_2=CHCH_2CH_2CO_2H$	$C_6H_6/AcOH$	30 (3500 Å)	$CH_2=CHCH_2CH_3$ (9), $CH_2=CHCH_2CH_2C_6H_5$ (6), mixture of 3-butenyl acetate, cyclopropylmethyl acetate, and cyclobutyl acetate (total 5)	59

376

Note: References 210–245 are on pp. 420–421.

TABLE VIII. DECARBOXYLATION OF UNSATURATED CARBOXYLIC ACIDS BY LEAD TETRAACETATE (*Continued*)

Reactant	Solvent	Temp, °C	Product(s) (% Yield)	Ref.

B. γ,δ-Unsaturated Acids (Continued)

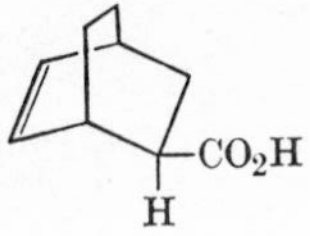

[structure: bicyclic with CO₂H, H]   AcOH   86–89   [structures: (16), (4), (6)]   114

[structures: (8), (2)]

C. δ,ε-Unsaturated Acids

C₈ [cycloheptadiene-CO₂H]   AcOH   70   [cycloheptadienyl-OAc] (70)   116b

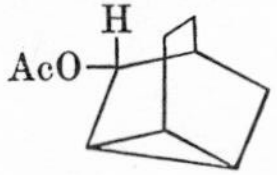

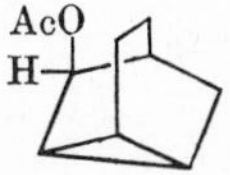

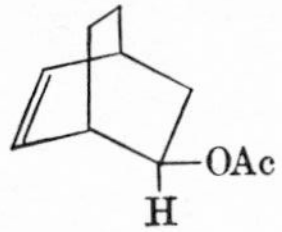

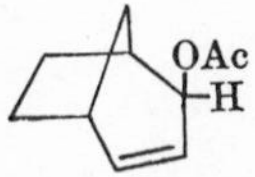

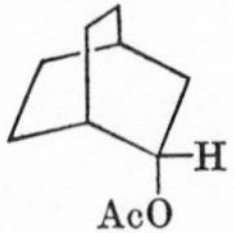

[cycloheptatriene-CO₂H]   AcOH   70   [tropylium Br⁻] (20)   117

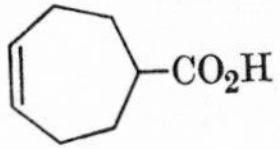

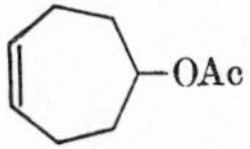

C₉ [cyclooctadiene-CO₂H]   —   —   [cyclooctadienyl-OAc] (—), [bicyclic-OAc] (—), [indene] (—)   86

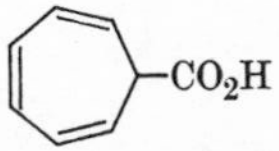

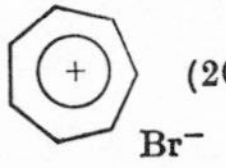

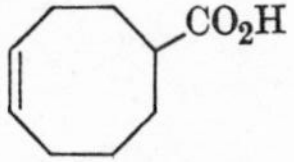

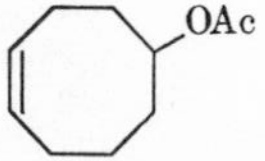

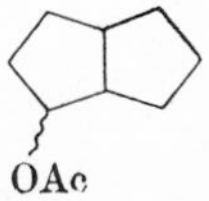

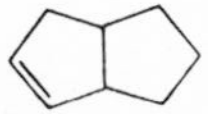

Note: References 210–245 are on pp. 420–421.

TABLE IX. DECARBOXYLATION OF o-SUBSTITUTED BENZOIC ACIDS BY LEAD TETRAACETATE (*Continued*)

Reactant	Solvent	Temp, °C	Product(s) (% Yield)		Ref.
C$_{13}$ 2-X-biphenyl-2'-CO$_2$H			6H-dibenzo[b,d]pyran-6-one		
X = H	C$_6$H$_6$	80	"	(85)	31
X = Cl	C$_6$H$_6$	80	"	(40)	31
X = NO$_2$	C$_6$H$_6$	80	"	(18)	42
X = OCH$_3$	C$_6$H$_6$	80	"	(99)	42
X = CO$_2$H	C$_6$H$_6$	80	"	(7)	42
X = CO$_2$H	CH$_3$CN	80	"	(50)	123
X = CO$_2$CH$_3$	C$_6$H$_6$	80	"	(37)	42

Note: References 210–245 are on pp. 420–421.

TABLE IX. Decarboxylation of o-Substituted Benzoic Acids by Lead Tetraacetate (*Continued*)

Reagent	Solvent	Temp, °C	Product(s) (% Yield)	Ref.
C_{14} 2-COC_6H_5-C_6H_4-CO_2H	C_6H_6	80	fluorenone (12), 2-COC_6H_5-biphenyl (5), $C_6H_5COC_6H_5$ (0.25)	121
2-$CH_2C_6H_5$-C_6H_4-CO_2H	C_6H_6	80	3-phenylphthalide (42)	121
C_{19} 2-C_6H_5-2'-CO_2H-biphenyl	C_6H_6	80	triphenylene (100)	42

Note: References 210–245 are on pp. 420–421.

TABLE X. Decarboxylation of Acids by Lead Tetraacetate/Cupric Acetate

	Reactant	Solvent	Temp, °C	Product(s) (% Yield)	Ref.
C_4	CH_2=$CHCH_2CO_2H$	C_6H_6/AcOH	80	CH_2=$CHCH_2OAc$ (87)	58
C_5	CH_2=$CHCH_2CH_2CO_2H$	C_6H_6/AcOH	30 (3500 Å)	1,3-Butadiene (24), C_4H_7OAc mixture (3), $C_8H_{13}OAc$ mixture (10)	59
	◁—CH_2CO_2H	C_6H_6/AcOH	30 (3500 Å)	1,3-Butadiene (38), C_4H_7OAc mixture (3), $C_8H_{13}OAc$ mixture (20)	59
	☐—CO_2H	C_6H_6	80	Cyclobutene (68)	60
		C_6H_6/AcOH	30 (3500 Å)	Cyclobutene (78), mixture of 3-butenyl, cyclopropylmethyl, and cyclobutyl acetates (total 7)	59
		C_2H_5OAc/AcOH	30 (3500 Å)	Cyclobutene (77), C_4H_7OAc mixture (11)	59
		CH_3CN/AcOH	30 (3500 Å)	Cyclobutene (3), C_4H_7OAc mixture (58)	59
C_6	$(C_2H_5)_2CHCO_2H$	C_6H_6	80	2-Pentene (cis 27; trans 45)	60
C_7	n-$C_6H_{13}CO_2H$	C_6H_5Cl	80	1-Hexene (72)	60
	"	"	30 (3500 Å)	" (70)	60
C_7	Cyclohexanecarboxylic acid	C_6H_6	80	Cyclohexene (100)	60
	3-Cyclohexenecarboxylic acid	C_6H_6/CH_3CN/AcOH	80	1,4-Cyclohexadiene (38), 1,3-cyclohexadiene (8), benzene (18)	60
C_8	n-$C_7H_{15}CO_2H$	C_6H_5Cl	80	1-Heptene (68)	60
		"	30 (3500 Å)	1-Heptene (71)	60

Note: References 210–245 are on pp. 420–421.

TABLE X. Decarboxylation of Acids by Lead Tetraacetate/Cupric Acetate (Continued)

Reactant	Solvent	Temp, °C	Product(s) (% Yield)	Ref.
$n\text{-}C_4H_9CH=CHCH_2CO_2H$	$C_6H_6/AcOH$	80	$n\text{-}C_4H_9CH=CHCH_2OAc$ (13), $n\text{-}C_4H_9CH(OAc)CH=CH_2$ (68)	58
⟨cyclohexyl⟩-CH_2CO_2H	C_6H_5Cl	80	Methylenecyclohexane (40)	60
"	"	30 (3500 Å)	Methylenecyclohexane (84)	60
$C_6H_5CH_2CO_2H$	$C_6H_6/AcOH$	80	$C_6H_5CH_2OAc$ (60)	58
$n\text{-}C_8H_{17}CO_2H$	C_6H_6	80	1-Octene (82)	60
"	"	30 (3500 Å)	1-Octene (87)	60
$C_6H_5CH_2CH_2CO_2H$	$C_6H_6/AcOH$	80	$C_6H_5CH=CH_2$ (47)	58
$C_6H_5CH_2CD_2CO_2H$	$C_6H_6/AcOH$	80	$C_6H_5CH=CD_2$ (79)	30
$C_2H_5OCO(CH_2)_4CH_2CH_2CO_2H$	C_6H_6/pyridine	80	$C_2H_5OCO(CH_2)_4CH=CH_2$ (60)	60
$C_6H_5CH(CH_3)CH_2CO_2H$	$C_6H_6/AcOH$	80	$C_6H_5C(CH_3)=CH_2$ (40)	58
$C_6H_5CH(C_2H_5)CO_2H$	$C_6H_6/AcOH$	80	$C_6H_5CH=CHCH_3$ (18), $C_6H_5CH(C_2H_5)OAc$ (69)	58
$C_6H_5C(CH_3)_2CO_2H$	$C_6H_6/AcOH$	80	$C_6H_5C(CH_3)=CH_2$ (21), $C_6H_5C(CH_3)_2OAc$ (54)	58
$C_6H_5\text{-}\triangle\text{-}CO_2H$ (cis and trans)	C_6H_6	80	$C_6H_5CH=CHCH_2OAc$ (cis and trans, total 30), $C_6H_5CH(OAc)CH=CH_2$ (14)[a]	66, 67

[a] For five other products the original references must be consulted.

C₁₁	[lactone with CH₃ and CO₂H]	C₆H₆	80	[bicyclic lactone with CH₃] (55)	214
C₁₃	C₆H₅(CH₂)₄CO₂H	C₆H₆	80	C₆H₅CH₂CH₂CH=CH₂ (65)	60
	[C₆H₅-cyclohexane-CO₂H] (cis and trans)	C₆H₆	80	1-Phenylcyclohexene (26), 3-Phenylcyclohexene (60)	30
C₁₅	(C₆H₅)₂CHCH₂CO₂H	C₆H₆	80	(C₆H₅)₂C=CH₂	30
C₂₀	[tricyclic diterpene with H₃C, CO₂H, isopropyl]	C₆H₆/pyridine	80	[tricyclic diterpene] (76)	101
	[tricyclic diterpene with H₃C, CO₂H, isopropyl, double bond]	C₆H₆/pyridine	80	[tricyclic diterpene] (76), [tricyclic diterpene with H₃C, OAc] (23)	101

Note: References 210–245 are on pp. 420–421.

TABLE X. Decarboxylation of Acids by Lead Tetraacetate/Cupric Acetate (Continued)

Reactant	Solvent	Temp, °C	Product(s) (% Yield)	Ref.
C_{22}	C_6H_6	80	(79)	215
C_{24}	C_6H_6/pyridine	80	(60)	90
C_{27}	C_6H_6/pyridine	80	(65)	216

C_{28}	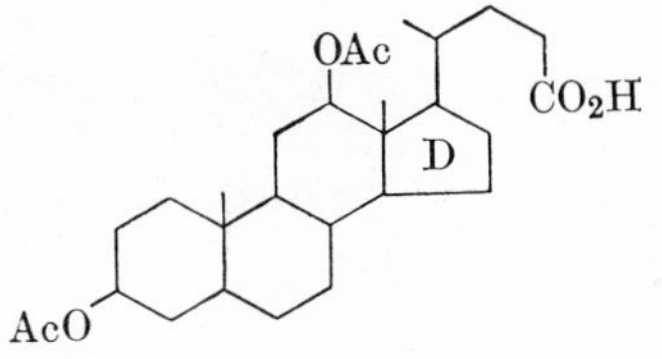 (structure with OAc, AcO, CO₂H, D ring)	$C_6H_6/$pyridine	80	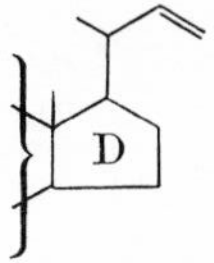 (—) (structure)	90
C_{31}	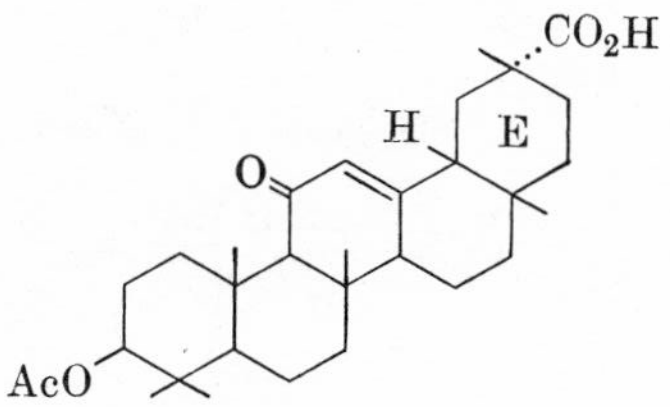 (structure with CO₂H, E ring, H, ketone, AcO)	C_6H_6	80	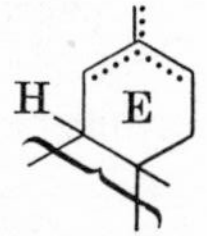 (88) (structure with H, E)	65
C_{32}	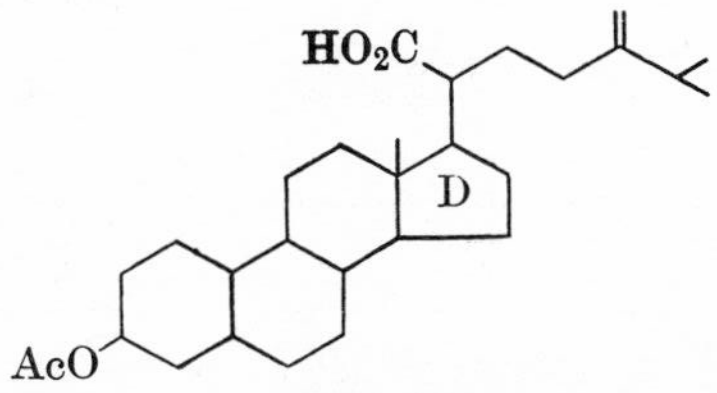 (structure with HO₂C, D ring, AcO)	$C_6H_6/$pyridine	80	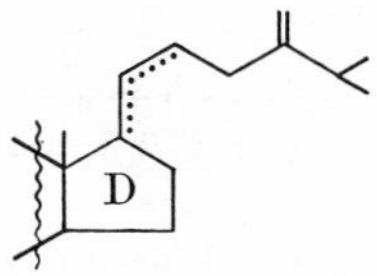 (>50) (structure with D)	217

Note: References 210–245 are on pp. 420–421.

TABLE XI. Decarboxylation of Malonic Acids, Half-Esters of Malonic Acids, β-Cyano Acids, and β-Keto Acids by Lead Tetraacetate

	Reactant	Solvent	Temp, °C	Product(s) (% Yield)	Refs.
			A. Malonic Acids		
C_6	◇$((CO_2H)_2)$	C_6H_6	80	▢=O (20)	93
C_7	⬠$(CO_2H)_2$	C_6H_6	80	⬠=O (45)	199
C_8	⬡$(CO_2H)_2$	C_6H_6	80	⬡=O (50)	199
C_9	$n\text{-}C_4H_9C(C_2H_5)(CO_2H)_2$	C_6H_6	80	$n\text{-}C_4H_9COC_2H_5$ (60)	199
	bicyclic$(CO_2H)_2$	C_6H_6	80	bicyclic=O (Low)	200
C_{10}	bicyclic$(CO_2H)_2$	C_6H_6	80	bicyclic=O (16)	153, 154
C_{11}	$i\text{-}C_5H_{11}C(C_2H_5)(CO_2H)_2$	C_6H_6	80	$i\text{-}C_5H_{11}COC_2H_5$ (70)	199
	$C_6H_5C(C_2H_5)(CO_2H)_2$	C_6H_6	80	$C_6H_5COC_2H_5$ (63)	199
	$(n\text{-}C_4H_9)_2C(CO_2H)_2$	C_6H_6	80	$(n\text{-}C_4H_9)_2CO$ (70)	199

	Substrate	Solvent	Temp.	Product	Ref.
C_{13}	(CH₃,CH₃-decalin)-CH(CO₂H)₂ structure	C_6H_6	80	(CH₃,CH₃-decalin)-(OAc)₂ (60)	201
C_{16}	biphenyl-CH₂CH(CO₂H)₂	C_6H_6	80	Phenanthrene (10), 9,10-dihydrophenanthrene (trace)	103

B. Half-Esters of Malonic Acids

	Substrate	Solvent	Temp.	Product	Ref.
C_6	$CH_3CH(CO_2H)CO_2C_2H_5$	C_6H_6	80, then distil	$CH_3CH(OAc)CO_2C_2H_5$ (40)	125
C_7	$C_2H_5CH(CO_2H)CO_2C_2H_5$	C_6H_6	80, then distil	$C_2H_5CH(OAc)CO_2C_2H_5$ (49)	125
C_8	$n\text{-}C_3H_7CH(CO_2H)CO_2C_2H_5$	C_6H_6	50–80, then distil with copper	$n\text{-}C_3H_7CH(OAc)CO_2C_2H_5$ (55)	124, 126
C_9	$CH_2=CHCH_2CH(CO_2H)CO_2C_2H_5$	C_6H_6	80, then distil	$CH_2=CHCH_2CH(OAc)CO_2C_2H_5$ (33)	125
	$n\text{-}C_4H_9CH(CO_2H)CO_2C_2H_5$	C_6H_6	80, then 200	$n\text{-}C_4H_9CH(OAc)CO_2C_2H_5$ (60)	124
	$i\text{-}C_4H_9CH(CO_2H)CO_2C_2H_5$	C_6H_6	—	$i\text{-}C_4H_9CH(OAc)CO_2C_2H_5$ (65)	124, 126
	$s\text{-}C_4H_9CH(CO_2H)CO_2C_2H_5$	C_6H_6	—	$s\text{-}C_4H_9CH(OAc)CO_2C_2H_5$ (37)	124, 126
C_{11}	cyclohexyl-CH(CO₂H)CO₂C₂H₅	C_6H_6	—	cyclohexyl-CH(OAc)CO₂C₂H₅ (35)	124, 126
C_{12}	$n\text{-}C_7H_{15}CH(CO_2H)CO_2C_2H_5$	C_6H_6	—	$n\text{-}C_7H_{15}CH(OAc)CO_2C_2H_5$ (82)	124, 126

Note: References 210–245 are on pp. 420–421.

TABLE XI. Decarboxylation of Malonic Acids, Half-Esters of Malonic Acids, β-Cyano Acids, and β-Keto Acids by Lead Tetracetate *(Continued)*

	Reactant	Solvent	Temp, °C	Product(s) (% Yield)	Ref.
B. Half-Esters of Malonic Acids (Continued)					
C$_{14}$	C$_6$H$_5$(CH$_2$)$_3$CH(CO$_2$H)CO$_2$C$_2$H$_5$	C$_6$H$_6$	80	C$_6$H$_5$(CH$_2$)$_3$C(OAc)(CO$_2$H)CO$_2$C$_2$H$_5$ (33), C$_6$H$_5$(CH$_2$)$_3$CH(OAc)CO$_2$C$_2$H$_5$ (10), [tetralin with CO$_2$C$_2$H$_5$] I (30), II (39), [tetralin with AcO, CO$_2$C$_2$H$_5$] (II, 7)	127
C$_{15}$	[structure: HO$_2$C–C(CH$_3$)(CO$_2$C$_2$H$_5$)–CH$_2$CH$_2$–C$_6$H$_4$]	C$_6$H$_6$	80	[tetralin with H$_3$C, CO$_2$C$_2$H$_5$] (66)	127
		AcOH	80	″ (73)	127
		CH$_3$CN	80	″ (80)	127
C. β-Cyano Acids and β-Keto Acids					
C$_6$	n-C$_3$H$_7$CH(CO$_2$H)CN	C$_6$H$_6$/toluene	0°, then distil with copper	n-C$_3$H$_7$CH(OAc)CN (25)	124, 127
	CH$_3$CH(CO$_2$H)COC$_2$H$_5$	C$_6$H$_6$/toluene	0°, then distil with copper	CH$_3$CH(OAc)COC$_2$H$_5$ (30)	124, 126
C$_8$	C$_2$H$_5$CH(CO$_2$H)COC$_3$H$_7$-n	C$_6$H$_6$/toluene	0°, then distil with copper	C$_2$H$_5$CH(OAc)COC$_3$H$_7$-n (21)	124, 126
	n-C$_4$H$_9$CH(CO$_2$H)COCH$_3$	C$_6$H$_6$/toluene	0°, then distil with copper	n-C$_4$H$_9$CH(OAc)COCH$_3$ (40)	124, 126

Note: References 210–245 are on pp. 420–421.

TABLE XII. DECARBOXYLATION OF γ-KETO ACIDS BY LEAD DIOXIDE

Reactant		Temp, °C	Products (% Yield)	Ref.
C_7	3-oxocyclohexane-CO₂H	250	cyclohex-2-enone (92)	130
C_9	2-acetylcyclohexane-CO₂H (COCH₃)	250	1-acetylcyclohexene —COCH₃ (45)	130
	2-acetyl-1-methylcyclopentane-CO₂H (CH₃, COCH₃)	150	1-methyl-2-acetylcyclopentene (CH₃, —COCH₃) (34)	129
C_{11}	$C_6H_5COCH_2CH(CH_3)CO_2H$	250	$C_6H_5COCH=CHCH_3$ (69)	130
C_{12}	$p\text{-}CH_3C_6H_4COCH_2CH(CH_3)CO_2H$	250	$p\text{-}CH_3C_6H_4COCH=CHCH_3$ (76)	130
C_{15}	$2\text{-}C_{10}H_7COCH_2CH(CH_3)CO_2H$	250	$2\text{-}C_{10}H_7COCH=CHCH_3$ (35)	130
C_{16}	$C_6H_5COCH_2CH(C_6H_5)CO_2H$	135	$C_6H_5COCH=CHC_6H_5$ (57)	130
		250	$C_6H_5COCH=CHC_6H_5$ (84)	130
C_{20}	$2\text{-}C_{10}H_7COCH_2CH(C_6H_5)CO_2H$	250	$2\text{-}C_{10}H_7COCH=CHC_6H_5$ (83)	130

Note: References 210–245 are on pp. 420–421.

TABLE XIII. HALODECARBOXYLATION OF ACIDS BY LEAD TETRAACETATE AND LITHIUM HALIDE OR IODINE

	Reactant	Solvent	Temp, °C	Product(s) (% Yield)	Refs.
		A. Chlorodecarboxylation Using Lithium Chloride			
C_2	CH_3CO_2H	C_6H_6	80	CH_3Cl (55)	62
C_4	$(CH_3)_2CHCO_2H$	C_6H_6	80	$(CH_3)_2CHCl$ (98)	61
C_5	$n\text{-}C_4H_9CO_2H$	C_6H_6	80	$n\text{-}C_4H_9Cl$ (92)	61
	$i\text{-}C_4H_9CO_2H$	C_6H_6	80	$i\text{-}C_4H_9Cl$ (99)	61
	$s\text{-}C_4H_9CO_2H$	C_6H_6	80	$s\text{-}C_4H_9Cl$ (96)	61
	$t\text{-}C_4H_9CO_2H$	C_6H_6	80	$t\text{-}C_4H_9Cl$ (72), $(CH_3)_2C{=}CH_2$ (9)	61
	$CH_2{=}CHCH_2CH_2CO_2H$	C_6H_6	80	$CH_2{=}CHCH_2CH_2Cl$ (2)	61
	Cyclobutanecarboxylic acid	C_6H_6	80	Cyclobutyl chloride (98)	61
C_6	$(CH_3)_3CCH_2CO_2H$	C_6H_6	80	$(CH_3)_3CCH_2Cl$ (92)	61
	$C_2H_5C(CH_3)_2CO_2H$	C_6H_6	80	$C_2H_5C(CH_3)_2Cl$ (91), $C_2H_5C(CH_3){=}CH_2$ (4), $CH_3CH{=}C(CH_3)_2$ (4)	61
	![Cl-cyclobutane-Cl / CO2H]	C_6H_6	80	![Cl-cyclobutane-Cl] (—)	142
C_7	Cyclohexanecarboxylic acid	C_6H_6	80	Cyclohexyl chloride (100)	61
	3-Cyclohexenecarboxylic acid	C_6H_6	—	3-Cyclohexenyl chloride (45)	218
	[deuterated cyclohexene with CO_2H]	C_6H_6	81	[deuterated cyclohexenyl chloride] (25)	193

	Substrate	Solvent	Yield	Product	Ref.
C_8	$C_6H_5CO_2H$	C_6H_6	80	C_6H_5Cl (8), CH_3Cl (58)	61
	$C_6H_5CH_2CO_2H$	$C_6H_6/AcOH$	80	$C_6H_5CH_2Cl$ (90–95)	58
	p-$ClC_6H_4CH_2CO_2H$	$C_6H_6/AcOH$	80	p-$ClC_6H_4CH_2Cl$ (90–95)	58
	p-$O_2NC_6H_4CH_2CO_2H$	$C_6H_6/AcOH$	80	p-$O_2NC_6H_4CH_2Cl$ (80)	58
	(norbornene-CO₂H, H)	C_6H_6	80	(lactone, AcO, H, C=O) (70)[a]	113
C_9	(3,5-dimethylcyclohexane-CO₂H, cis,cis)	C_6H_6	75	(3,5-dimethylchlorocyclohexane) (65) cis,cis:trans,trans = 38:62	138
	(bicyclo[2.2.2]octane-CO₂H)	—	—	(bicyclo[2.2.2]octyl-Cl) (—)	219
C_{10}	(1-isopropylbicyclo[2.2.2]octane-CO₂H)	C_6H_6	80	(1-isopropylbicyclo[2.2.2]octyl-Cl) (—)	219

Note: References 210–245 are on pp. 420–421.

[a] No halodecarboxylation was observed.

TABLE XIII. HALODECARBOXYLATION OF ACIDS BY LEAD TETRAACETATE AND LITHIUM HALIDE OR IODINE
(Continued)

A. Chlorodecarboxylation Using Lithium Chloride

Reactant	Solvent	Temp, °C	Product(s) (% Yield)	Refs.
C₁₅ (HO₂C-tricyclic structure)	C₆H₆	80	(Cl-tricyclic structure) (—)	220
(decalone with CH(CH₃)CO₂H, CH₃, CH₃, O)	C₆H₆	80	(decalone with CH(CH₃)Cl, CH₃, CH₃, O) (80)	221
(H₃C-cage structure with CH(CH₃)CO₂H, CH₃)	—	—	(H₃C-cage with CH(CH₃)Cl, CH₃) (—)	212
C₁₆ (octahydrophenanthrene with CO₂H, CH₃)	C₆H₅CH₃	111	(octahydrophenanthrene with Cl, CH₃) (65)	222
(octahydrophenanthrene with H₃C, CO₂H)	C₆H₆	80	(octahydrophenanthrene with H₃C, Cl) (62)	222

392

B. Bromodecarboxylation Using Lithium Bromide

C_4	$(CH_3)_2CHCO_2H$	C_6H_6	80	$(CH_3)_2CHBr$ (50–60)	62
C_9	H, HO_2C (structure)	C_6H_6	80	H, Br (structure)	143

C. Iododecarboxylation Using Iodine

C_2	CH_3CO_2H	b	150–180	$CH_3CO_2CH_3$ (20)[a]	155
C_4	$(CH_3)_2CHCO_2H$	b	150–180	$(CH_3)_2CHCO_2CH(CH_3)_2$ (18)[a]	155
	$C_2H_5CHClCO_2H$	CCl_4	77 (hv)	C_2H_5CHClI (12)	223
	$C_2H_5CHBrCO_2H$	CCl_4	77 (hv)	C_2H_5CHBrI (18)	223
C_5	$t\text{-}C_4H_9CO_2H$	b	150–180	No ester formed[a]	155
		CCl_4	77 (hv)	$t\text{-}C_4H_9I$ (10), $(CH_3)_2C{=}CH_2$ (20)	144, 146
	$HO_2C(CH_2)_3CO_2H$	—	150–180	γ-Butyrolactone (19)	155
		CCl_4	77 (hv)	$I(CH_2)_3I$ (12)	144, 146
C_6	$n\text{-}C_5H_{11}CO_2H$	b	150–180	$n\text{-}C_5H_{11}CO_2C_5H_{11}\text{-}n$ (34)[a]	155
		CCl_4	77 (hv)	$n\text{-}C_5H_{11}I$ (100)	144, 146
	$HO_2C(CH_2)_4CO_2H$	CCl_4	77	$I(CH_2)_4I$ (33)	144, 146

Note: References 210–245 are on pp. 420–421.

[a] No halodecarboxylation was observed.
[b] The solvent was 1,1,2,2-tetrachloroethane, o-dichlorobenzene, or mineral oil.

TABLE XIII. Halodecarboxylation of Acids by Lead Tetraacetate and Lithium Halide or Iodine (Continued)

C. Iododecarboxylation Using Iodine (Continued)

Reactant	Solvent	Temp, °C	Product(s) (% Yield)	Refs.
C₇ n-C$_6$H$_{13}$CO$_2$H	b	150–180	n-C$_6$H$_{13}$CO$_2$C$_6$H$_{13}$-n (50)[a]	155
Cyclohexanecarboxylic acid	CCl$_4$	77 ($h\nu$)	Cyclohexyl iodide (91)	144, 146
C$_6$H$_5$CO$_2$H	b	150–180	C$_6$H$_5$I (62)	155
	CCl$_4$	77 ($h\nu$)	C$_6$H$_5$I (56)	144, 146
C₈ ⟨bicyclic⟩–CO$_2$H	CCl$_4$	77 ($h\nu$)	⟨bicyclic⟩–I (74)	153, 154
⟨spiro⟩–CO$_2$H	CCl$_4$	77 ($h\nu$)	⟨spiro⟩–I (—)	224
cyclohexane-1,2-di-CO$_2$H	CCl$_4$	77 ($h\nu$)	cyclohexene (34)[a]	144, 146
C₉ C$_2$H$_5$CH(CH$_3$)CH$_2$CH(CH$_3$)CH$_2$CO$_2$H	CCl$_4$	77 ($h\nu$)	C$_2$H$_5$CH(CH$_3$)CH$_2$CH(CH$_3$)CH$_2$I (85)	225
(CH$_3$)$_2$C=C(CH$_3$)$_2$ cyclopropane–CO$_2$H	CCl$_4$	77 ($h\nu$)	iodo-lactone (CH$_3$)$_2$ (52)[a]	107

394

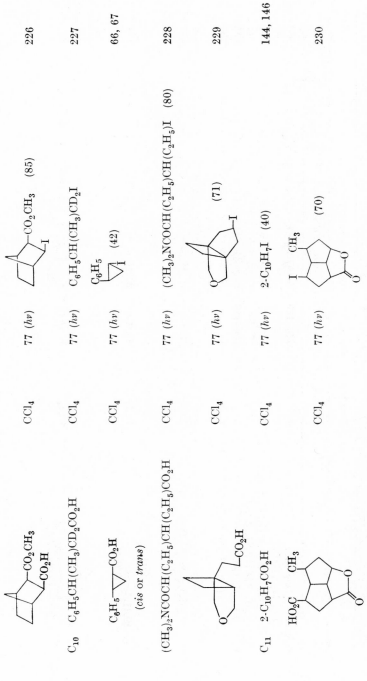

Note: References 210–245 are on pp. 420–421.

[a] No halodecarboxylation was observed.

[b] The solvent was 1,1,2,2-tetrachloroethane, *o*-dichlorobenzene, or mineral oil.

TABLE XIII. HALODECARBOXYLATION OF ACIDS BY LEAD TETRAACETATE AND LITHIUM HALIDE OR IODINE (*Continued*)

	Reactant	Solvent	Temp, °C	Product(s) (% Yield)	Ref.
			C. Iododecarboxylation Using Iodine (Continued)		
C_{12}	$n\text{-}C_{11}H_{23}CO_2H$	b	150–180	$n\text{-}C_{11}H_{23}CO_2C_{11}H_{23}\text{-}n$ (50)	155
	$n\text{-}C_5H_{11}CH(CH_3)\text{-}CH_2CH(CH_3)CH_2CO_2H$	CCl_4	77 ($h\nu$)	$n\text{-}C_5H_{11}CH(CH_3)CH_2CH(CH_3)CH_2I$ (80)	231
C_{13}	$C_6H_5CONH(CH_2)_5CO_2H$	CCl_4	77 ($h\nu$)	$C_6H_5CONH(CH_2)_5I$ (63)	144, 146
C_{16}	$n\text{-}C_{15}H_{31}CO_2H$	b	150–180	$n\text{-}C_{15}H_{31}CO_2C_{15}H_{31}\text{-}n$ (54)	155
	Cyclopentadecanecarboxylic acid	C_6H_6	—	Mixture of cyclopentadecyl iodide and acetate (total 80)	232
	$(C_6H_5)_2\triangleleft\text{—}CO_2H$	CCl_4	77 ($h\nu$)	$(C_6H_5)_2\triangleleft\text{—}I$ (57)	66, 67
C_{17}	$(C_6H_5)_2\underset{CH_3}{\triangleleft}CO_2H$	CCl_4	77 ($h\nu$)	$(C_6H_5)_2\underset{CH_3}{\triangleleft}I$ (45)	150
	[triptycene-CO$_2$H structure]	CCl_4	77 ($h\nu$)	[triptycene-I structure] (63)	233

C_{18}	n-C$_6$H$_{13}$CO(CH$_2$)$_{10}$CO$_2$H	CCl$_4$	77 ($h\nu$)	n-C$_6$H$_{13}$CO(CH$_2$)$_{10}$I (79)	144, 146
C_{20}	n-C$_6$H$_{13}$CH(OAc)(CH$_2$)$_{10}$CO$_2$H	CCl$_4$	77 ($h\nu$)	n-C$_6$H$_{13}$CH(OAc)(CH$_2$)$_{10}$I (82)	144, 146
C_{21}	n-C$_{20}$H$_{41}$CO$_2$H	b	150–180	n-C$_{20}$H$_{41}$CO$_2$C$_{20}$H$_{41}$-n (52)	155

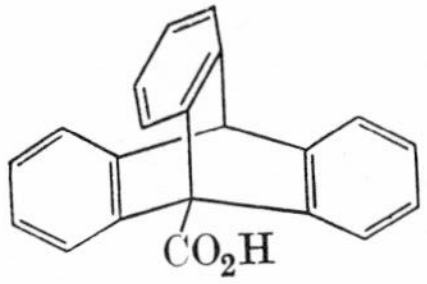

C_{22}

[Triptycene-type structure with CO$_2$H] CCl$_4$, 77 ($h\nu$), [Triptycene-type structure with I] (23), 233

[Steroid with CH(CH$_3$)CO$_2$H] CCl$_4$, 77 ($h\nu$), [Steroid with CH(CH$_3$)I] (27), 234

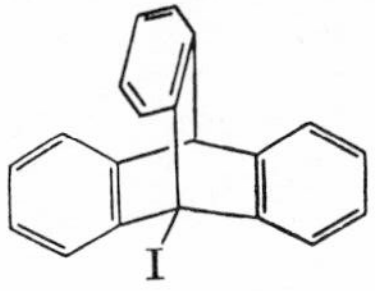

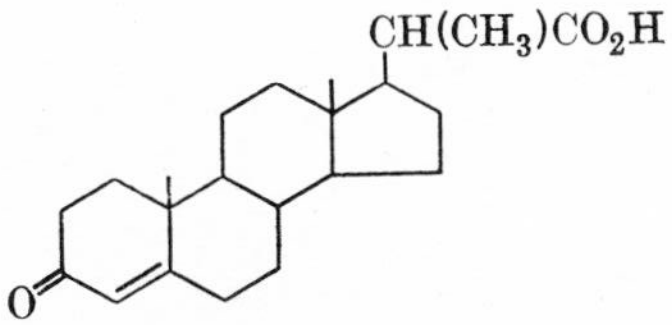

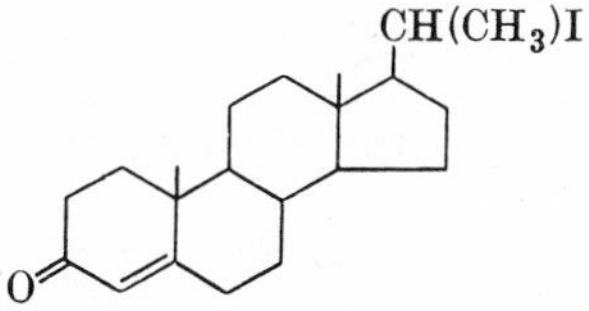

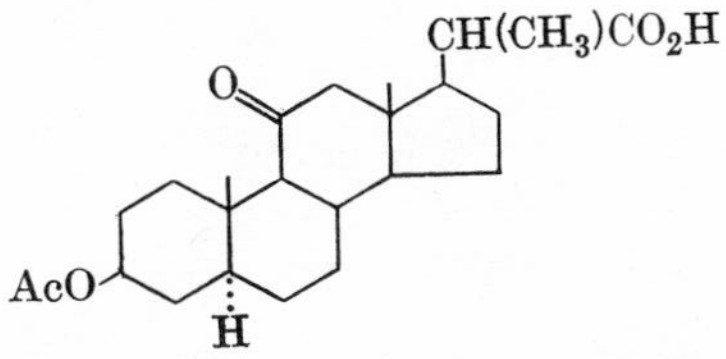

C_{24}

[Steroid with CH(CH$_3$)CO$_2$H, AcO, H] CCl$_4$, 77 ($h\nu$), [Steroid with CH(CH$_3$)I, AcO, H] (85), 144, 146

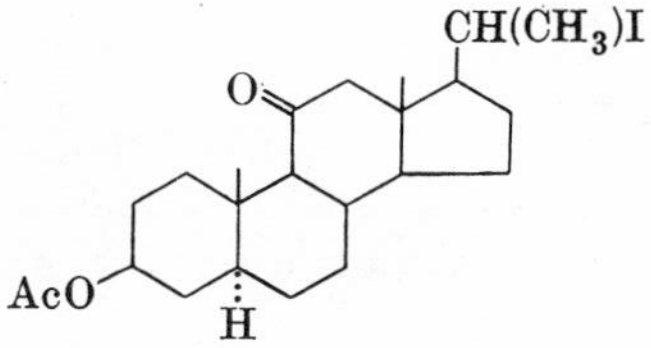

Note: References 210–245 are on pp. 420–421.

[a] No halodecarboxylation was observed.
[b] The solvent was 1,1,2,2-tetrachloroethane, o-dichlorobenzene, or mineral oil.

TABLE XIV. COMPARISONS OF HALODECARBOXYLATION PROCEDURES

	Reactant	Method	Products (% Yield)	Refs.
C_5	n-$C_4H_9CO_2H$	$Pb(OAc)_4$/LiCl	n-C_4H_9Cl (92)	61, 62
	$(CH_3)_3CCO_2H$	$Pb(OAc)_4$/LiCl	$(CH_3)_3CCl$ (72)	61, 62
		$Pb(OAc)_4/I_2$	$(CH_3)_3CI$ (10), $(CH_3)_2C{=}CH_2$ (20)	144, 145
		Ag(I) salt/Br_2	No definite products	235
	◇-CO_2H	$Pb(OAc)_4$/LiCl	◇-Cl (98)	61, 62
		Ag(I) salt/Br_2	◇-Br (57)	236
C_6	n-$C_5H_{11}CO_2H$	$Pb(OAc)_4/I_2$	n-$C_5H_{11}I$ (100)	145
		$(CH_3)_3COI$	n-$C_5H_{11}I$ (91)	145
	$HO_2C(CH_2)_4CO_2H$	$Pb(OAc)_4/I_2$	$I(CH_2)_4I$ (33)	144, 145
		$(CH_3)_3COI$	$I(CH_2)_4I$ (62)	145
		Ag(I) salt/Br_2	$Br(CH_2)_4Br$ (58)	237

C_7	cyclohexyl-CO_2H	$Pb(OAc)_4$/LiCl	Cyclohexyl chloride (100)	61, 62
		$Pb(OAc)_4$/I_2	Cyclohexyl iodide (91)	144, 145
		$(CH_3)_3COI$	Cyclohexyl iodide (70)	145
		Ag(I) salt/Cl_2	Cyclohexyl chloride (70)	238
		Ag(I) salt/Br_2	Cyclohexyl bromide (73–80)	238
C_{10}	(H$_3$C–CH$_3$ bicyclic–CO_2H)	$(CH_3)_3COI$	(H$_3$C–CH$_3$ bicyclic–I) (62)	145
C_{18}	n-$C_{17}H_{35}CO_2H$	HgO/I_2	'' (72)	233
		Ag(I) salt/Br_2	n-$C_{17}H_{35}Br$ (73–86)	239
		Hg(II) salt/Br_2	n-$C_{17}H_{35}Br$ (93)	133
		Ag(I) salt/I_2	n-$C_{17}H_{35}I$ (65)	240

Note: References 210–245 are on pp. 420–421.

TABLE XV. DECARBOXYLATION OF 1,2-, 1,3-, AND 1,4-DICARBOXYLIC ACIDS

A. 1,2-Dicarboxylic Acids

	Reaction	Reagent	Solvent	Temp, °C	Products (% Yield)	Refs.
C_4	$HO_2CCH_2CH_2CO_2H$	PbO_2	None	200	Ethylene (1)	159
C_6	cyclobutane-1,2-dicarboxylic acid	$Pb(OAc)_4$/ 1–2 eq pyridine	$(CH_3)_2SO$	35	Cyclobutene (9)	191
	cyclobutane-1,2-d_2-dicarboxylic acid	$Pb(OAc)_4$/ 1–2 eq pyridine	C_6H_6	80	cyclobutene-d_2 (—)	192
C_8	cyclohexane-1,2-dicarboxylic acid	PbO_2	Decalin	190	Cyclohexene (12)	159
	cyclohexane-d_8-1,2-dicarboxylic acid	$Pb(OAc)_4$	Pyridine	50–60	cyclohexene-d_8 (20)	241

Substrate	Reagent	Solvent	Temp	Product (Yield %)	Ref
cyclohexane-1,2-dicarboxylic acid (4-ene)	Pb(OAc)$_4$/O$_2$	Pyridine	67	cyclohexadiene (76)	83
cyclobutane-1,2-dicarboxylic acid	Pb(OAc)$_4$	Pyridine	50–53	cyclobutene (30–38)	179
cyclobutane-dicarboxylic anhydride	Pb(OAc)$_4$	Pyridine	50–53	cyclobutene (15)	179
cyclobutane-dicarboxylic anhydride	Pb(OAc)$_4$	Pyridine	43–45	cyclobutene (20)	171
bicyclic diacid	PbO$_2$	Decalin	180	bicyclic enone (7)	196

Note: References 210–245 are on pp. 420–421.

TABLE XV. Decarboxylation of 1,2-, 1,3-, and 1,4-Dicarboxylic Acids
(Continued)

A. 1,2-Dicarboxylic Acids (Continued)

Reaction	Reagent	Solvent	Temp, °C	Products (% Yield)	Refs.
C_9 (norbornane-2,3-dicarboxylic acid)	PbO_2	Decalin	190	(Low)	159
(norbornene-dicarboxylic acid)	$Pb(OAc)_4$	—	—	(—)	194
(oxo-cyclopentane dicarboxylic acid)	$Pb(OAc)_4/O_2$	Pyridine	67	(46)	83
(methoxy-norbornane dicarboxylic acid)	PbO_2	Decalin	180	(3)	196

402

C_{10} (CH₃, CH₃, CO₂H, CO₂H cyclohexane)	PbO₂	Decalin	190	(cyclohexene with 2 CH₃) (16)	159
(bicyclic lactone with CH₃)	PbO₂	None	200	(bicyclic alkene with CH₃) (15)	163
(bicyclic CO₂H, CH₂CO₂H)	PbO₂	None	210	(bicyclic =CH₂) (17)	162
(bicyclic anhydride)	PbO₂	None	250	(bicyclic alkene) (20)	159

Note: References 210–245 are on pp. 420–421.

TABLE XV. Decarboxylation of 1,2-, 1,3-, and 1,4-Dicarboxylic Acids (*Continued*)

A. *1,2-Dicarboxylic Acids (Continued)*

Reaction	Reagent	Solvent	Temp, °C	Products (% Yield)	Refs.
(structure with CO₂H, CO₂H and C=O)	Pb(OAc)₄/O₂	C₆H₆	80	(bicyclic enone) (60)	161
	Pb(OAc)₄/O₂	Pyridine	67	" (52)	83
(structure with two C=O and CO₂H, CO₂H)	Pb(OAc)₄/ 1–2 eq pyridine	CH₃CN	50	(bicyclic dione with ene) (30–38)	81
C₁₁ (structure with OCH₂OAc, CO₂H, CO₂H)	Pb(OAc)₄/ 1–2 eq pyridine	C₆H₆	100	(OCH₂OAc bicyclic ene) (2)	196

404

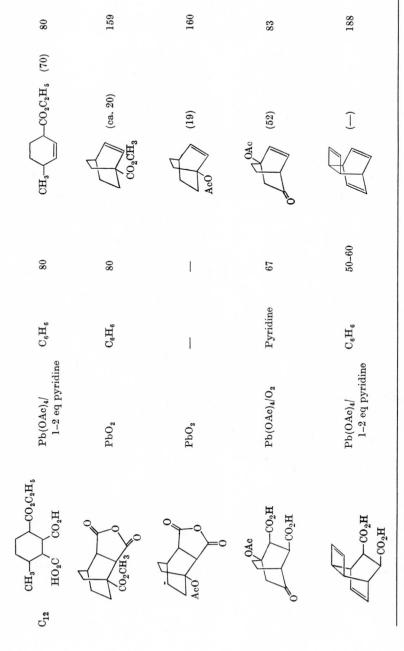

TABLE XV. DECARBOXYLATION OF 1,2-, 1,3-, AND 1,4-DICARBOXYLIC ACIDS
(*Continued*)

A. 1,2-Dicarboxylic Acids (Continued)

Reaction	Reagent	Solvent	Temp, °C	Products (% Yield)	Refs.
C_{12} (Cont.)					
(structure with CO_2H, CO_2H)	$Pb(OAc)_4$	Pyridine	60–65	(30)	180–182
(structure with CO_2H, CO_2H)	$Pb(OAc)_4$/ 1–2 eq pyridine	—	—	(28)	184
C_{13} (structure with CO_2H, CO_2H, $C_2H_5O_2C$)	$Pb(OAc)_4$/ 1–2 eq pyridine	C_6H_6	Room temp	(65)	165
	″	$(CH_3)_2SO$	″	(45)	
	″	C_6H_6 or CH_3CN	80 (C_6H_6) 55 (CH_3CN)	(63–88)	82, 79, 80
(lactone structure with $C_2H_5O_2C$)	PbO_2	None	170–190	(30–38)	79, 80

Substrate	Reagent	Solvent	Temp	Product(s) (Yield %)	Refs.
(lactone with OAc, C14)	PbO₂	—	—	(bicyclic OAc) (—)	Quoted in 160
H₃C—C(CH₃)—...CO₂H, CO₂H, CH₃ (keto diacid)	Pb(OAc)₄	Pyridine	67	H₃C—C(CH₃)—...CH₃ (ketone) (80)	83
CH₃, C₂H₅O₂C, CO₂H, CO₂H	Pb(OAc)₄ / 1–2 eq pyridine	C₆H₆ or CH₃CN	50	CH₃, C₂H₅O₂C (50–70)	79, 80
CO₂CH₃, CH₃O₂C, CO₂H, CO₂H	Pb(OAc)₄ / 1–2 eq pyridine	C₆H₆	80	CO₂CH₃, CH₃O₂C (I, 45)	168, 169
		Dioxane[a]	Room temp	CO₂CH₃, CH₃O₂C (lactone-dione) I (45), (16)	165, 168

Note: References 210–245 are on pp. 420–421.

[a] Similar results were obtained in other solvents (Ref. 165).

TABLE XV. Decarboxylation of 1,2-, 1,3-, and 1,4-Dicarboxylic Acids (*Continued*)

Reaction	Reagent	Solvent	Temp, °C	Products (% Yield)	Refs.
A. 1,2-Dicarboxylic Acids (Continued)					
C_{14} (Cont.)					
[structure with CH_3O_2C, CO_2H, CO_2H, CH_3O_2C]	$Pb(OAc)_4$/ 1–2 eq pyridine	C_6H_6	80	[structure with CH_3O_2C, CH_3O_2C] (50)	242
[structure with OAc, CO_2H, CO_2H, AcO]	$Pb(OAc)_4$	C_6H_6/pyridine/ CH_3CN	80	[structure with OAc, AcO] (49)	243
[structure with H_3C, CH_3, OAc, CO_2H, CO_2H, O]	$Pb(OAc)_4/O_2$	Pyridine	67	[structure with H_3C, CH_3, OAc, O] (69)	83

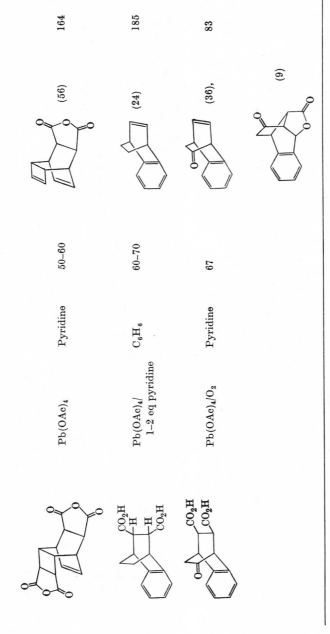

TABLE XV. Decarboxylation of 1,2-, 1,3-, and 1,4-Dicarboxylic Acids
(*Continued*)

A. *1,2-Dicarboxylic Acids* (*Continued*)

Reaction	Reagent	Solvent	Temp, °C	Products (% Yield)	Refs.
C_{15} [structure with H$_3$C, CH$_3$, CO$_2$H, CO$_2$H, OAc]	Pb(OAc)$_4$/O$_2$	Pyridine	67	[structure with H$_3$C, CH$_3$, OAc] (Low)	83
[structure with CO$_2$H, CO$_2$H, N-C$_6$H$_5$SO$_2$]	Pb(OAc)$_4$	Pyridine	80	[structure with N-C$_6$H$_5$SO$_2$] (50)	58
C_{16} *dl*-C$_6$H$_5$CH(CO$_2$H)-CH(C$_6$H$_5$)CO$_2$H	Pb(OAc)$_4$/ 1–2 eq pyridine	C$_6$H$_6$	80	*trans*-C$_6$H$_5$CH=CHC$_6$H$_5$ (44)	26
meso-C$_6$H$_5$CH(CO$_2$H)-CH(C$_6$H$_5$)CO$_2$H	Pb(OAc)$_4$/ 1–2 eq pyridine	C$_6$H$_6$	80	*trans*-C$_6$H$_5$CH=CHC$_6$H$_5$ (41)	26

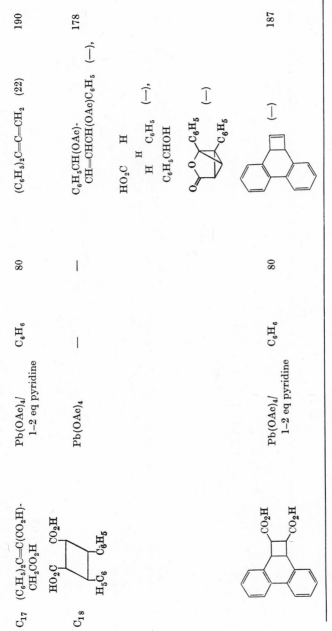

Note: References 210–245 are on pp. 420–421.

TABLE XV. Decarboxylation of 1,2-, 1,3-, and 1,4-Dicarboxylic Acids (*Continued*)

A. *1,2-Dicarboxylic Acids (Continued)*

Reaction	Reagent	Solvent	Temp, °C	Products (% Yield)	Refs.
C$_{18}$ (Cont.)	Pb(OAc)$_4$/ 1–2 eq pyridine	C$_6$H$_6$	80	(—)	187
	Pb(OAc)$_4$/ 1–2 eq pyridine	C$_6$H$_6$ or CH$_3$CN	50	(50–70)	79, 80
	"	(CH$_3$)$_2$SO	Room temp	(20)	82
	"	CH$_3$CN	60	(18)	82
	Pb(OAc)$_4$/O$_2$	Pyridine	67	(21)	83
	Pb(OAc)$_4$/ 1–2 eq pyridine	(CH$_3$)$_2$SO	Room temp	(15)	82

C_{19}	$(p\text{-}CH_3C_6H_4)_2C{=}C(CO_2H)\text{-}CH_2CO_2H$	,,	CH_3CN	60	,, (4) 82
		$Pb(OAc)_4/O_2$	Pyridine	67	,, (21) 83
		$Pb(OAc)_4/$ 1–2 eq pyridine	C_6H_6	80	$(p\text{-}CH_3C_6H_4)_2C{=}C{=}CH_2$ (17) 190
	$(p\text{-}CH_3OC_6H_4)_2C{=}C(CO_2H)\text{-}CH_2CO_2H$	$Pb(OAc)_4/$ 1–2 eq pyridine	C_6H_6	80	$(p\text{-}CH_3OC_6H_4)_2C{=}C{=}CH_2$ (15) 190
C_{20}	[structure with HO₂C, HO₂C, OAc]	$Pb(OAc)_4/O_2$	Pyridine	67	[structure with OAc] (16) 83
C_{22}	[anhydride structure]	PbO_2	—	—	[structure] (45) 245

Note: References 210–245 are on pp. 420–421.

TABLE XV. DECARBOXYLATION OF 1,2-, 1,3-, AND 1,4-DICARBOXYLIC ACIDS (*Continued*)

Reaction	Reagent	Solvent	Temp, °C	Products (% Yield)	Refs.
A. 1,2-Dicarboxylic Acids (Continued)					
C₂₃	Pb(OAc)₄	Pyridine	50		(—), 197
					(—)
	Pb(OAc)₄	Pyridine	Reflux		(—) 100

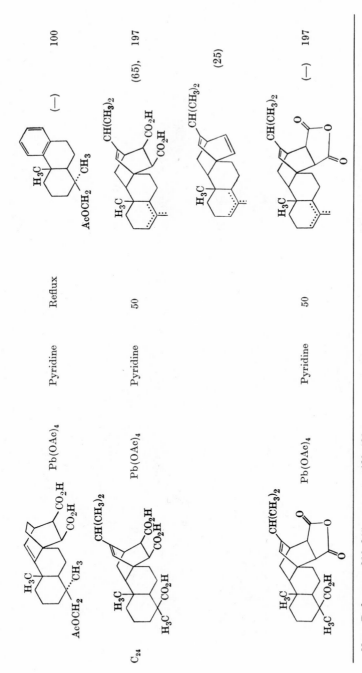

TABLE XV. Decarboxylation of 1,2-, 1,3-, and 1,4-Dicarboxylic Acids (*Continued*)

A. *1,2-Dicarboxylic Acids* (*Continued*)

Reaction	Reagent	Solvent	Temp, °C	Products (% Yield)	Refs.
C_{24} (Cont.)	PbO_2	—	—	(45)	245
C_{25}	$Pb(OAc)_4$	Pyridine	50	(I, —)	197
	$Pb(OAc)_4$	Pyridine	Room temp	I(7), (10) (and other products)	92

B. 1,3- and 1,4-Dicarboxylic Acids

C$_{10}$	(cyclopentane with CO$_2$H, CH$_3$, CH$_3$, CH$_3$, HO$_2$C, CH$_3$)	Pb(OAc)$_4$/ 1–2 eq pyridine	C$_6$H$_6$	80	(bicyclic lactone with H$_3$C, CH$_3$, O, CH$_3$) (70)	102
C$_{14}$	(benzene ring with C(CH$_3$)$_2$CO$_2$H, C(CH$_3$)$_2$CO$_2$H)	Pb(OAc)$_4$/ 1–2 eq pyridine	Dioxane	75	(isochromanone: H$_3$C, CH$_3$, O, H$_3$C, CH$_3$, O) (—)[,b] (benzofused 7-membered dilactone: H$_3$C, CH$_3$, O, O, H$_3$C, CH$_3$) (—)	203
C$_{17}$	C$_6$H$_5$CH(CO$_2$H)CH(C$_6$H$_5$)- CH$_2$CO$_2$H	Pb(OAc)$_4$/ 1–2 eq pyridine	C$_6$H$_6$	80	(γ-butyrolactone with C$_6$H$_5$, C$_6$H$_5$) (72)	102

Note: References 210–245 are on pp. 420–421.
[b] A 1:1 mixture of the products was obtained.

TABLE XVI. Decarboxylation of α-Amino and α-Hydroxy Acids by Lead Tetraacetate

	Reactant	Solvent	Temp, °C	Product(s) (% Yield)	Ref.
A. α-Amino Acids					
C_6	$CH_3CONHCH_2CO_2H$	$HCON(CH_3)_2$	25	CH_3CONH_2 (75), CH_2O (70)	76
C_7	$CH_3CONHCH(C_3H_7\text{-}i)CO_2H$	$HCON(CH_3)_2$	25	CH_3CONH_2 (40), $i\text{-}C_3H_7CHO$ (43), $CH_3CONHCH{=}C(CH_3)_2$ (34)	76
	$CH_3CONHCH(CH_2CH_2SCH_3)CO_2H$	$HCON(CH_3)_2$	25	CH_3CONH_2 (41), $CH_3SCH_2CH_2CHO$ (56)	76
C_9	$C_6H_5CH_2NHCH_2CO_2H$	AcOH	100	$C_6H_5CH_2NHCH_2OAc$ (20)	16–18
	$C_6H_5CONHCH_2CO_2H$	AcOH	100	$C_6H_5CONHCH_2OAc$ (—)	16–18
C_{11}	$CH_3CONHCH(CH_2C_6H_5)CO_2H$	$HCON(CH_3)_2$	25	CH_3CONH_2 (51), $C_6H_5CH_2CHO$ (33)	76
B. α-Hydroxy Acids					
C_2	CH_2OHCO_2H	—[a]	—	CH_2O (—)	73
C_4	$HO_2CCH_2CHOHCO_2H$	—[a]	—	HO_2CCH_2CHO (—)	73
C_6	$(HO_2CCH_2)_2C(OH)CO_2H$	—[a]	—	$(HO_2CCH_2)_2CO$ (—)	73

C_8	$C_6H_5CHOHCO_2H$	—[a]	—	C_6H_5CHO (—)	73
C_{10}	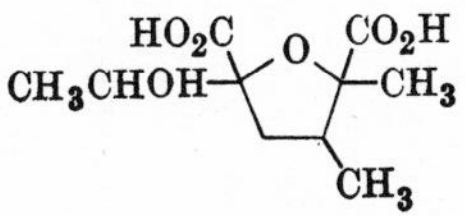	AcOH	100	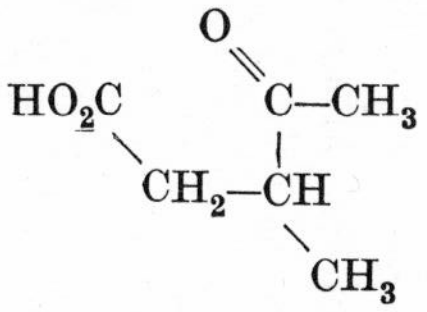 (—)	204
C_{14}	$(C_6H_5)_2C(OH)CO_2H$	—	—	$(C_6H_5)_2CO$ (—)	73
C_{21}	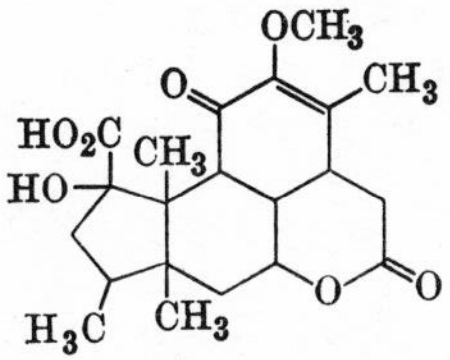	AcOH	Room temp	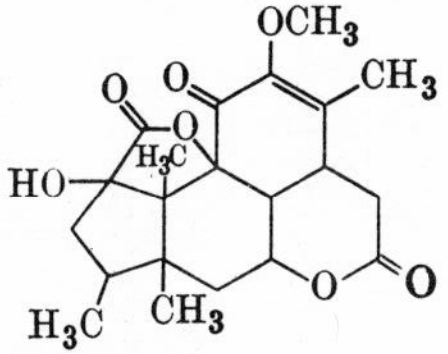 (—)	116a

Note: References 210–245 are on pp. 420–421.

[a] A variety of solvents and temperatures was used.

TABLE XVII. Decarboxylation of α-Keto Acids by Lead Tetraacetate[a]

	Reactant	Solvent	Product(s) (% Yield)
C_3	CH_3COCO_2H	C_6H_6	CO_2 (100), AcOH (—)
		$C_6H_5CH_2OH$	$C_6H_5CH_2OAc$ (—)
C_6	$(CH_3)_3CCOCO_2H$	AcOH	$(CH_3)_3CCO_2H$ (—)
C_8	$C_6H_5COCO_2H$	AcOH	$C_6H_5CO_2H$ (—)
C_9	$C_6H_5CH_2COCO_2H$	AcOH	$C_6H_5CH(OAc)CO_2H$ (—)

[a] All reactions are at room temperature. Details for all are in Ref. 75.

REFERENCES TO TABLES IV–XVII

[210] M. Julia and F. LeGoffic, *Bull. Soc. Chim. Fr.*, **1965**, 1555.
[210a] R. E. Partch, private communication.
[211] M. Julia, F. LeGoffic, and L. Katz, *Bull. Soc. Chim. Fr.*, **1964**, 1122.
[212] F. Kido, R. Sakuma, H. Uda, and A. Yoshikoshi, *Tetrahedron Lett.*, **1969**, 3169.
[213] I. Kitagawa, K. Kitazawa, and I. Yosioka, *Tetrahedron Lett.*, **1970**, 1905.
[214] P. R. Bruck, R. D. Clark, R. S. Davidson, W. H. H. Günther, P. S. Littlewood, and B. Lythgoe, *J. Chem. Soc., C*, **1967**, 2529.
[215] W. Herz, R. N. Mirrington, and H. Young, *Tetrahedron Lett.*, **1968**, 405; W. Herz, R. N. Mirrington, H. Young, and Y. Yeng Lin, *J. Org. Chem.*, **33**, 4210 (1968).
[216] W. Herz and M. G. Nair, *J. Org. Chem.*, **34**, 4016 (1969).
[217] D. H. R. Barton, D. Giacopello, P. Manitto, and D. L. Struble, *J. Chem. Soc., C*, **1969**, 1047.
[218] F. R. Jensen and C. H. Bushweller, *J. Amer. Chem. Soc.*, **91**, 5774 (1969).
[219] E. Müller and U. Trense, *Tetrahedron Lett.*, **1967**, 2045.
[220] P. K. Freeman and D. M. Balls, *Tetrahedron Lett.*, **1967**, 437.
[221] T. Nozoe, T. Asao, M. Ando, and K. Takase, *Tetrahedron Lett.*, **1967**, 2821.
[222] H. Christol, D. Moers, and T. Pietrasanta, *Bull. Soc. Chim. Fr.*, **1969**, 962.
[223] R. C. Neuman, Jr. and M. L. Rahm, *J. Org. Chem.*, **31**, 1857 (1966).
[224] B. Weinstein, A. H. Fenselau, and J. G. Thoene, *J. Chem. Soc.*, **1965**, 2281.
[225] G. Odham, *Ark. Kemi*, **23**, 431 (1965).
[226] U. Scheidegger, J. E. Baldwin, and J. D. Roberts, *J. Amer. Chem. Soc.*, **89**, 894 (1967).
[227] A. Streitwieser and P. J. Stang, *J. Amer. Chem. Soc.*, **87**, 4953 (1965).
[228] D. H. R. Barton, L. D. S. Godinho, and J. K. Sutherland, *J. Chem. Soc.*, **1965**, 1779.
[229] E. Buchta and S. Billenstein, *Ann.*, **702**, 38 (1967).
[230] T. Sakan and K. Abe, *Tetrahedron Lett.*, **1968**, 2471.
[231] G. Odham, *Ark. Kemi*, **21**, 379 (1964).
[232] H. Nozaki, T. Mori, and R. Noyori, *Tetrahedron Lett.*, **1967**, 779.
[233] J. B. Dence and J. D. Roberts, *J. Org. Chem.*, **33**, 1251 (1968).
[234] B. Krieger and E. Kaspar, *Chem. Ber.*, **100**, 1169 (1967).
[235] M. Heintzeler, *Ann.*, **569**, 102 (1950).
[236] J. D. Roberts and H. E. Simmons, *J. Amer. Chem. Soc.*, **73**, 5487 (1951).
[237] H. Schmid, *Helv. Chim. Acta*, **27**, 127 (1944).
[238] C. Hunsdiecker, H. Hunsdiecker, and E. Vogt, Ger. Pat. 730,410 (1942) [*C.A.*, **38**, 374 (1944)].
[239] T. N. Mehta, V. S. Mehta, and V. B. Thosar, *J. Indian Chem. Soc., Ind. News Ed.*, **4**, 170 (1941) [*C.A.* **36**, 4486 (1942)]; J. W. H. Oldham, *J. Chem. Soc.*, **1950**, 100.
[240] J. W. H. Oldham and A. R. Ubbelohde, *J. Chem. Soc.*, **1941**, 368.

[241] F. A. L. Anet and M. Z. Haq, *J. Amer. Chem. Soc.*, **87**, 3147 (1965).
[242] J. Kazan and F. D. Greene, *J. Org. Chem.*, **28**, 2965 (1963).
[243] J. Kopecky and J. Smejkal, *Tetrahedron Lett.*, **1967**, 3889.
[244] A. C. Oehlschlager and L. H. Zalkow, *Can. J. Chem.*, **47**, 461 (1969).
[245] D. Ginsburg, *Bull. Soc. Chim. Fr.*, **1960**, 1348.

BIS(CHLOROMETHYL) ETHER IN CHLORO-
METHYLATIONS—WARNING

Bis(chloromethyl) ether is a potent carcinogen, so chloromethylations should be run with great caution.

Chloromethylation of aromatic compounds is a useful reaction that is widely used. Fuson and McKeever reviewed the reaction in the first volume of *Organic Reactions*.[1] Bis(chloromethyl) ether is a by-product and/or intermediate. Recently Van Duuren and associates found this ether carcinogenic to rats when injected subcutaneously or repeatedly applied to their skins.[2] In subsequent studies, rats and hamsters exposed to 0.1 ppm of bis(chloromethyl) ether in the air 5 hours a day for 2 years had a high incidence of lung cancer.[3] The ether is also an acute lung irritant. Methyl chloromethyl ether, which is occasionally used as a reagent for chloromethylation, is also quite toxic, although less so than bis(chloromethyl) ether.[2,3]

Accordingly, avoid chloromethylation if alternative synthetic routes are satisfactory. If you must chloromethylate, use great caution. Run chloromethylations only in the back of a well-ventilated hood. Wear neoprene gloves when pouring and handling reaction mixtures. In distillations, be careful with foreruns, which may contain bis(chloromethyl) ether (bp 105–106°). Avoid procedures in which bis(chloromethyl) ether is prepared and isolated. Destroy the ether in foreruns or reaction mixtures by treatment with aqueous sodium hydroxide (hydrolysis to formaldehyde and hydrogen chloride occurs readily). If you get bis(chloromethyl) ether on your skin, wash the affected area with water promptly and thoroughly. If you inhale bis(chloromethyl) ether, seek medical attention at once.

[1] R. C. Fuson and C. H. McKeever, *Org. Reactions*, **1**, 63 (1942).

[2] B. L. Van Duuren, A. Sivak, B. M. Goldschmidt, C. Katz, and S. Melchionne, *J. Nat. Cancer Inst.*, **43**, 481 (1969).

[3] R. T. Drew, V. Cappiello, S. Laskin, and M. Kuschner, 1970 Conference, American Industrial Hygiene Association, Abstract 164.

AUTHOR INDEX, VOLUMES 1–19

Adams, Joe T., 8
Adkins, Homer, 8
Albertson, Noel F., 12
Angyal, S. J., 8
Archer, S., 14

Bachmann, W. E., 1, 2
Baer, Donald R., 11
Behr, Lyell C., 6
Bergmann, Ernst D., 10
Berliner, Ernst, 5
Blatchly, J. M., 19
Blatt, A. H., 1
Blicke, F. F., 1
Bloomfield, Jordan J., 15
Brand, William W., 18
Brewster, James H., 7
Brown, Herbert C., 13
Brown, Weldon G., 6
Bruson, Herman Alexander, 5
Bublitz, Donald E., 17
Buck, Johannes S., 4
Butz, Lewis W., 5

Carmack, Marvin, 3
Carter, H. E., 3
Cason, James, 4
Cope, Arthur C., 9, 11
Corey, Elias J., 9
Cota, Donald J., 17
Crounse, Nathan N., 5

Daub, Guido H., 6
Dave, Vinod, 18
DeTar, DeLos F., 9
Djerassi, Carl, 6
Donaruma, L. Guy, 11
Drake, Nathan L., 1
DuBois, Adrien S., 5

Eliel, Ernst L., 7

Emerson, William S., 4
England, D. C., 6

Fieser, Louis F., 1
Folkers, Karl, 6
Fuson, Reynold C., 1

Geissman, T. A., 2
Gensler, Walter J., 6
Gilman, Henry, 6, 8
Ginsburg, David, 10
Govindichari, Tuticorin R., 6
Gutsche, C. David, 8

Hageman, Howard A., 7
Hamilton, Cliff S., 2
Hamlin, K. E., 9
Hanford, W. E., 3
Harris, Constance M., 17
Harris, J. F., Jr., 13
Harris, Thomas M., 17
Hartung, Walter H., 7
Hassall, C. H., 9
Hauser, Charles R., 1, 8
Heldt, Walter Z., 11
Henne, Albert L., 2
Hoffman, Roger A., 2
Holmes, H. L., 4, 9
Houlihan, William J., 16
House, Herbert O., 9
Hudson, Boyd E., Jr., 1
Huyser, Earl S., 13

Ide, Walter S., 4
Ingersoll, A. W., 2

Jackson, Ernest L., 2
Jacobs, Thomas L., 5
Johnson, John R., 1
Johnson, William S., 2, 6
Jones, G., 15

Jones, Reuben G., 6
Jorgenson, Margaret J., 18

Kende, Andrew S., 11
Kloetzel, Milton C., 4
Kochi, Jay K., 19
Kornblum, Nathan, 2, 12
Kosolapoff, Gennady M., 6
Kreider, Eunice M., 18
Krimen, L. I., 17
Kulka, Marshall, 7

Lane, John F., 3
Leffler, Marlin T., 1

McElvain, S. M., 4
McKeever, C. H., 1
McOmie, J. F. W., 19
Maercker, Adalbert, 14
Magerlein, Barney J., 5
Manske, Richard H. F., 7
Martin, Elmore L., 1
Martin, William B., 14
Moore, Maurice L., 5
Morgan, Jack F., 2
Morton, John W., Jr., 8
Mosettig, Erich, 4, 8
Mozingo, Ralph, 4

Nace, Harold R., 12
Newman, Melvin S., 5
Nielsen, Arnold T., 16

Pappo, Raphael, 10
Parham, William E., 13
Parmerter, Stanley M., 10
Pettit, George R., 12
Phadke, Ragini, 7
Phillips, Robert R., 10
Pine, Stanley H., 18
Posner, Gary H., 19
Price, Charles C., 3

Rabjohn, Norman, 5
Rinehart, Kenneth L., Jr., 17
Roberts, John D., 12
Roe, Arthur, 5

Rondestvedt, Christian S., Jr., 11
Rytina, Anton W., 5

Sauer, John C., 3
Schaefer, John P., 15
Schulenberg, J. W., 14
Schweizer, Edward E., 13
Semmelhack, Martin F., 19
Sethna, Suresh, 7
Sharts, Clay M., 12
Sheehan, John C., 9
Sheldon, Roger A., 19
Shirley, David A., 8
Shriner, Ralph L., 1
Simonoff, Robert, 7
Smith, Lee Irvin, 1
Smith, Peter A. S., 3, 11
Spielman, M. A., 3
Spoerri, Paul E., 5
Stacey, F. W., 13
Struve, W. S., 1
Suter, C. M., 3
Swamer, Frederic W., 8
Swern, Daniel, 7

Tarbell, D. Stanley, 2
Todd, David, 4
Touster, Oscar, 7
Truce, William E., 9, 18
Trumbull, Elmer R., 11

van Tamelen, Eugene E., 12

Walling, Cheves, 13
Wallis, Everett S., 3
Warnhoff, E. W., 18
Weston, Arthur W., 3, 9
Whaley, Wilson M., 6
Wilds, A. L., 2
Wiley, Richard H., 6
Wilson, C. V., 9
Wolf, Donald E., 6
Wolff, Hans, 3
Wood, John L., 3

Zaugg, Harold E., 8, 14
Zweifel, George, 13

CHAPTER AND TOPIC INDEX, VOLUMES 1-19

Many chapters contain brief discussions of reactions and comparisons of alternative synthetic methods which are related to the reaction that is the subject of the chapter. These related reactions and alternative methods are not usually listed in this index.
In this index the volume number is in BOLDFACE, the chapter number in ordinary type.

Acetoacetic ester condensation, **1**, 9
Acetic anhydride, reaction with quinones, **19**, 3
Acetoxylation of quinones, **19**, 3
Acetylenes, synthesis of, **5**, 1
Acid halides, reactions with organometallic compounds, **8**, 2
Acrylonitrile, addition to (cyanoethylation), **5**, 2
α-Acylamino acid mixed anhydrides, **12**, 4
α-Acylamino acids, azlactonization of, **3**, 5
α-Acylamino carbonyl compounds, in preparation of thiazoles, **6**, 8
Acylation of ketones to diketones, **8**, 3
Acyl hypohalites, reactions of, **9**, 5
Acyloins, **4**, 4; **15**, 1
Aldehydes, synthesis from carboxylic acids, **4**, 7; **8**, 5
synthesis of, **4**, 7; **5**, 10; **8**, 4, 5; **9**, 2
Aldol condensation, **16**
Aliphatic and alicyclic nitro compounds, synthesis of, **12**, 3
Aliphatic fluorides, **2**, 2
Alkali amides, **1**, 4
Alkylating agents for esters and nitriles, **9**, 4
Alkylation, of aromatic compounds, **3**, 1
of esters and nitriles, **9**, 4
of metallic acetylides, **5**, 1
γ-Alkylation of dianions of β-dicarbonyl compounds, **17**, 2
Alkylations with amines and ammonium salts, **7**, 3
Alkylidenesuccinic acids, preparation and reactions of, **6**, 1
Alkylidene triphenylphosphoranes, preparation and reactions of, **14**, 3
Alkynes, synthesis of, **5**, 1
π-Allylnickel complexes, **19**, 2

Allylphenols, preparation by Claisen rearrangement, **2**, 1
Aluminum alkoxides, in Meerwein-Ponndorf-Verley reduction, **2**, 5
in Oppenauer oxidation, **6**, 5
α-Amidoalkylations at carbon, **14**, 2
Amination, of heterocyclic bases by alkali amides, **1**, 4
of hydroxyl compounds by Bucherer reaction, **1**, 5
Amine oxides, **11**, 5
Amines, preparation by reductive alkylation, **4**, 3; **5**, 7
reactions with cyanogen bromide, **7**, 4
Anhydrides of aliphatic dibasic acids in Diels-Alder reaction, **5**, 5
Anthracene homologs, synthesis of, **1**, 6
Anti-Markownikoff hydration, **13**, 1
Arndt-Eistert reaction, **1**, 2
Aromatic aldehydes, preparation of, **5**, 6
Aromatic compounds, chloromethylation of, **1**, 3
Aromatic fluorides, preparation of, **5**, 4
Aromatic hydrocarbons, synthesis of, **1**, 6
Arsinic acids, **2**, 10
Arsonic acids, **2**, 10
Arylactic acids, synthesis of, **1**, 1
β-Arylacrylic acids, synthesis of, **1**, 8
Arylamines, preparation and reactions of, **1**, 5
γ-Arylation, **17**, 2
Arylation of unsaturated compounds by diazonium salts, **11**, 3
Arylglyoxals, condensation with aromatic hydrocarbons, **4**, 5
Arylsulfonic acids, preparation of, **3**, 4
Aryl thiocyanates, **3**, 6

Azides, preparation and rearrangement of, **3**, 9
Azlactones, **3**, 5

Baeyer-Villiger reaction, **9**, 3
Bart reaction, **2**, 10
Bechamp reaction, **2**, 10
Beckmann rearrangement, **11**, 1
Benzils, reduction of, **4**, 5
Benzoin condensation, **4**, 5
Benzoins, synthesis of, **4**, 5
Benzoquinones, synthesis of, **4**, 6
  acetoxylation of, **19**, 3
Benzylamines, from Sommelet-Hauser rearrangement, **18**, 4
Biaryls, synthesis of, **2**, 6
Bicyclobutanes, from cyclopropenes, **18**, 3
Bischer-Napieralski reaction, **6**, 2
von Braun cyanogen bromide reaction, **7**, 4
Bucherer reaction, **1**, 5

Cannizzaro reaction, **2**, 3
Carbenes, **13**, 2
Carbon alkylations with amines and ammonium salts, **7**, 3
Carboxylic acids, reaction with organolithium reagents, **18**, 1
Catalytic hydrogenation of esters to alcohols, **8**, 1
Chapman rearrangement, **14**, 1; **18**, 2
Chloromethylation of aromatic compounds, **1**, 3; see Warning, 19
Cholanthrenes, synthesis of, **1**, 6
Chugaev reaction, **12**, 2
Claisen condensation, **1**, 8
Claisen rearrangement, **2**, 1
Cleavage, of benzyl-oxygen, benzyl-nitrogen, and benzyl-sulfur bonds, **7**, 5
  of carbon-carbon bonds by periodic acid, **2**, 8
  of non-enolizable ketones with sodium amide, **9**, 1
Clemmensen reaction, **1**, 7
Conjugate addition of organocopper reagents, **19**, 1
Copper-catalyzed Grignard conjugate additions, **19**, 1
Coumarins, preparation of, **7**, 1
Coupling of diazonium salts with aliphatic compounds, **10**, 1, 2
  of allyl ligands, **19**, 2
Curtius rearrangement, **3**, 7, 9
Cyanoethylation, **5**, 2

Cyanogen bromide, reactions with tertiary amines, **7**, 4
Cyclic ketones, formation by intramolecular acylation, **2**, 4
Cyclization of aryl-substituted aliphatic acids, acid chlorides and anhydrides, **2**, 4
Cycloaddition reactions, thermal, **12**, 1
Cyclobutanes, preparation from thermal cycloaddition reactions, **12**, 1
Cyclopentadienylchromium nitrosyl dicarbonyl, **17**, 1
Cyclopentadienylmanganese tricarbonyl, **17**, 1
Cyclopentadienylrhennium tricarbonyl, **17**, 1
Cyclopentadienyltechnetium tricarbonyl, **17**, 1
π-Cyclopentadienyltransition metal compounds, **17**, 1
Cyclopentadienylvanadium tetracarbonyl, **17**, 1
Cyclopropane carboxylates, from diazoacetic esters, **18**, 3
Cyclopropenes, preparation of, **18**, 3

Darzens, glycidic ester condensation, **5**, 10
Deamination of aromatic primary amines, **2**, 7
Debenzylation, **7**, 5; **18**, 4
Decarboxylation of acids, **19**, 4
Dehalogenation of α-haloacyl halides, **3**, 3
Dehydrogenation, in preparation of ketenes, **3**, 3
  in synthesis of acetylenes, **5**, 1
Demjanov reaction, **11**, 2
Desoxybenzoins, conversion to benzoins, **4**, 5
Desulfurization with Raney nickel, **12**, 5
Diazoacetic esters, reactions with alkenes, alkynes, heterocyclic and aromatic compounds, **18**, 3
Diazomethane, in Arndt-Eistert reaction, **1**, 2
  reactions with aldehydes and ketones, **8**, 8
Diazonium fluoroborates, preparation and decomposition, **5**, 4
Diazonium ring closure reactions, **9**, 7
Diazonium salts, coupling with aliphatic compounds, **10**, 1, 2
  in deamination of aromatic primary amines **2**, 7
  in Meerwein arylation reaction, **11**, 3
  in synthesis of biaryls and aryl quinones, **2**, 6

Dieckmann condensation, **1**, 9; **15**, 1
  for preparation of tetrahydrothiophenes, **6**, 9
Diels-Alder reaction, with acetylenic and olefinic dienophiles, **4**, 2
  with cyclenones, **5**, 3
  with maleic anhydride, **4**, 1
Diene additions to quinones, **5**, 3
3,4-Dihydroisoquinolines, preparation of, **6**, 2
Diketones, pyrolysis of diaryl-, **1**, 6
Doebner reaction, **1**, 8

Elbs reaction, **1**, 6
Electrophilic reagents, preparation of, **14**, 2
Epoxidation with organic peracids, **7**, 7
Esters, acylation with acid chlorides, **1**, 9
  reaction with organolithium reagents, **18**, 1
Exhaustive methylation, Hofmann, **11**, 5

Favorskii rearrangement, **11**, 4
Ferrocenes, **17**, 1
Fischer indole cyclization, **10**, 2
Fluorination, **2**, 2
Formation of carbon-carbon bonds by alkylations with amines and ammonium salts, **7**, 3
  by π-allylnickel complexes, **19**, 2
  by amidoalkylation, **14**, 2
  by free radical additions to olefins, **13**, 3
  of carbon-germanium bonds by free radical addition, **13**, 4
  of carbon-heteroatom bonds by free radical chain additions to carbon-carbon multiple bonds, **13**, 4
  of carbon-nitrogen bonds by free radical addition, **13**, 4
  of carbon-phosphorus bonds by free radical additions, **13**, 4
  of carbon-silicon bonds by free radical addition, **13**, 4
  of carbon-sulfur bonds by free radical addition, **13**, 4
Formylation of aromatic hydrocarbons, **5**, 6
Free radical additions, to olefins and acetylenes to form carbon-heteroatom bonds, **13**, 4
  to olefins to form carbon-carbon bonds, **13**, 3
Friedel-Crafts cyclization of acid chlorides, **2**, 4

Friedel-Crafts reaction, **2**, 4; **3**, 1; **5**, 5; **18**, 1
Fries reaction, **1**, 11

Gattermann aldehyde synthesis, **9**, 2
Gattermann-Koch reaction, **5**, 6
Germanes, addition to olefins and acetylenes, **13**, 4
Glycidic esters, reactions of, **5**, 10
Gomberg-Bachmann reaction, **2**, 6; **9**, 7
Grundmann synthesis of aldehydes, **8**, 5

Haller-Bauer reaction, **9**, 1
Halocarbenes, preparation and reaction of, **13**, 2
Halocyclopropanes, reactions of, **13**, 2
Halogenated benzenes in Jacobsen reaction, **1**, 12
Halogen-metal interconversion reactions, **6**, 7
Heterocyclic bases, amination of, **1**, 4
Heterocyclic compounds, synthesis by Ritter reaction, **17**, 3
Hoesch reaction, **5**, 9
Hofmann elimination reaction, **11**, 5; **18**, 4
Hofmann exhaustive methylation, **11**, 5
Hofmann reaction of amides, **3**, 7, 9
Hunsdiecker reaction, **9**, 5; **19**, 4
Hydration of olefins, dienes, and acetylenes, **13**, 1
Hydrazoic acid, reactions and generation of, **3**, 8
Hydroboration, **13**, 1
Hydrogenation of esters, **8**, 1; *see also* Reduction, hydrogenation of esters with copper chromite and Raney nickel
Hydrogenolysis of benzyl groups attached to oxygen, nitrogen, and sulfur, **7**, 5
Hydrogenolytic desulfurization, **12**, 5
Hydrohalogenation, **13**, 4
Hydroxylation with organic peracids, **7**, 7

Imidates, rearrangement of, **14**, 1
Isoquinolines, synthesis of, **6**, 2, 3, 4

Jacobsen reaction, **1**, 12
Japp-Klingemann reaction, **10**, 2

Ketenes and ketene dimers, preparation of, **3**, 3
Ketones, comparison of synthetic methods, **18**, 1
  preparation from acid chlorides and organometallic compounds, **8**, 2; **18**, 1

synthesis from organolithium reagents and carboxylic acids, **18**, 1
Kindler modification of Willgerodt reaction, **3**, 2
Knoevenagel condensation, **1**, 8; **15**, 2
Koch-Haaf reaction, **17**, 3
Kostaneck; synthesis of chromones, flavones, and isoflavones, **8**, 3

β-Lactams, synthesis of, **9**, 6
β-Lactones, synthesis and reactions of, **8**, 7
Lead tetraacetate, in oxidative decarboxylation of acids, **19**, 4
Leuckart reaction, **5**, 7
Lithium aluminum hydride reductions, **6**, 10
Lossen rearrangement, **3**, 7, 9

Mannich reaction, **1**, 10; **7**, 3
Meerwein arylation reaction, **11**, 3
Meerwein-Ponndorf-Verley reduction, **2**, 5
Metalations with organolithium compounds, **8**, 6
Michael reaction, **10**, 3; **15**, 1, 2; **19**, 1

Nitro compounds, preparation of, **12**, 3
Nitrosation, **2**, 6; **7**, 6

Oligomerization of 1,3-dienes, **19**, 2
Oppenauer oxidation, **6**, 5
Organoboranes, isomerization and oxidation of, **13**, 1
Organolithium compounds, **6**, 7; **8**, 6; **18**, 1
Organometallic compounds, of copper, **19**, 1
 of germanium, phosphorus, silicon, and sulfur, **13**, 4
 of magnesium, zinc, and cadmium, **8**, 2; **18**, 1; **19**, 1, 2
Osmocene, **17**, 1
Oxidation, of alcohols and polyhydroxyl compounds, **6**, 5
 of aldehydes and ketones, Baeyer-Villiger reaction, **9**, 3
 of amines, phenols, aminophenols, diamines, hydroquinones, and halophenols, **4**, 6
 of α-glycols, α-amino alcohols, and polyhydroxyl compounds by periodic acid, **2**, 8
 with selenium dioxide, **5**, 8
Oxidative decarboxylation, **19**, 4

Oximes, formation by nitrosation, **7**, 6

Pechman reaction, **7**, 1
Peptides, synthesis of, **3**, 5; **12**, 4
Peracids, preparation of epoxidation and hydroxylation with, **7**, 7
Periodic acid oxidation, **2**, 8
Perkin reaction, **1**, 8
Phosphinic acids, synthesis of, **6**, 6
Phosphonic acids, synthesis of, **6**, 6
Phosphonium salts, preparation and reactions of, **14**, 3
Phosphorus compounds, addition to carbonyl group, **6**, 6
Pictet-Spengler reaction, **6**, 3
Polyalkylbenzenes, in Jacobsen reaction, **1**, 12
Pomeranz-Fritsch reaction, **6**, 4
Prévost reaction, **9**, 5
Pschorr synthesis, **2**, 6; **9**, 7
Pyrazolines, intermediates in diazoacetic ester reactions, **18**, 3
Pyrolysis, of amine oxides, phosphates, and acyl derivatives, **11**, 5
 of ketones and diketones, **1**, 6
 of xanthates, **12**, 2
 for preparation of ketenes, **3**, 3
π-Pyrrolylmanganese tricarbonyl, **17**, 1

Quaternary ammonium salts, rearrangements of, **18**, 4
Quinolines, preparation by Skraup synthesis, **7**, 2
Quinones, acetoxylation of, **19**, 3
 diene additions to, **5**, 3
 synthesis of, **4**, 6

Reduction, by Clemmensen reaction, **1**, 7
 hydrogenation of esters with copper chromite and Raney nickel, **8**, 1
 with lithium aluminum hydride, **6**, 10
 by Meerwein Ponndorf-Verley reaction, **2**, 5
 by Wolff-Kishner reaction, **4**, 8
Reductive alkylation for preparation of amines, **4**, 3; **5**, 7
Reductive desulfurization of thiol esters, **8**, 5
Reformatsky reaction, **1**, 1, 8
Reimer-Tiemann reaction, **13**, 2
Resolution of alcohols, **2**, 9
Ritter reaction, **17**, 3
Rosenmund reaction for preparation of arsonic acids, **2**, 10

Rosenmund reduction, **4,** 7
Ruthenocene, **17,** 1

Sandmeyer reaction, **2,** 7
Schiemann reaction, **5,** 4
Schmidt reaction, **3,** 8, 9
Selenium dioxide oxidation, **5,** 8
Silanes, addition to olefins and acetylenes, **13,** 4
Simonini reaction, **9,** 5
Skraup synthesis, **7,** 2
Smiles rearrangement, **18,** 4
Sommelet reaction, **8,** 4
Sommelet-Hauser rearrangement, **18,** 4
Stevens rearrangement, **18,** 4
Stobbe condensation, **6,** 1
Sulfonation of aromatic hydrocarbons and aryl halides, **3,** 4

Tetrahydroisoquinolines, synthesis of, **6,** 3
Tetrahydrothiophenes, preparation of, **6,** 9
Thiazoles, preparation of, **6,** 8
Thiele-Winter acetoxylation of quinones, **19,** 3

Thiocarbamates, synthesis of, **17,** 3
Thiocyanation of aromatic amines, phenols, and polynuclear hydrocarbons, **3,** 6
Thiocyanogen, substitution and addition reactions of, **3,** 6
Thiophenes, preparation of, **6,** 9
Thorpe-Ziegler condensation, **15,** 1
Tiemann reaction, **3,** 9
Tiffeneau-Demjanov reaction, **11,** 2

Ullmann reaction, in synthesis of diphenylamines, **14,** 1
in synthesis of unsymmetrical biaryls, **2,** 6

Willgerodt reaction, **3,** 2
Wittig reaction, **14,** 3
Wolff-Kishner reduction, **4,** 8

Xanthates, preparation and pyrolysis of, **12,** 2

Ylides, in Stevens rearrangement, **18,** 4
structure and properties of, **14,** 3

SUBJECT INDEX, VOLUMES 1–19

Since the table of contents provides a quite complete index, only those items not readily found from the contents pages are listed here. Numbers in BOLDFACE type refer to experimental procedures.

3β-Acetoxy-20-iodoallopregnan-11-one, 331
11-Acetoxy-1-iodoheptadecane, **359**
Acetoxylation of quinones, 199-277
   application to structures of natural products, 207-208, 219
   effect of methoxyl group, 209, 211-212
   failures, 208-210, 212-213
   formation of tetraacetates, 203-204, 208
   steric effects, 205-206, 214
Acyloin condensation, 142
Alkanes, from carboxylic acids, 300, 332
   from peresters, 300
Alkylation of amines with 1,3-butadiene and zero-valent nickel, 139-140
Alkyl chlorides, synthesis by halodecarboxylation, 328
Allylation, 147, 150-159
3-Allylcyclohexanone, **61**
Allyl N,N-dimethylsulfamate, 157
Allyl pyrrolidinedithiocarbamate, 157
Aminoalkylation via π-allylnickel compounds, 138-139
Ammonitriles from carboxyamidenes, 298
Aposantene, 335
Asymmetric induction in conjugate additions of organocopper reagents, 62

Basketene, 339, 341
Benzobarralene, 340
3,4-Benzocoumarin by oxidation of 2′-substituted-2-carboxybiphenyls, 322-323
Bicyclohexene, 339
Bicyclo[2,2,2]oct-2-en-5-one, **362**
Bigeranyl, 156
Bis-(acetylacetonato)nickel(II), 129, 145
Bis-(π-allyl)nickel, 119, 122-126, 128

Bis-(π-allyl)palladium, 141
Bis-(π-2-carbethoxyallyl)nickel, **176**
Bis-(π-crotyl)nickel, 125-127
Bis-(1,5-cyclooctadiene)nickel, 161, 178-179
Bis-(cyclooctatetraene)iron, 141
Bis-(π-cyclooctatrienyl)nickel, 124
Bisdecarboxylation of 1,2-dicarboxylic acids, 298-299, 334-335, 361
Bis-(π-2-methylallyl)nickel, **176**
α-Bromocrotonic acid, 23
cis- and *trans*-4-*t*-Butylcyclohexanecarboxylic acid, chlorodecarboxylation, 328
2-Butynonitrile, 44

π-(2-Carbethoxyallyl)nickel bromide, 179
Carbon-carbon bond formation via π-allyl-nickel compounds, 115-198
   carbonylation, 146-147, 170-174
   synthesis of, allyl-containing compounds, 147, 150-159
      aminopolyenes, 139-140
      butenoic acid derivatives, 146, 171-173
      carbocyclic compounds, 129-138, 165-169
      hexadienoic acid derivatives, 147, 174
      ketones, 124-126
      linear polyenes, 124-126, 136, 138-142, 162-163, 165
      pentenoic acid derivatives, 173
Carbon monoxide insertion during coupling of π-allyl ligands, 127, 147, 170-171
Carbonylation of π-allylnickel halides, 146-147, 170-174
Carminic acid, structure of, 108
Chlorodecarboxylation with Pb(OAc)$_4$-LiCl, **359**

Conjugate addition, of Grignard reagents, 1-113
  of methyl, using lithium dimethylcopper, 59
    using *tris*(trimethyl phosphite)-methyl copper complex, 60
  of organocopper reagents, 1-113
    for introduction of axial alkyl groups in steroids, 52
    inverse addition, definition, 10
    normal addition, definition, 10
    1,6 and 1,8 addition, 40-42
    side reactions, 27-29
    to *cis-trans* isomers, 30
    to isopropylidenemalonates, 29
    to isopropylidenecyanoacetate, 29
    to octalones, 4, 33-39
    in synthesis, of hydroazulenes, 53
      of insect juvenile hormone, 53
      of terpenes, 52-53
  of phenyl using lithium diphenylcopper, 60
Co-oligomerization of 1,3-butadiene with alkenes and alkynes, 136-137
Copper-catalyzed Grignard conjugate addition, 1-113
Cristol-Firth halodecarboxylation, 326
Cubane, 301
Cyclobutene, 304-305, 356
cis, trans-1,5-Cyclodecadiene, 129, 133, 139
1,5,9-Cyclododecatriene, 129
trans, trans, cis-1,5,9-Cyclododecatriene, 140
1,5-Cyclooctadiene, 129, 132-133, 182
cis-1-(*cis*-Cyclooct-2-en-1-yl)cyclooct-2-ene, 181
Cyclopropanation of benzene, 286

trans-1,4,9-Decatriene, 136
1-Dialkylamino-2,6,10-dodecatrienes, 140
1-Dialkylamino-2,6-octadienes, 139
Dibenzocyclooctatetraene, 340
Diethylaluminum cyanide, 50
Diethyl phosphorochloridate, 18
1,4-Dimethylallylbenzene, 151
3,5-Dimethyl-1-chlorocyclohexane, 360
1,2-Dimethyl-*cis, cis, trans*-1,4,7-cyclodecatriene, 133
4,5-Dimethyl-*cis, cis, trans*-1,4,7-cyclodecatriene, 137, 177

3,5-Dimethylcyclohexanone, 60
2,11-Dimethyl-1,11-dodecadiene, 151
2,5-Dimethyl-1,5-hexadiene, 151
Dermocybin, 207
Dermoglaucin, structure of, 207
2,5-Dimethyl-1,5-hexadiene, 151
9α, 10α-Dimethyl-1-isopropenyl-*trans*-3-decalone, 59
1,1-Diphenylhydrindene, 307
Disproportionation of π-allylnickel halide, 128
1,2-Divinylcyclobutene, 129, 133
1,3,6,10-Dodecatetraene, 141

(±)-Eremophil-11-en-3-one, 59
1-Ethoxycarbonylbicyclo[2,2,2]oct-2-ene, 362
Ethyl 2-acetoxyhexanoate, 359
Ethyl 6-chloro-2-methylenehexanoate, 152
Ethyl 2-cyano-3-isopropyl-3-methylpentanoate, 58
Ethyl isopropylidenecyanoacetate, 10

β-Farnesene, 156-157
Fremy's salt, 206
Fumigatin, structure of, 207, 219

cis and *trans*-Geranylcyclohexane, 180
Grignard reagents, in conjugate addition reactions, 1-113

Halodecarboxylation, in lead tetraacetate oxidative decarboxylations, 326-334
6-Heptenoic acid, from suberic acid, 357
1-Hexene by Pb^{IV}/Cu^{II} decarboxylation of 1-heptanoic acid, 356
Hummulene, 166
Hunsdiecker reaction, 326, 333
4-Hydroxy-2-methyl-4-phenyl-1-butene, 180
2-Hydroxy-1,4-naphthoquinone, 222

Iododecarboxylation with $Pb(OAc)_4$-I_2, 359
Isophorone, 4
7β-Isopropenyl-4-methyloctalone, 37
7β-Isopropenyl-10-methyloctalone, 37

Kolbe electrolysis, 286